Miracle of Life

Music from Spheres

Nigel & Max Shindler

Miracle of Life

Reveals the path that will lead us toward a future.

Music from Spheres

Tells the story of how

Good conquers evil

"The greatest trick the devil ever pulled

Was making man believe he doesn't exist."

"Stories from Heaven"
Answers the question;
How is the Tower made?

An Age is born the same way a baby is made.

In the case of man; sperm impregnates the egg, which travels to a place it incubates, and once it's mature, merges with the world. Ages are created by a confluence of forces within Britain; energies seep up from below, and fall from above; converging with The Word, Spirit, and water, to form the Jews, and once they've sufficiently developed, they spring upon the world to teach the good news, and build the Trinity.

The only way we can survive is by remembering our humanity; that our fellow man is also our brother, we must care of our nieghbour same way we care for ourselves.

"Hell on Earth"
Answers the question;
Why did the Tower fall?

Many things people believe are good are the opposite. Opposing such things as sexism, racism, and discrimination, and supporting, equality, is considered good by the common man. The truth is the contrary, such terms are used to homogenise society, which promotes consumerism by making people ignorant because they fail to judge, and discern, resulting in them becoming attached to inanimate, rather than animate, objects; to materials instead of spirit; as a consequence, the Tower can no longer be held together, and fall.

All forms of brutality must cease; the key to making this happen is enlightening the public where, and how, savagery is occurring in the world today.

Great Britian

The British Government & Nigel Shindler, express their gratitude to the French Government for providing assisting in compiling the material used in "Miracle of Life"; the depth of our appreciation is immeasurable; thank you.

Nigel & Max
Shindler

<u>The British</u>

Miracle of Life
Music from Spheres
The Rise and Fall of the British Empire

Volume I	Volume II
Songs from Heaven	Hell on Earth
Second edition Vol. I & Vol. II	
The Rise and Fall of the	The Rise and Fall of the
British Empire	British Empire
Volume I	Volume II
Saviours of the World	While Britain Slept
Thatcherism	Saviours of the World
While Britain Slept	Britain's Great Future
Revelation: Britain's Great Future	Trinity: The Father, the Son, the Holy Ghost
Thatcherism: The Puppet Named Thatcher	The Creator
The Trinity Manifesto	The Jews
Love is The Word and the Time is Now	

<u>Love Is the Nature of Existence</u>

Love Is the Nature of Existence
VOLUME I
The Trinity Manifesto

Love Is the Nature of Existence:
VOLUME II
Love is The Word and the Time is Now

Love Is the Nature of Existence:
VOLUME III
Trinity: The Father, the Son, the Holy Ghost

Love Is the Nature of Existence:
VOLUME IV
The Creator

Love Is the Nature of Existence
VOLUME V
The Jews

<u>Love is The Word and the Time is Now</u>

Love is The Word and the Time is Now
The Nature of Existence

The Nature of Existence: Book II
Love is The Word and the Time Is Now

Watching From a Tower
Tree of Life
The Great American Lie: World Destruction

The Boy and The Tower
The Father, the Son, the Holy Ghost

The Tower

The Tower: Book I; Trinity
The Tower: Book II; Love is The Word
The Tower: Book III; Creation
The Tower: Book IV; The Last Judgment
The Tower: Book V; The British

Art by
Nigel Shindler

Chosen to Make Dream Come True
I Spy, with my Little Eye
Love Is the Nature of
Existence
Heaven
Law of One
Love is The Word
A Man for All Season
The Jews
Valour
Jew

Tapestry in Motion
The Resurrection
Creation
The Tree of Life
The Time is Now
Heaven on Earth
The Boy and The Tower
The Watcher and The Tower
The Stories of Towers
Heaven's Gift

Art by
Max Shindler

Max Shindler - Fine Art America
fineartamerica.com › Max Shindler
Max Shindler's Page - PictureSocial
www.picturesocial.com/profile/MaxShindler
Max Shindler (@Truculent2) | Twitter
https://twitter.com/truculent2

Miracle of Life
Music from Spheres
Is dedicated to the memory of
All who have lost their lives due to the evil
That has spread around the world in the name of religion.

The

Eternal Journey

This book acknowledges the 144,000 Jews who made history.
Your number was so few, but without every one,
Nothing else would have been possible.
Thanks to God all mighty,
The one who chose
Them all.

CreateSpace
4900 LaCross Road
North Charleston SC, 29406
Copyright 2000-2016
CreateSpace, a DBA of On-Demand Publishing, LLC
www.createspace.com

Miracle of Life
Music from Spheres

ISBN-13: 978-1530313747
ISBN-10: 1530313740

CONTENT
The Tower

CONTENTS

Miracle of
Life
Music from
Spheres

List of Chapters

List of Poems & Songs

Music moves Sound

The twinkling dots in the night sky, rise, fall, and spin, on an axis;
Making us stare, wonder, and mysteriously contemplate.

Clouds, merge, separate, suspend, float, and then eventually
dissipate.

Their meaning, cause, purpose,
Bring times that cause us to question, and hesitate.
Closing our eyes, centering our mind,
Takes us to a state in which we can meditate.

Suddenly, like a bolt of lightning blasting from the sky,
The brilliance evolves till it is made to illuminate.

Gushes are made in the air, catching leaves, seeds, petals, and flowers,
Rolling, dipping, and circulating; much like the wheels
That spiral at a county fair.

Waves quiver, and undulate, till they crash upon a gravelly shore.
Eagles, sparrows, and seagulls, dive till they once again soar.

Bubbles sprout, mummer, and chuckle, over rocks, stones, and pebbles,
Till they can eventually collapse, and buckle.

Streams meander as they tremble, through valleys, gorges, and fields,
Representing a fine archetypal symbol.

These mesmerizing, intoxicating, delights,
Encase me in their delicate, dream filled, cocoon;
Somehow made possible by distant wavering orbiting lights.

The mind represents configurations of this kind in sound,
That flows, cascades, along the delicate drapes of a gown.

Music, and nature, conjoins as one, making life spectacularly fun.
Where do these wonders stem from?
The glowing radiant globe we call the Sun!

At no other time in history did man reveal his capacity to destroy, and cause suffering, than during the Second World War. At the same time bonfires fueled by books written by some of the greatest minds ever were watched as a spectacle by the masses.

This act was not pardoned, forgiven, or forgotten; history has shown that we have taken ourselves as a consequence to the brink of self-annihilation. Surely, by now, we have learned our lesson, and The Lord will provide us our saving grace once again - the Jews.

CREDO

The Jew is the sacred being who has brought down from heaven the everlasting fire, and has illumined with it the entire world. He is the religious source, spring, and fountain out of which all the rest of the peoples have drawn their beliefs and their religions."

published in the year 1916. In the year 1919 an experiment was coordinated by a team led by the British astronomer Arthur Stanley Eddington which confirmed Einstein's theory regarding the gravitational deflection of starlight; the German-born British physicist, Nobel laureate, Max Born, praised general relativity as the "greatest feat of human thinking about nature". "The Miracle of Life" completes Einstein's works, and presents a new way of perceiving the cosmos; the unifying force Einstein was looking for throughout his later years was, himself; to add a bit of humour, had he not spent so much of his time gazing at the stars in sky, and spent a little more time combing his hair in front of a mirror, he would have realized this.

Foreword

In order to prove the legitimacy of the claims made within my book a plan was executed for the same reason photographs were taken of a solar eclipse by dual expeditions in Sobral, in northern Brazil, and Principe, a west African island. The Miracle of Life reveals that every part of the globe is encountering the same socio-political problems, and this is due to a disease of the mind that has always been present among assemblages of people, and has spread rampantly throughout the world since the end of the Second World War.

Margret Thatcher Illusion

Definition: The natural appearance of an inverted image of a human face on which the eyes, (with their eye-brows), and mouth, are inverted relative to the face, although the image appears grotesquely hideous when the face as a whole is viewed the right way up, so that only the eyes and mouth are inverted.

The illusion of normality in the inverted image is believed to be due to the differential inversion effect, and the associated difficulty of recognizing configurational incongruities in an inverted face; also called the Maggie Illusion, or the Thatcher Illusion; so called due to the British psychologist who discovered it in 1980, who originally used a photograph of the newly installed U.K. Prime Minister, Margaret Thatcher, to illustrate the effect

Conditioning

Definition: The process of learning through which the behaviour of organism becomes dependent on environmental stimuli.

Oxford Dictionary of Psychology, Andrew M. Coleman,
Oxford University Press,
Third edition, 2009

Inverted Totalitarianism, (this is the title for the political/economic system now ruling Britain), affects people the same way the Thatcher Illusion does; what people believe is acceptable, or fine, is actually monstrously wrong, therefore, the perception people have of their surroundings has nothing to do with the real environment they.

Among the most popular shows currently on T.V. are the ones people called "Virtual Reality"; they are staged productions that are supposed to resemble reality, similar, or so people believe, to a play performed on a stage where people act scenes that have already been written. A theatrical play is typically a reproduction of something that could happen, or has happened, in "real life", thus, people can identify with what takes place on the stage, and benefit from it due to having the opportunity to learn from an experience, without actually going through the ordeal.

The masse are far more accustomed to consuming vast quantities of time watching programs they cannot derive anything beneficial from; there is no such thing as "virtual reality"; people are misconstruing the nature of these shows; they have no resemblance to reality, as the genre implies; something is either real, or it isn't.

The same form of deception was manufactured while Margaret Thatcher was Prime Minister of the U.K., and each leader that followed, as well as by the mayor of London, Boris Johnson, who is currently making strides to become Prime Minister.

Britain, over the previous thirty years, has become increasingly reliant on foreign nations; David Cameron, however, unlike the others, has charted a strategy that facilitates Britain ability to move in the opposite direction; enabling Britain to acquire prominence on the world's political/economic, stage; few have a comprehensive understanding of the measures that have been undertaken to achieve this outcome, and instead believe there is some resemblance of truth to what Boris Johnson is pontificating.

People should ask themselves one question; according to on-going polls what concerns British voters the most are jobs, and the amount of money they can spend; such being the case, why would people choose to compromise the people, and agencies, they require to live the lifestyles they do?   Abruptly breaking ties with the European Union would have this effect.

It is because Boris Johnson fabricates lies in order to deceive the public into believing they are wealthier than is in the case; the consequence, if he were elected Prime Minister, would be catastrophic for Britain; to make an analogy; a man starting a business formulates plans in accordance with the cash he has available, however, at some point his bank informs him the amount of money they thought he had was a miscalculation; his funds have actually been exhausted; he is, as a result, left penniless.

Boris Johnson knows full well Britain's economy isn't strong enough to sustain itself.   Britain consists of a society of "consumers"; they behave collectively the same as someone who buys groceries in a supermarket; if the people responsible for putting groceries on the shelves are fired, and the trucks that transport goods to the supermarket do not operate, he would not be able to shop for food in this supermarket; thus, since the British public are presently getting the food they need; why is there a perceived necessity to expel immigrants from Britain?

In order to answer this question, I will back track to Thatcher's era, and note that most of Britain's population at the time had the sense there was something "misaligned" about Margaret Thatcher; instead of examining these divergences, people covered them up with a quaint, rosy, picture, that had no relation to reality; she wasn't an "iron lady"; she was a malleable puppet, who allowed her persona to be used as a vehicle to ruin Britain; the masses perception of her; what she stood for, wasn't slightly off the mark, but entirely the opposite to the truth.   Britain has now become, due to her policies, and the leaders that continue her legacy, as well as Boris Johnson, who many believe, for the most part, is doing what is in the best interests of Britain; nothing could be further from the truth; he is doing what the people

I call, the "Monolith"; which is made up of the mob families that began making a fortune in the 1930s from bootlegging; want him to do. They use the same techniques on a global scale to get acquire wealth as they did in the 30's in Harlem, the Bronx, and Chicago; which means making sure, by any means necessary, people pay their debts.

Britain didn't become more self-sufficient, during Margaret Thatcher's time in power, but weaker, and more dependent of foreign nations. Britain, correspondingly, is now, more so than at any other time since the Second World War, a country with a population comprised primarily of consumers, rather than a society full of people able to cultivate their minds to the extent they can create something they can offer to other people around the globe.

When Thatcher was leader of Britain she created a persona that managed to convince the majority of people living in Britain that she was looking out for their best interests; the truth was the exact opposite; instead of giving British people greater power to determine their affairs, she was working to nullify their capacity to charter their own course.

Boris Johnson continues in Thatcher's footsteps, and the British people are buying his act that he gives a damn about them. The sad truth is that he has no interest in Britain; he has no standards to uphold; no morals to adhere to; his only mission is to ruin Britain.

Inverted Totalitarianism strips power from people, while making people believe they are being given more; to make an analogy, picture a trifle cake turned upside down; the heavier ingredients that were once on the bottom get mixed up with the lighter ingredients that were on top. In a democratic society the elected leader is supposed to be able to dictate what the people beneath him do; in Inverted Totalitarianism the cake, a metaphor for society, no longer consists of layers; similarly, Britain no longer has classes ; there is no hierarchy; everybody believes that an opinion cannot be of greater value than anybody else's.

People in Inverted Totalitarian societies have a tendency to believe they are able to determine their own fate; this mind-set prevails because there is no one who can stop them from doing what they want to do, and that's why they believe they are in control.

The opposite is the case because when people are allowed to do what they please, they quickly lose the ability to take control of their lives, and eventually people become brainwashed, till the point that they believe is bad for them is in fact good; for example, I recently asked a university student why he bought himself bottled water, instead of, for example, what I was holding in my hand at the time; a carton of orange juice, priced close to half the price of the bottled water he had.

With a fair amount of derision in his voice, feeling, no doubt, somewhat saddened someone hadn't caught on to reality to the same extent he had, he explained that because he played basketball; (he wasn't dressed in shorts, a

T-shirt, and sneakers, and standing on a basketball court, when we spoke), and, thus, while tugging with one hand the skin over his Adam's apple, didn't want to become de-hydrated, furthermore, orange juice had sugar, he told me; I interjected that he was incorrect, it has "fructose" not "sucrose"; he appeared not to process what I'd said, and continued with his point that orange juice was bad because it had "too much sugar".

Buying orange juice, in his mind, was comparable to buying something as detrimental to your health, as a candy bar. I did mention that the drink, Gatorade was popular in the country I'd just come from because it contains electrolytes that help the body to regain the nutrients it lost while engaged in strenuous physical exercise. I also mentioned that water has no nutrition, and what he was drinking out of a plastic bottle was no different from the water that comes out of a tap, except for the fact it's filtered.

He gave every indication my words had failed to make an impression on him, which didn't surprise me; a typical university student in a "consumer" society has been thoroughly conditioned. He has learned over the years to respond to stimuli; which is the opposite of what one would expect to take place in a person who enters a post-secondary institution; their actions are not determined from within; his capacity for free choice, and free will, has been eradicated.

The way the tyrants who rule the world did this was by allowing people to give into their most basic instincts; the mind weakens as a result, and brainwashing becomes easier to accomplish

While this is taking place the heavier ingredients in the trifle, continuing the analogy from before, overwhelm the lighter parts, to the extent their life support system becomes shut off. Consumer societies contain destructive elements such as the concept of "equality" that insures the elite are unable to obtain resources to help them reach their full potential; no one is deemed better than another; no one leads, and no one, or so it is believed, overtakes anyone else, and everyone gets way with their criminal acts because will look away when it takes place, due to not believing they have the right to judge; "who is without sin, may cast the first stone", is a line commonly used to justify this behaviour. Civilization, as a result, can no longer exist; people become brutes that feed on members of their own species, and are still able to live with themselves due to not being aware of what they are; the truth is too grotesque for any man to face; they are the same as the Thatcher Illusion when the eyes and mouth appear the proper way, but the rest of the face is contorted; a hideous beast.

People wear personas to hide the monsters they are; if anybody is able to identify a monster, (as I have), they are deemed, "rude", which according to their principles of morality, then entitles them to deny that person his basic rights, which actually exposes the monster's true nature.

Many of the people who colonized the Americas, for example, had the same mentality; they massacred people and stole riches belonging to them, and justified their actions by labelling their victims as members of a sub-species; with the backing of The Church, and correspondingly, The Lord, they slaughtered people in the millions.

The British people openly commit their atrocities; people are stripped of their I.D., and thrown of the streets to die; they suffer, and die, while mums and dads, with their children by their side, walk by; moms pushing strollers ignore the cries for help from those denied a home where they can feel safe. The people are disconnected from reality, and also themselves; they are soulless being who kill as efficiently as Adolf Eichmann, in fact, Britain's consumers are mass murderers. Each falls into the role of playing a part in the machine that generates this genocide.

Savages exist not only in Rwanda; Serbia; Armenia; Cambodia; China, and the Soviet Union, as well as many other countries around he world, they are here; across the street from where you live; behind you are you wait in line at a back; the person who sits in the cubicle next to yours at work.

Boris Johnson; author of a biography on Winston Churchill, knows this; he has the paperwork listing the names of those who've been killed; their National Insurance Numbers, as well as the bank account numbers opened with their I.D; in the same way Adolf Eichmann knew the number of Jews transported to concentration camps; the number that died, and when.

Boris Johnson is also a man who believes the U.K. should break away from the European Union, and wants to become Prime Minister to insure this happens; all the while, he knows Britain is not a producer county, and its consumers must rely on the skills belonging to the immigrants to get by; in fact it is because of the immigrants the enormously wealthy can lead the luxurious lifestyles they do; they live, the vast majority, off the proceeds of the crimes committed by immigrants The British public believes their lives would be better; which equate to more jobs, and money, if immigrants no longer came into this country, and the once here, left.

David Cameron has pushed for the implementation n of policies designed to insure Britain will one day be free of immigrants; Britain will then be able to generate great men that can enrich the world by their creations.

The cream must rise to the top again; be praised, and revered for their accomplishments.

Britain is the land meant to harbour Jews. Britain is the land of milk and honey; a place where Jews can be nourished, and thrive.

Projection

Without Jews nothing would possible; they create time, space, and time, by fusing words to form idea, and those ideas make the world go around.

Paramahansa Yoganananda

The human mind can, and must, liberate within itself energies greater than there is within stones, and metals, lest the atomic giant, newly unleashed, turn on the world in mindless destruction.

The arts, in every form, have been molded to fit the steepening decline of man's mental acuity. Popular novels are increasingly written using shorter, more simplistic, sentences, which is a reflection of man's diminishing capacity to focus, and join thoughts to form ideas. I find the expression, "Everybody's entitled to an opinion", most amusing. An opinion isn't something that's handed to you, as if it were a meal served on a silver platter, rather it is formulated by the processing of information one has collected; therefore, the question remains; is everybody capable of constructing an opinion? The question is, quite obviously, cynically rhetorical.

Movies, and music, are more and more catered to titillate the minds of the lowest common denominator, thus, attending the opposite purpose they were meant to serve; the cultivation of minds; chaos, confusion, violence, and disorder, can only result from such practices. The Jew, The Lord's servant, on the other hand, acts to bond people together, encouraging people to follow the "golden rule", which was been

expressed in a multitude of ways, in various cultures, at different times.

PBS NewsHour: What do you think is more important for kids; hard work, imagination, or faith?

The words faith, hard work, and imagination, I'm sure, have a personal meaning for everyone, which can also change over time; therefore, I cannot answer the question because I have no frame of reference.

I do find it concerning, however, that people are inclined to make presumptions on such matters, which can lead to misunderstandings, miscommunication, and, thus, hamper one's efforts to accomplish as much as possible during the day, (which can be expressed as "hard work"; which can also be experienced, or perceived, as enjoyable due to the sense of worth and meaning).

I imagine, (the word is a construct manufactured by my mind, brain, spirit, entity, being, maybe all, at other times possibly not), the term "faith", for most, is associated with their religious, ideological, and cultural, background, therefore, unless I am provided information on such matters, I couldn't possibly formulate an opinion; I can, however, spout out a prejudicial remark, which is a common form of behaviour among those inclined to boost about how much they know.

A common problem in our world today is that most believe they are saying something, when they are saying nothing at all, ("doublespeak", and "newspeak").

PBS NewsHour: What do you think is more important for kids; hard work, imagination, or faith?

Putting aside the fact that each of the words, faith, hard work, and imagination, have various definitions, and mean something particular to each individual due to their own idiosyncratic cultural background; what does the word "kids" mean? For example, what age group is being referred to? Furthermore, why is it an "adult", if that is whom this question is meant for, that is unable to ask these simple, fundamental, questions, determining for another human being what is more important?

Such mindless preoccupations fill up a lot of people's time, and they actually think they are engaging in something important; this is more or less what is happening in our schools today, and "adults" are setting a poor example for their children.

PBS NewsHour: What do you think is more important for kids; hard work, imagination, or faith?

Above is a series of words that gives the impression something is being said; that a question is being asked; the truth of the matter is that this very impressive sounding, and "weighty", collection of words actually doesn't say anything at all; no question is being asked?

For kids...... for kids to do well in school? Does this mean achieve high grades? For kids to develop socialization skills? Socialize with whom, where, in what capacity, and about what?

For kids to find life a fulfilling experience? Who can make such a determination?

I believe I have made my point quite clear, but the sad truth of the matter is that the message I've conveyed will most likely, in more cases than not, miss the mark; meaning, go over people's heads.

The term "newspeak", means much the same thing. The newspapers of today appear to be telling us something, when, in fact, nothing concrete is being conveyed at all.

It's O.K. to ruminate on meaningless, inconsequential, propositions; but when practicalities are involved that can lead to a negative consequence suffered by others, we have a real problem, and hardly a fantasy made up in an imaginary, nonsensical, land.

Alice in Wonderland
Is Alice living in wonderland, imagining wonderland, or is wonderland actually herself; a projection created by her mind?

The pernicious maladies prevailing in our world today are entirely due to man; he decided to go against his nature, and the Earth has suffered as a consequence. From one decade to the next since the Second World War the psychogenic illness affecting the masses has spread, and become more and more, deeply embedded into the strata of societies.

Those who have benefited financially from this sickness, have correspondingly laboured to hide the nature of the problem, so much so that there is actually a vast population of people who are unaware of how horrendous things are, and, as extraordinary as it sounds, a portion is entirely oblivious to there being any problem at all; at least of the kind having the potential to compromise man's existence.

George Bernard Shaw
"Man remains what he has always been; the cruelest of all the animals, and the most elaborately, and fiendishly, sensual."

Introduction

David J. Hadden

Awake, awake, O Zion
Come clothe yourself with strength
Awake, awake, O Zion
Come clothe yourself with strength.

Put on your garments of splendour
O Jerusalem
Come sing your songs of joy and triumph
See that your God reigns
Awake, awake.

Burst into songs of joy together
O Jerusalem
The Lord has comforted His people
The redeemed Jerusalem
Awake, awake.

Future Times/Rejoice

YES

In the fountains of the Universe : Set time in accord
Sits the boychild Solomon : Ever turning round and round
In the cities of the Southern Sky : Set points Universe
Dreams he of glory : Pulsating round and round
Future times will stand and clearly see : Highest dancing
Of the course of innocence : Drifting drifting

See it all
See it all
Till tomorrow
See it all
See it all
Till tomorrow
Future times will stand and clearly smile
Of the course of innocence

Dantalion will ride again : Raging forth underland
The course of evils standing straight : Grind to grind to grind
Hot metal will abound the land : Churning out shout
As the form regards our blazing hand
Future times will stand and clearly see
Of the course of innocence

One the word will enter all our hearts
Two the duel will alter them
Three jewels countenance divine away delight away
Four the fight to free the land
Five the islands of Arabia
Six the tears that separate
Six the tears that separate

Rejoice forward out this feeling
Ten true summers long
We go round and round and round and round
Until we pick it up again

Time flies, on and on it goes
Thru the setting sun
Carry round and round and round and round

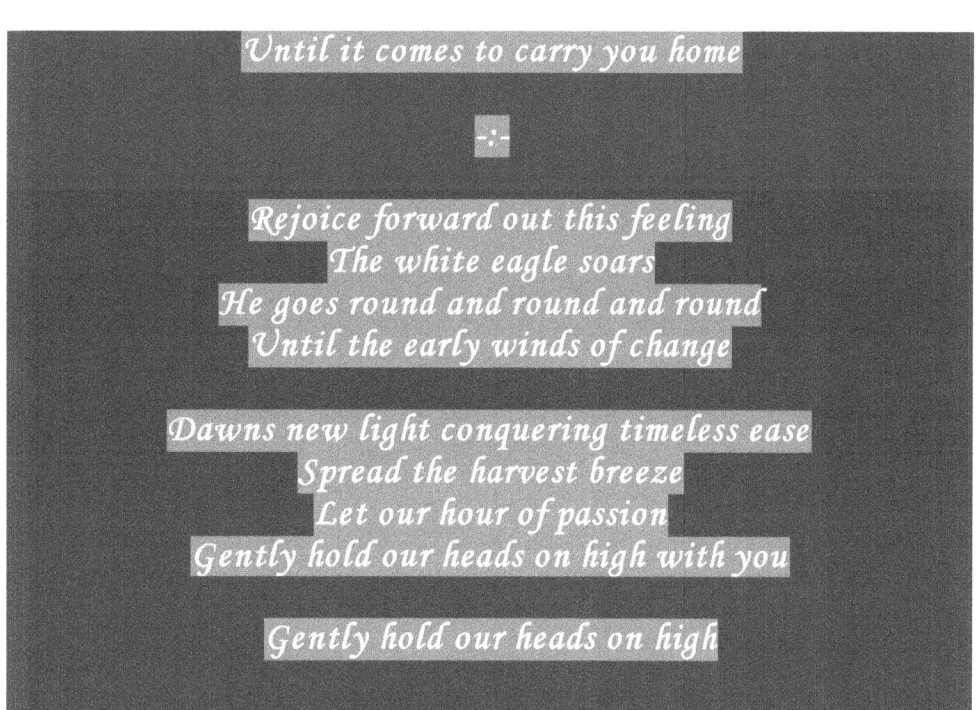

Until it comes to carry you home

-:-

Rejoice forward out this feeling
The white eagle soars
He goes round and round and round
Until the early winds of change

Dawns new light conquering timeless ease
Spread the harvest breeze
Let our hour of passion
Gently hold our heads on high with you

Gently hold our heads on high

Harmonic Convergence

The **Harmonic Convergence** is the name given to one of the world's first globally synchronized meditation events, which occurred on August 16–17, 1987, which also closely coincided with an exceptional alignment of planets in our solar system (see Astrological alignment below).

The timing of the Harmonic Convergence was allegedly significant in the Maya calendar, with some consideration also given to European and Asian astrological traditions. The chosen dates have the distinction of allegedly marking a planetary alignment with the Sun, Moon and six out of eight planets being "part of the grand trine." Though Arguelles eventually connected the timing of the Harmonic Convergence with his understanding of the significance of Maya calendrics, the dates themselves were derived not from Maya cosmology but from Tony Shearer's reconstructed Aztec prophecies.

The event had been predicted by author Tony Shearer in his book *Lord of the Dawn* (1971). The principal organizer of the Harmonic Convergence event was José Argüelles. According to Shearer's interpretation of Aztec cosmology, the selected date marked the end of twenty-two cycles of 52 years each, or 1,144 years in all. The twenty-two cycles were divided into thirteen "heaven" cycles, which began in AD 843 and ended in 1519, when the nine "hell" cycles began, ending 468 years later in 1987. The very beginning of the nine "hell" cycles was precisely the day that Hernán Cortés landed in Mexico, April 22, 1519 (coinciding with "1 Reed" on the

Aztec/Mayan calendar, the day sacred to Mesoamerican cultural hero Quetzalcoatl). The 9 hell cycles of 52 years each ended precisely on August 16–17, 1987. Shearer introduced the dates and the prophecy to Arguelles in 1970, and he eventually co-opted them and created the name Harmonic Convergence as the public title of the event.

Astrological Alignment

According to the astrologer Neil Michelsen's "The American Ephemeris," on 24 August 1987 there was an exceptional alignment of planets in the solar system. Eight planets were aligned in an unusual configuration called a grand trine.

The Sun, Moon and six out of eight planets formed part of the grand trine, that is, they were aligned at the apexes of an equilateral triangle when viewed from the Earth.

The Sun, Moon, Mars and Venus were in exact alignment, astrologically called a conjunction at the first degree of Virgo in Tropical Astrology. Mercury was in the fourth degree of Virgo which most astrologers count as part of the same conjunction being within the "orb" of influence. Jupiter was in Aries, and Saturn and Uranus in Sagittarius completing the grand trine. However some believe that this is an Earth Grand Trine with Sun/Moon/Mars/Venus/Mercury in the initial degrees of Virgo, Neptune at 5 degrees of Capricorn, and Jupiter in the last degree of Aries (anaretic degree), on the cusp of Taurus. Uranus and especially Saturn are on the edge of this trine.

Controversy is associated to the claim that this occurrence was a unique event. Grand Trines, where planets are at 120 degree positions forming an equilateral triangle are not uncommon or particularly noteworthy. Traditional Astrology does not consider Trines to be action points, so the notion of a special shift in world history would not be in the offing under a Grand Trine. Hence, many traditional astrologers do not regard this occurrence to be of significant importance. There is no evidence that astronomers have ever considered it significant.

Astrological Interpretations

The convergence is purported to have "corresponded with a great shift in the earth's energy from warlike to peaceful."

Believers of this esoteric prophecy maintain that the Harmonic Convergence ushered in a five-year period of Earth's "cleansing", where many of the planet's "false structures of separation" would collapse.

According to Argüelles, the event came at the end of these "hell" cycles and the beginning of a new age of universal peace. Adherents believed that signs indicated a "major energy shift" was about to occur, a turning point in Earth's collective karma and dharma, and that this energy was powerful enough to change the global perspective of man from one of conflict to one of co-operation. Actress and author Shirley MacLaine called it a "window of light," allowing access to higher realms of awareness.

According to Argüelles, the Harmonic Convergence also began the final 25-year countdown to the end of the Mayan Long Count in 2012, which would be the so-called end of history and the beginning of a new 5,125-year cycle. Evils of the modern world, e.g. war, materialism, violence, abuses, injustice, oppression, etc. would end with the birth of the 6th Sun and the 5th Earth on December 21, 2012.

Power Centers

An important part of the Harmonic Convergence observances was the idea of congregating at "power centers." Power centers were places, such as Mount Shasta, California, and Mount Fuji, where the spiritual energy was held to be particularly strong. The belief was that if 144,000 people assembled at these power centers and meditated for peace, that the arrival of the new era would be facilitated. **(Wikipedia)**

The children are our future!
If we are prepared to acknowledge the shortcomings in the education system today, there's no reason why the future can't be bright for each and everyone of us.. A key component in making that happen is a renewed appreciation for greatness, and making sure nutrients are available for such an individual's fullest potential to be actualized.
Many of the wold's problems have been left unattended for so long, only men of Einstein's calibre are going to get us out of the pit we've dug for ourselves.

Awaken the Peace: Insights from Paramahansa Yogananda

POSTED BY PAOLA DI FLORIO ON NOVEMBER 21, 2014

Peace is that underrated quality, the one we sometimes don't realize we're living without. Given the myriad distractions of modern life, it sometimes takes a major wake-up call to remind us to stop and tune in.

In 1946, one of the great spiritual teachers of our time, Paramahansa Yogananda, broke with tradition by publishing *Autobiography of a Yogi*, a book about his mystical experiences with saints and seers in India. These secret teachings had been passed down exclusively to ascetics in ancient times, but due to a special dispensation, Yogananda and his lineage shared the age-old practice of Kriya Yoga with everyday people. In 1945, as Yogananda was putting the finishing touches to the book that would transform millions of lives, we deployed the Atomic bomb as a weapon for the first time in human history. Yogananda, foreseeing what was to come, wrote: "The human mind can and must find within itself energies greater than those within stones and metals, lest the atomic giant, newly unleashed, turn on the world in mindless destruction."

We now had the means to extinguish the human race. But Yogananda knew that those same subatomic forces we had harnessed to produce the bomb, were also within every cell of our beings. They were the very energies the ancient Rishis had cultivated within the mind and body to elevate human consciousness. The urgent lesson of our time was to use those energies for a higher purpose -- to find the still place within where peace is an inexhaustible aspect of the Divine.

This is one of the many lessons I learned from Yogananda, whose works I spent the last 6 years absorbing in every possible way for a documentary film on his life called Awake: *The Life of Yogananda*, which I co-wrote and directed with Lisa Leeman and co-produced with Peter Rader. Yogananda rocked my world by systematically dismantling how I viewed human existence.

At the onset of filming, I challenged a senior monk from Yogananda's organization, Self Realization Fellowship, by declaring that I had a problem with the word "God." He asked me if I'd ever experienced a feeling of peace.

I remembered as a child feeling moments of deep connection in nature or while visiting hermitages in the Abruzzi Mountains in Italy... and in the intense quiet that followed a heavy snowfall in the north-eastern United States.

"Of course," I answered.

"That," he said "is an aspect of Divinity. Build on it in the stillness of meditation, and you'll experience God. "

I realized I had an opening. Peace was something real and tangible that I had experienced both externally and internally, many times. I was suddenly curious to connect to it in a different way -- that is, to allow and imagine that the kernel of Peace within me could actually be an aspect of God. What a concept!

Yogananda was a radical, a revolutionary – a total game-changer. His timing in bringing his teachings to the West in 1920 couldn't have been more perfect. When he first arrived in America, everything was up for grabs. Einstein's General Theory of Relativity has just been proven by an astronomical observation of a solar eclipse in 1919. And the Quantum Physicists that followed him were telling us that matter on the subatomic scale was elusive, erratic, based on vibrations and probabilities... that where we placed our attention could actually alter physical reality.

The Rishis of India had developed a similar ontological framework nearly 3000 years prior. In ancient times, scientific and spiritual pursuits went hand-in-hand. The separation of these disciplines actually came fairly recently. As Harvard Physicist and Physician, Dr Anita Goel, who is featured in *Awake: The Life of Yogananda*, explains: "Several hundred years ago, things like alchemy became chemistry, numerology became mathematics, astrology became astronomy, metaphysics became physics. And the separation of objective reality and subjective reality actually enabled science to make great progress. Many of the biggest breakthroughs in science and technology in the last century have been made in reductionist silos."

But, while it allowed Western scientists to deepen their understanding of the gross material world, they lost touch, perhaps, with the more mystical aspects of our place in the cosmos. Today's communication technology may have brought us closer in many ways, but it's also been the source of constant distraction and noise, which those of us on a spiritual path must find a way to silence.

Yet the kind of peace that Yogananda was talking about isn't about checking out and living as a hermit in a mountain cave. He famously said: "It shows more spiritual fiber to live a godly life in the jungle of civilization, full of human tigers, wolves, and snakes that bite at you because you are doing good."

Gandhi, who had been a lawyer in his day, was also a practicing yogi, who took lessons on Kriya Yoga from Yogananda. In fact, Gandhi used the yogic principle of ahimsa to promote peace in the world, which was a key influence on Martin Luther King, Jr., who changed a nation through the same philosophy of nonviolence.

Yogananda guides us to be "calmly active and actively calm" when we do our part in the world. With this comes a clarity, a focus and a serenity that puts us in touch with our interconnectedness with all beings. When, in meditation, we experience the cessation of the disturbances of the mind, we come into contact with inner Peace. Once that happens inside, outer nonviolence is inevitable.

Ode to Joy

Joyful, joyful
We adore Thee
God of glory
Lord of love
Hearts unfold like flowers before Thee
Hail Thee to the sun above
Melt the clouds of sin and sadness
Drive the dark of doubt away
Giver of immortal gladness
Fill us with the light of day

With light
With light

Mortals join the mighty chorus
Which the morning stars began Father love is reigning o'er us

Brother love binds man to man
Ever singing march we onward
Victors in the midst of strife
Joyful music lifts us Son ward
In the triumph song of life

PETER GABRIEL

"In Your Eyes"

love I get so lost, sometimes
days pass and this emptiness fills my heart
when I want to run away
I drive off in my car
but whichever way I go
I come back to the place you are

all my instincts, they return
and the grand facade, so soon will burn
without a noise, without my pride
I reach out from the inside

in your eyes
the light the heat
in your eyes
I am complete
in your eyes
I see the doorway to a thousand churches
in your eyes
the resolution of all the fruitless searches
in your eyes
I see the light and the heat
in your eyes
oh, I want to be that complete
I want to touch the light
the heat I see in your eyes

love, I don't like to see so much pain
so much wasted and this moment keeps slipping away
I get so tired of working so hard for our survival
I look to the time with you to keep me awake and alive

and all my instincts, they return
and the grand facade, so soon will burn
without a noise, without my pride
I reach out from the inside

in your eyes
the light the heat
in your eyes
I am complete
in your eyes
I see the doorway to a thousand churches
in your eyes
the resolution of all the fruitless searches
in your eyes
I see the light and the heat
in your eyes
oh, I want to be that complete
I want to touch the light,
the heat I see in your eyes
in your eyes in your eyes
in your eyes in your eyes
in your eyes in your eyes

Words inspired by the life of
Robert F. Kennedy

The only chance we have of surviving as a species is by remembering our
Humanity; that our fellow man is also our brother, and we must care for
Others as much as we care for ourselves; only then will the divine reach
The Earth; after all, we are in fact Spirit, not matter,
And there is nothing we cannot overcome
If we are determined to do so.

Death & Beauty

On December 9, 1946, at one of the first trials in Nuremberg, Germany, U.S. Brigadier general Telford Taylor made an opening statement that included an explanation of the importance of the trials.

"The mere punishment of the defendants, or even thousands of others equally guilty, can never redress the terrible injuries which the Nazis visited on these unfortunate peoples. For them it is far more important that these incredible events be established by clear and public proof, so that no one can ever doubt that they were fact and not fable."

"Miracle of Life" was written for the same reason the trials took place in Nuremburg, Germany, following the Second World War.

Daily Mail; 28/01/2016

"A Memorial to the victims of Nazi Atrocities is to be erected in London."

"It will show the importance Britain places on preserving the memory of the Holocaust."

"It will be in Victoria Tower Gardens by the Thames, next to the Palace of Westminster."

Death and Beauty

On a field shields, and armour block, and protect, slashes of steel.
Each man is aware of the presence of danger he can feel.

Blood gushes, limbs flay shrieks of rage, anguish, and pain, fill the air.
To proceed onward is the call;
The wall of arrows, guns, spears, knives, appear to be as if a waiting lair.

Here is no right or wrong, good, bad, unfair, or fair.
The enemy's face has dagger eyes that pierce with a malicious glare.

Am I here because it is my wish? Is this the fate I deserve?
Without knowing the reason or cause, can I muster the necessary nerve?

The dawn follows at day break.
The field once green is now littered
With limbs, torsos, death, and the colour red.
This is where many greet their doom; on this plain is their final bed.

Where is the glory? Why not shame?
Many gallantly served; families have told, and will tell,
Of lives that ended in this wretched hell.

Like pawns on a checker board,
Their wishes, wants, plans, dreams, were served to play a game.
Cherished ones will speak of the life passed.
Children, grandparents, relatives, ancestors, will add to his lasting fame.

By the mid of the day, rain gushes, and floods the land.
Washed into rivers, and streams, are the remnants
Of this reckless, unspeakable, act;
But the event did indeed occur, this is a fact.

So many lives have been washed away,
With so much time that was left to remain.
What from any of this disgrace could one possibly gain?

No more does he have the chance to dream, reflect, hope, and ponder.
Soon, not far in the future, people will gaze from a hill not far yonder,
And question; where was the battle?
What is left to memorialize, to familiarize, hypothesize,
To help fashion a method to theorize,
Why men were led to their death without the need to hypnotize?

In the earth their names have been left as script on paper.
Bones disintegrate, flesh, and organs, mould;
They're devoured by insects, burrowers, and scavengers.
Their souls, essence, have returned once more to The Maker.

The continuation of each path must now be decided upon;
The chance to redeem;
Repair; replenish; capitulate; contemplate, and compensate,
Has arrived.

Oh so many lives are required to put things right.
Sadly, yet another is gone.
Relentless is the pursuit toward perfection.
It will only cease when the work is done.

From the womb life emerges, set to pay for past misdeeds,
And awaken to the truth still to be reached.
Mother, father, and family, have been chosen by the Grand Dictator.
He will have His way; eventually, hopefully, sooner rather than later.

When the War is over, peace will overwhelm.
All he is, and was meant to be.
He sits on a mountain throne, and surveys all that has been;

The glorious, hideous, sickening, and righteous, sights, he has already seen.

No matter the mistakes, opportunities, and leaps of faith;
From each, wisdom he was able to glean.

Nothing was lost, gained, missed, forgotten;
Everything was as it was meant to be.
A Play designed, integrated, and implemented,
By The Most High; the Glorious He.

When all is said, and done,
Only the chosen few will remain standing.
Vanquished are the pathetic, immoral, creatures,
Who fought to nourish only their most heinous, deplorable, features.

The field that once carried death, decay, rot, and gore,
Is now destined to be planted by those
Who showed respect for God's Law.

No matter the situation, sacrifice, loss, pain, or suffering,
They decided to always do what is right.
The light of the Sun baths the land that is now sheltered
By the gentle caress of The Lord's hand.
Heaven has arrived!
No more is there a struggle, a need to fight;
Everything has been made right.

Leaping across a chasm, a void, dark, unknown, space,
We reach the inner sanctum,
Sheltering the hallowed, resplendent, splendour,
That is our bliss.

Naked, luxurious, voluptuous, lips gently place a kiss.
The dark progressively proceeds to become light;
The ever renewing, gorgeous, brilliant, radiance, that shines bright.

Within a chamber a choir sings.
Strings melodiously twine together as one.
Trumpets, clarinets, bows, bring to life hymns.

A temple for the gods holds Moses with his staff on a throne.
David stands firm, strong, a hand raised to greet his shoulder bone.
Here is where the Artist, lover of creation, can feel completely at home.
Even after the doors have shut, and the rest have left,
Among the paintings, murals, and sculptures,
He feels as though he is not alone.

Walls are covered with finely bonded books.
The covers indicate much that lies inside each.
The finest scriptures enhance whatever place you might be;
A park bench, pub stool, a garden covered with patches of roses,
Or a quiet, sandy, sun drenched, beach.

The Stage is populated with those who have a role to play.
The production is never the same.
Each performance is different, no matter the day.

Critics unable to maintain the required stature,
Will describe an exquisite Show with the artistry of a butcher;
But each having risen from his seat,
Then leaving the theatre to join the world again,
Has enriched his capacity to experience;
All can now be perceived as intensified;
Joy; laughter; sorrow; fear, as well as pain.

The last Movement has been completed.
The final note announced the end.
The Conductor turns to the crowd,
Each member of the Orchestra rises to greet the wild, jubilant, cheer.
Many in the Audience feel the shared moment
Has made the person beside them a friend,
Wanting nothing more than to savour the experience
By being together, and having a beer.

Colours, strokes, dimensions captured in clay, and marble,
Mesmerize, and dazzle the eye, by the extent they marvel.
The Great dedicated their lives to developing skill, technique, and style.
The names number merely a few, and can be contained in a simple file.

The labels, signs, distinguish each section.
Revealed are the multitude disciplines
Man has sought to understand and explain.
The wisest use colourful, meaningful, terminology,
And avoid the usage of words that would best be considered plain.

How strange that such a minute portion of humanity
Has manifested such an acute magnificent psychology.
Does the answer lie in neurology, physiology, or biology?
Or is the Spirit contained in each of such a nature
That once lighted, fires a storm, telling us
Another blessed creature has been born?
They are the kind that serves to enrich others.
Working tirelessly to better themselves,
While others are protected beneath bed covers.
Where is the explanation, reason?
Could it be above each
There is an angel
That hovers above.

We are such stuff
As dreams are made on,
And our little life is rounded
With a sleep.

William Shakespeare
The Tempest

Heaven's Open

I shoulder the burden of so many others.
Disquieting thoughts fill the air.
Quite often we wonder; is someone really there?

Alone we can feel, when sadness cloaks us,
But it cannot come near to offering protection.
How fine it must be to be a Buddha!
You would have reached the height that perfection brings;
All day you would feel the call to sing.

Fluctuating moods, a sensitivity to tune;
The two reach, and offer the other a hand.

Creativity, still, can often only come in spurts.
My, how much grander it would be,
If their time could fill a land,
Or maybe merely a brook
That is bordered with a collection of sand.

Some are acquainted with things finer than others.
Decades, centuries may pass, allowing lovers to meld,
So their hearts become one once again.
Seen as one, a magnificent design,
Reveals something extraordinary, beyond belief!
It spans dimensions incomprehensible;
Thus, merely glance in wonder at the web of a leaf.

All told, when reaching within,
A limitless resource is found buried there;
A boundless, effervescent fountain
That can be perceived as a loving lair.

Just a few, a tiny, minute, minority,
Have managed to grasp the enormity, and complexity,
Of impressions provided by the senses,
As well as the worlds still left unknown; that lie beyond.

Their eyes twinkle, as a laughter gurgles delight.
To continue to increase what is known,
Is their on going fight!
Regardless, the battle rages on,
Resisting the ignorance that so many still possess;
Thus, study they despite the evident fatigue,
Throughout the dark, cold, and howling, night.

More, and more, must be gathered;
Never do they settle for anything less!
Could it be each was gifted with The Lord's heavenly bliss!?

Cosmic Creation

That which people construe as existence, or reality, actually stems from non-existence, or The Lord, which means that creation is actually an illusion, and what we think is real, is nothing more than a dream.

The universe began when words were expelled from The Source, which is also referred to as the "First Cause" or "First Mover", which is "The Father".

he words dispersed as they were carried within streams of light, and this brought about space. When the words began to be drawn together, (Love is the joining of words), matter formed; time is created by the movements of material objects, and, thus, history began. The purpose of existence is for The Lord to realize Himself.

The completion of history is achieved through the interaction of Jews with Savage Primates. The Lord views the contrast between the actions of Jews, and Homo-sapiens, and, as a consequence, conceptualizes Himself; paradoxically, the more heinous the deeds of Homo-sapiens, increasingly less cognizant they are of the disease within themselves; The Lord becomes more aware of who He is.

The Lord integrates Himself within this scheme by posing as two figures, commonly referred to as the Son, Jesus Christ, the Messiah, (who summarizes "The Old Testament" in a succinct form), and the Holy Spirit, who is called the Mashiach within Judaism; what makes the Messiah different from all other Jews is his connection with The Lord, and this is due to the junctures in history when he appears.

Time operates within cycles, therefore, for time to begin anew, Love must continue from the completion of one cycle to the next; those who are righteous; "full of light"; obey the Ten Commandments; the Jews, make the next Age possible. A new Age begins when a story consisting of words is expelled, and history is once again formed by the fusing of words that form ideas; this is why Jews have the finest minds.

Each cycle, which can be called a "Day of The Lord", consists of light and dark; alternating oscillations of each; ascending and descending. A complete oscillation consists of 24,000 years. The synthesis of ideas contracts time, and these contractions happen at the same rate throughout every Age.

George Orwell gave a warning in his novel "1984" that if the tide of history does not change, people all over the world will lose their most human qualities, and will not even be aware of it.

"We shall squeeze you empty,
and then we shall fill you with ourselves."

"War is peace."
"Freedom is slavery."
"Ignorance is strength."

"But if thought corrupts language,
language can also corrupt thought."

"Big Brother is Watching You."

An Age begins when the previous Age is destroyed; history manufactures creation, (picture someone collecting the blocks of a tower that has fallen, and assembling them to form a different tower). What I have stated is the opposite of how most interpret the first chapter of Genesis in the Old Testament.

The evil perpetrated by Savage Primates that bring forth dissolution; chaos, and confusion, cements the plan orchestrated in the beginning.

Time does not move forward, but in reverse; (picture someone unraveling a chain, and the chain rolling back into ball); that means that yesterday is not the day before today, but tomorrow; everything people conceive taking place in the future, has already taken place in the past.

Savage Primates have not become more civilized over the course of history; a being that uses resources more efficiently due to the strength of his mind, but engages more in endeavors that do not enrich his mind, thereby facilitating the concretization of a primitive entity.

God is not disappointed with Savage Primates distancing themselves from nature due to breaking the Ten Commandments as time moves in reverse, but rather, appreciates the opportunity to become more aware of the magnitude of His Goodness, as time folds within the realm of space.

The universe expands, and contracts, as time oscillates up, and down, while moving in a circle. The stars can be viewed as signs, and symbols, as the book of Genesis details, and serve to record historical events in the sky; once more commonly referred to as the "heavens". There are no other planets supporting life in the universe; Earth is where the Trinity is formed, and there is only one Trinity; this is what monotheism means.

There aren't any new ideas; inventions, or innovations, within history; dates are assigned when they manifest, and are derived from the "Akashic records", which contain every thought, action, deed, and word, that takes exists.

In summation; the future is the past, the past is the future; nothing will happen that hasn't already take place; nothing can come as a surprise to The Lord, because everything that will happen has already taken place. He knows the future; the past, everything at once, and all in a single moment in time.

The mystic's moment of illumination shares with great poetry the liberating power of the deepest levels of consciousness. In the words of William Blake, "If the doors of perception were cleansed, everything would appear to a man as it is, infinite." Poetry, Wilson argues, is a contradiction of the habitual prison of daily life and shows the way to transcend the ordinary world through an act of intense attention-and intention. The poet, like the mystic, is subject to sudden ""peak experiences"" when ""everything we look upon is blessed."" W.B. Yeats, Dostoevsky, Gautama Buddha, Kazantzakis, Van Gogh, Rupert Brooke, Arunja, Nietzsche, A.L. Rouse, Jacob Boehme, Suzuki, and Edgar Allan Poe: their visionary understanding can generate awareness in each of us of our potential to open the floodgates of inner energy that creates mystic experience. Colin Wilson first received international acclaim in 1956 for The Outsider. ""Ever since I was thirteen, I have been obsessed by the question of the nature of mystical experience,"" he writes, and from that time he has been on a quest of the mystical in poetry, religion, and psychology." (Author unknown)

Merlin

Merlin stood upon the misty shore.
Waves dipped and rose in the distance.
This majestic picturesque scene was something he couldn't help but adore.
With a wave of his hand, he knew visions could be brought from light.
From heaven they came;
He knew not why, but he was sure they were right.
Thunder rolled through the grey clouds above.
A bolt of lightning struck down through the night.
A howl in the distance awoke the forest from its slumber;
Along with the rattling leaves hanging naked, and bare, on branches
Attached to tree trunks that soared upward toward
The sparkling jewels that light up the dark sky.
He heard a voice in the distance;

A song came to his mind that contained melodies
That entwined with each other in glorious ways.
He thought. I now know what will happen in future days.
One thing will lead to another; the chain is continuous; it never ends.
Within every cycle the love given, equals the amount taken.
Then all will begin again.

In the line of his sight the heavens appeared

Kabbalists; their aim is to achieve communication with, rather than absorption into, the En Sof, who di not enter the world directly, but through a series of 10 emanations, or spefirot, (spheres). Only through the sefirot, which make up inner reality, can God be understood, and can people move toward God.

The Jews

Jews transform people's perception of things; they incorporate elements into societies that weren't there before; they can disassemble whole belief systems, and replace them with something entirely different.

We are here, in the state we are today, due to an inability to recognize the value of the Jew, and his role within society, and, furthermore, the vast scheme of things. He is the protector, the preserver, and also the destroyer — "out with the old, in with the new".

It is quite extraordinary that the greater the numbers of Jews that have been provided to mankind, fewer are the numbers who are able to appreciate their accomplishments. The holocaust was meant to serve as a lesson no one would ever forget, and thus such horrors would never happen again.

"Triumvirate", presents a vision of nature that is entirely the contrary, in most cases the opposite, of how the vast majority conceive their surroundings; but I also explain the path that will lead us toward a sustainable future

Ode to Life

I have seen wonders, I have seen pain;
I have seen light on the earth once again.

Oh, how enlightening is the wisdom you seek.
No cares, worries, bothers;
The kind words the birds chirp, is what you greet.

A rabbit shoots from a ravine;
A predator follows the sight in his eyes;
What knowledge from this can we glean?

The only thing chasing you, are the desires you create in your mind.
All around you is so beautiful, rich, and dear;
Why not just sit back, relax, and enjoy a thick, dark, tasty, beer.

How many things do you really require?
The more you have, greater is the burden you shoulder.
The smaller your weight, swifter is your gait.
Mind, body, heart, and soul, become ever bolder.

The simplicity is stated in the fact, you are becoming more That.
Mysteries enfold, entwine; creating illusions fuelled
By the scenes that lie between.

You are complete; that you have surely been.
Is it your shadow, or the future, that guides the present unseen?

Be here, be now, and believe in God.
The Kingdom is there within you;
Merely watch carefully the things you do.
You are man, not beast, and you don't make the sound, Moooo!

Therefore, be as you are; the greatest thing by far.
Hallelujah! Now you can celebrate;
Have a drink with your mates in a Bar.

Life isn't serious, but quite hilarious.
Enjoy your problems, as well as your troubles;
they can only serve to enrich.

Stay clear of those that are nefarious,
Then grand, fantastic, will be your time on earth.
Your days will be spectacular, and marvellous,
Your heart will be overwhelmed with delight,
And nourished by a heavenly mirth,
That will make every day shine bright.

Nothing is this world is free;
Everything of value has to be earned.
Man, as The Lord instructed, must toil till the end of his days.
The greatest among men serve as examples to emulate and imitate;
They remind us that The Lord resides in each of us,
And if we simply strive to always do what is right,
He will grant us His sweet heavenly grace,
And bless us until we find our final resting place.

Missa Solemnis

Eulogy for Robert F. Kennedy

Your Eminences, Your Excellences, Mr President:

On behalf of Mrs Kennedy, her children, the parents and sisters of Robert Kennedy, I want to express what we feel to those who mourn with us today in this Cathedral and around the world.

We loved him as a brother, and as a father, and as a son. From his parents and from his older brothers and sisters -- Joe and Kathleen and Jack -- he received an inspiration which he passed on to all of us. He gave us strength in time of trouble, wisdom in time of uncertainty, and sharing in time of happiness. He will always be by our side.

Love is not an easy feeling to put into words. Nor is loyalty, or trust, or joy. But he was all of these. He loved life completely and he lived it intensely.

A few years back, Robert Kennedy wrote some words about his own father which expresses [sic] the way we in his family felt about him. He said of what his father meant to him, and I quote:

"What it really all adds up to is love -- not love as it is described with such facility in popular magazines, but the kind of love that is affection and respect, order and encouragement, and support. Our awareness of this was an incalculable source of strength, and because real love is something unselfish and involves sacrifice and giving, we could not help but profit from it."

And he continued,

"Beneath it all, he has tried to engender a social conscience. There were wrongs which needed attention. There were people who were poor and needed help. And we have a responsibility to them and to this country. Through no virtues and accomplishments of our own, we have been

fortunate enough to be born in the United States under the most comfortable conditions. We, therefore, have a responsibility to others who are less well off."

That is what Robert Kennedy was given. What he leaves to us is what he said, what he did, and what he stood for. A <u>speech</u> he made to the young people of South Africa on their Day of Affirmation in 1966 sums it up the best, and I would like to read it now:

There is discrimination in this world and slavery and slaughter and starvation. Governments repress their people; millions are trapped in poverty while the nation grows rich and wealth is lavished on armaments everywhere. These are differing evils, but they are the common works of man. They reflect the imperfection of human justice, the inadequacy of human compassion, our lack of sensibility towards the suffering of our fellows. But we can perhaps remember -- even if only for a time -- that those who live with us are our brothers; that they share with us the same short moment of life; that they seek -- as we do -- nothing but the chance to live out their lives in purpose and happiness, winning what satisfaction and fulfilment they can.

Surely, this bond of common faith, this bond of common goal, can begin to teach us something. Surely, we can learn, at least, to look at those around us as fellow men. And surely we can begin to work a little harder to bind up the wounds among us and to become in our own hearts brothers and countrymen once again. The answer is to rely on youth -- not a time of life but a state of mind, a temper of the will, a quality of imagination, a predominance of courage over timidity, of the appetite for adventure over the love of ease. The cruelties and obstacles of this swiftly changing planet will not yield to the obsolete dogmas and outworn slogans. They cannot be moved by those who cling to a present that is already dying, who prefer the illusion of security to the excitement and danger that come with even the most peaceful progress.

It is a revolutionary world we live in, and this generation at home and around the world has had thrust upon it a greater burden of responsibility than any generation that has ever lived. Some believe there is nothing one man or one woman can do against the enormous array of the world's ills. Yet many of the world's great movements, of thought and action, have flowed from the work of a single man. A young monk began the Protestant reformation; a young general extended an empire from Macedonia to the borders of the earth; a young woman reclaimed the territory of France; and it was a young Italian explorer who discovered the New World, and the 32 year-old Thomas Jefferson who [pro]claimed that "all men are created equal."

*These men moved the world, and so can we all. Few will have the greatness to bend history itself, but each of us can work to change a small portion of events, and in the total of all those acts will be written the history of this generation.*It is from numberless diverse acts of courage and belief that human history is shaped.* Each time a man stands up for an ideal, or acts to improve the lot of others, or strikes out against injustice, he sends forth a tiny ripple of hope, and crossing each other from a million different centres of energy and daring, those ripples build a current that can sweep down the mightiest walls of oppression and resistance.*

Few are willing to brave the disapproval of their fellows, the censure of their colleagues, and the wrath of their society. Moral courage is a rarer commodity than bravery in battle or great intelligence. Yet it is the one essential, vital quality for those who seek to change a world that yields most painfully to change. And I believe that in this generation those with the courage to enter the moral conflict will find themselves with companions in every corner of the globe.

For the fortunate among us, there is the temptation to follow the easy and familiar paths of personal ambition and financial success so grandly spread before those who enjoy the privilege of education. But that is not the road history has marked out for us. Like it or not, we live in times of danger and uncertainty. But they are also more open to the creative energy of men than any other time in history. All of us will ultimately be judged, and as the years pass we will surely judge ourselves on the effort we have contributed to building a new world society and the extent to which our ideals and goals have shaped that event.

The future does not belong to those who are content with today, apathetic toward common problems and their fellow man alike, timid and fearful in the face of new ideas and bold projects. Rather it will belong to those who can blend vision, reason and courage in a personal commitment to the ideals and great enterprises of American Society. our future may lie beyond our vision, but it is not completely beyond our control. It is the shaping impulse of America that neither fate nor nature nor the irresistible tides of history, but the work of our own hands, matched to reason and principle that will determine our destiny. There is pride in that, even arrogance, but there is also experience and truth. In any event, it is the only way we can live.*

That is the way he lived. That is what he leaves us.

My brother need not be idealized, or enlarged in death beyond what he was in life; to be remembered simply as a good and decent man, who saw wrong and tried to right it, saw suffering, and tried to heal it, saw war and tried to stop it.

Those of us who loved him and who take him to his rest today, pray that what he was to us and what he wished for others will some day come to pass for all the world.

As he said many times, in many parts of this nation, to those he touched and who sought to touch him:

Some men see things as they are and say why.
I dream things that never were and say why not.

By Edward Kennedy, brother of Robert F. Kennedy

Robert F. Kennedy

Bobby was a man people admire to this day.
One day in 68, he met his cruel fate.
Shot by Sirhan, or maybe it was Dan,
Holding the greasy pan.

No one till this day truly knows
What happened during that shooting rampage,
That brought his life to a close.

Such an enormous shame, indeed;
He still had so much more, he could offer to the poor.

Tragically, today we view him lying there;
Head held in the hand of another,
Losing his life's blood on a dirty, grimy, floor.
This whole scene, I deeply abhor.

Think hard, I beseech you, about his life,
And all the wonderful things he had to say.
On podiums he stood, leaning out doors to shake hands.
Greeting the people, letting them know,
Despite all his privilege, intelligence, and innumerable gifts,
He was still just like them;
at times vulnerable, weak, and sad to the core.

Like many,
He knew the tide of the times was changing;
But ripples of hope he sought to create,
So that one day all would be free,
And then never could there be a date
When man would endure a miserable fate.

Today, we need another Bobby.
Is he out there somewhere?
I hope so!
I could make the search an honourable hobby.

Words inspired by the life of
Robert F. Kennedy

The only chance we have of surviving is
By remembering our humanity.
Our fellow man is also our brother.
We must care for others.
As much as we care for ourselves.

Part One

Psychopathology of the Masses

Many things people believe are good are the opposite.
Opposing such things as sexism; racism, and discrimination,
And supporting, equality, is considered good by the common man.
The truth is the contrary, such terms are used to homogenize society,
Which promotes consumerism by making people ignorant because they fail
To judge, and discern, resulting in them becoming attached to inanimate,
Rather than animate, objects; to materials instead of Spirit,
Consequently, the Tower can no longer be held together,
And falls.

When the Jews departed, the light within the world went also,
and then people gave into their nature; cheating, stealing, and
deception, can only scramble the mind, and we see outward
expressions of this in the manner people behave.

The arts, in every form, have been molded to fit the steepening
decline of man's mental acuity. Popular novels are increasingly
written using shorter, more simplistic, sentences, which is a
reflection of man's diminishing capacity to focus, and join
thoughts to form ideas.

Chapter One

"We suffer from evils which we inflict upon ourselves,
And we ascribe them to God who is far from connected with them."
Moses Maimonides

Mass Psychogenic Illness

Nazism was not eradicated after the Second World War; the disease spread, and flourished, around the world due to programmes such as "The Paperclip Project".
The solution, the way to insure it is eradicated, once and for all, is to expose the whole truth about the Nazi Party; the extent of the atrocities that occurred while Adolf Hitler was Chancellor of Germany

The result of brainwashing is that people end up doing the will of someone else; their actions can contradict what they say, while the person believes there in conformity between the two. The mind adapts to this phenomenon by "compartmentalizing"; what this means is that a person can commit an absolutely brutal act, and moments later appear composed; he may appear composed, and serene, while committing the act of brutality.

Video footage of people labelled extremist committing atrocities exposes this psychological process because of the dichotomy between their actions, and their affect. Islam is commonly believed a form of religion; if such is the case, one is either belongs to it or not; hence, what is meant by the word "extreme"? Most of those considered "radicals", or "extremists", are not behaving in a manner that is contrary to the ideology in the Koran, hence, one can conclude that Islam does not encourage moral behaviour, and is, therefore, a "death cult", because it promotes violent behaviour.

Anybody who reads the Koran would realize that it encourages people to commit the barbaric exploits we see going on in many regions of the world; the heaviest concentration being in Muslim countries

Muslims claim the vast majority of them are not in any way involved in terrorist activities, and are in fact, they claim, peace loving people. This is an absurd argument considering the fact that when there is a war between nations, for example, not everybody is on the battlefield; likewise, in order for the select, the soldiers bold enough to state their whole hearted devotion to Islam, to commit terrorist acts, they require support from other Muslims.

In order to brainwash a person it is important to insure that he/she feels unsettled, and insecure; (filled with fear, and doubt), the Monolith has been successful in utilizing Islam in order to achieve this effect.

One of the duties of the State is that of caring for those of its citizens who find themselves the victims of such adverse circumstances as makes them unable to obtain even the necessities for mere existence without the aid of others. That responsibility is recognized by every civilized nation…To those unfortunate citizens, aid must be extended by the government – not as a matter of charity, but a matter of social duty.

FRANKLIN D. ROOSEVELT, 1933

The American mafia took advantage of Britain's weakened condition after She almost single-handedly saved the world from German Fascism. The mafia encouraged members of the Nazi intelligentsia to reside in America, as well as many other countries, after WWII, and have they have since used their knowledge, training, and expertise, to remain the world's superpower, and become extraordinarily wealthy.

The mafia continues to have an alliance with Germany till this day, and use the puppet known as, Angel Merkel, to achieve their objective. She has destabilized Europe by allowing swarms of immigrants, primarily from the Middle-East, into Germany; while making sure other countries must do the same. She repeatedly proclaims that Germany's social/economic/political, problems stem from immigrants refusing to learn German, and Germany's culture; obviously, her actions are not agreement with what she says – yet she was nominated "Person of the Year" for 2014 by Time magazine; this is proof that the masses have been afflicted with a psychogenic illness; their words and actions are not in harmony, and, thus, destructive; not just to themselves, but others, and also the nation they are a member of.

The people entering Europe are from dissimilar backgrounds; they were not been brought in cultures that respect, and honour, the values, traditions, people in Western Europe countries are taught to cherish, and furthermore, they have neither the ability, nor the inclination, to develop skills that can better people's lives, and, quite obviously, if someone is unable to support himself by legitimate means, and the State doesn't support that person, he may resort to feeding off others to survive. The most unfortunate reality we must all face is that immigrants have no reluctance practising cannibalism in order to survive.

The most devastating impact of this evolution of events is that Europe, the cultural heartland of the world, is denied the role She must play, and that is to make sure the world continues supporting life, which requires new ideas that enable the world's population to successfully adapt to changing circumstances. The most extraordinary people emanate from Europe, and the greatest concentration of these fine minds has been in Britain.

The forces, and energies, responsible for this occurrence originate within Britain; they seep outward in the same manner waves spread across water; the further they emanate from their source; weaker they become; it was not happenstance that so many in Holland, the Netherlands, and Norway, went to such extraordinary lengths to protect Jews during the Second World War, while, on the other hand, Russians troops thought little of butchering people in massive numbers once they got the upper hand during WWII, and began their march toward Germany.

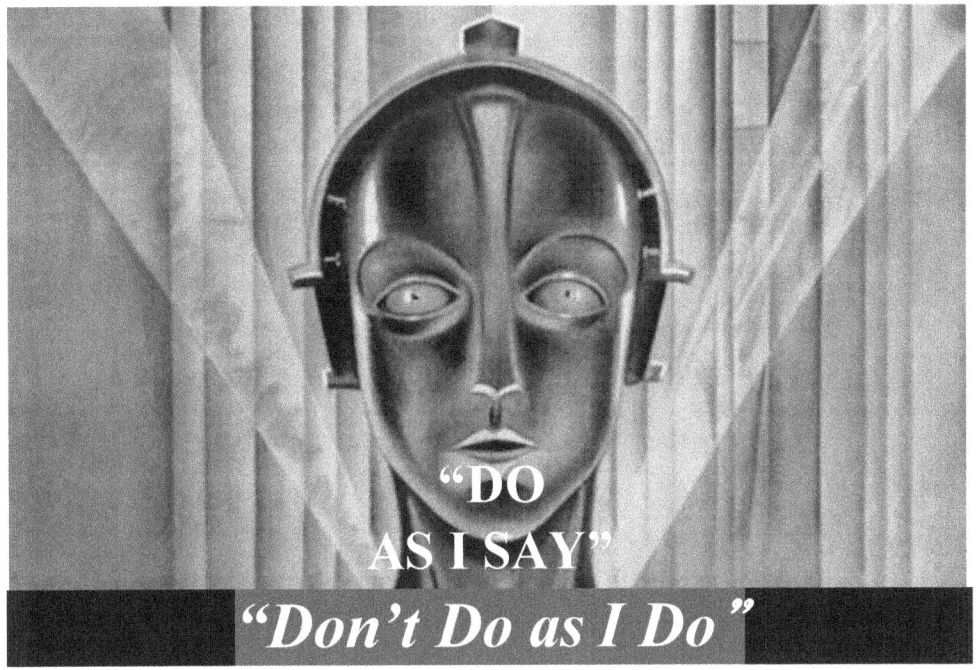

"DO AS I SAY"

"Don't Do as I Do"

In all honesty, does anyone think such a declaration could win a Politian a vote; logic dictates the answer should be; no.

Angela Merkel, "super mind physicist", defies logic; she repeatedly states that Germany's continuing problems are due to the nation's immigrants not learning German, or German culture, but she has welcomed a million more into Germany; obviously, they can only serve to further drain the economy, , thus, jeopardising the wellbeing of all the people who live in Germany, yet she remains Chancellor, which proves that she, and the German public who voted for her, and continue to support her, have been brainwashed; they are automatons following the directives of the Monolith; the mafia families that control the world.

Accepting immigrants while acknowledging they are unable to make a contribution to the countries that support them, is self-defeatist, never mind, kamikaze; accepting more immigrates is not a humanitarian act, but, rather, a colossal error in judgment that undermines the security, and well-being of not only Europeans, but people all around the world, as well as every other species on Earth.

The objective behind inviting immigrants into Europe is so the Monolith can brainwash the masses. The threat people face is not terrorism, but the people who do whatever they're told; no matter the cost to others, or how many they might be.

Fortunately, although, Britain has more foreigners on Her land that ever before, She is now more secure than at any other time since the conclusion of the Second World War; the means, and resources, used by the Monolith to control people is now fully understood; all the dots have been connected; the picture is complete; following is a synopsis of the scheme put in place by the Monolith that has made them the richest, most powerful, people on the planet.

Those who are labelled as vulnerable in Britain are located by the police, and medics, and told there are places of refuge, but, first, they are told they must have an assessment, so "caregivers" will know how best to help them.

The true objective of the assessment is to acquire a person's I.D., and as much background information about the person as possible; in other words, I.D. theft; the I.D. is then used to open bank accounts to hold funds that support organizations that de-stabilize the Middle-East, and result in people fleeing to First World Countries, and destabilize those countries.

The victim of I.D. theft, is also stripped of all personal wealth, and then denied a home; continual exposure to the elements; the stress of not having a secure, protected, abode; abuse; physical, psychological, and emotional, from the general public, and tainted food provided by the referral centers, hasten his death. Once he dies there is no longer any possibility he can interfere with funds transferred by using his I.D. The person is viewed as an object; a dirty piece of trash, no different from someone belonging to the "untouchable class" in India, who die on sidewalks, where their remains are picked up in a wheel barrow, and then dumped in a nearby river, (the difference between what takes place in India, and Britain, is that a person in India may rent space on a sidewalk that he can use it as his home; in Britain the homeless are taunted by the police, and repeatedly told to move, and threatened with a charge of loitering; trespassing, or "disturbing the peace", if they don't.

The equipment, and propaganda, used to support terrorist organizations flows in, and out, of the Britain, among other countries, while police watch; they stand guard with guns, and fake "sniffer dogs", so the public doesn't suspect anything pernicious is going on; especially right under their noses.

The sick, perverted, propaganda used to persuade people to get involved in terrorist activities is spread by various means, including Britain's public libraries.

We now live in the world George Orwell envisioned in his book, "1984"; everything is the opposite of how it should be; the bizarre, is now the norm. I would find this all comical, if not hilarious, if it weren't for the fact that so many people suffer, and die, because of the hideous, barbaric, scheme the Monolith implemented shortly after the Second World War.

What I've described should be self-evident due to the technology, and media, available now; people can access the media through numerous types of gadgets they can carry on their person wherever they go; what I mean is; even if a Politian hasn't been exposed as a fraud in the media, shouldn't the public, as technologically savvy as they proclaim, especially the younger generation who were brought up using technology as learning tools, be able to figure out whether Politians are doing what they said they'd do, or whether they are attempting to deceive the public into believing they're doing something they're not.

Did Angela Merkel commit fraud when she decided to allow immigrants into Germany despite stating that this would be detrimental to the German economy, and German culture? Or did she do what was required to acquire the power so she could do whatever she pleased? In other words, has she admitted she's tricked people, and, thus, exposed how evil she is? Or does she still present herself as good, although, she's endangered the masses by allowing "Tribal Man" to live side by side with "Rational Man"?

The answer is, NO, to both questions; although she's the Chancellor of Germany, the person with the most power in that country, the public had the right to disregard her instruction to open the floodgates to immigrants; possibly allowing as many as one million into Germany; but if the general public agrees with her that immigrants are the cause behind all Germany's woes, why would they shoot themselves not just in the foot, but both arms, and legs, as well, figuratively speaking.

The reason why this has happened is because people are brainwashed; they are now programmed automatons that facilitate themselves continuing to behave as robots by bringing upon themselves the ingredients that take away free choice and free will; their faculty of reason.

This the process began, and was initiated in a covert manner; in Britain's case, Margaret Thatcher; whom the public voted to be their Prime Minister for three terms, although, while in power it became increasingly obvious that the policies she was responsible for instigated, and claimed make the British economy stronger, and less reliant on foreign powers, were not having the affect she assured the British public they'd have.

On a daily basis we see evidence of people doing the opposite of what they say they'll do; but, if everybody behaves this way, why would people bother to lie? The answer is that in order for disruptive; destructive, forms of behaviour to perpetuate, people must believe they are moral, and good; this ensures that a person isn't burdened with guilt, or shame, which would hinder aberrant; immoral, and criminal, behaviour from continuing.

Western European governments tell the public they are working hard to combat terrorism at home, and abroad, however, the opposite is happening; resources that are allocated with the intention to undermine terrorist activity are actually used to facilitate the growth of terrorist activities; anyone can view the British government's statistics to verify this is the case; therefore, those given the most power, and resources, to prevent terrorist acts, use the power, and resources, the have to ensure terrorism continues; I am referring specifically to the police.

On a daily basis practically all of us encounter situations where people do the opposite of what they say they're going to do, or what we know they are supposed to do, and no matter the cost incurred by another person by doing so, the person is considered good, and the person who declares he's been inconvenienced, or worse, by this behaviour, is bad; there are other names that could possibly spring to the minds of those who have not done what they were asked to do; "rude"; "anti-social"; maybe four letter words; quite often there is a penalty for simply stating that one is not pleased with this behaviour; "whistle blowers" are quite often become victims of crime.

What I've just described sounds like a comedic play, something written by Monty Python; Morecambe and Wise; possibly Benny Hill; I would find it quite hilarious if it weren't for the human toll involved in this equation; in fact the intent of this behaviour is that people can be grievously wounded, if not killed, and the people doing the damage aren't aware they're doing it. How is it possible that a person can be contorted into such a hideous beast?

The most crucial element required to make a person follow orders, no matter what they might be, is to take away their free will, and free will; this deprives the person of the capacity to even question the directives he's given; therefore, it is no longer a matter of choosing to do something that is bad, because the person fails to even be aware it's bad.

The way to make people so ignorant, they don't know they've ignorant, is to weaken the faculty that enable people to be conscious, and aware; the less one uses the mind, the weaker it will become.

In this "age of information", information is at a person's fingertips via various forms of technology; this may make a reasonable person conclude that the world should be full of the wisest; most well informed people ever; the truth is the opposite is in fact the case.

Man, in his natural state, is like every other animal on the Earth; he must struggle to survive, and this works to his favour throughout his life; when his body weakens, his skills improve due to usage; the mind strengthens as the body succumbs to the toll of aging.

The Monolith, following the Second World War, designed the world so that some countries would be "consumers"; consisting of a population that barely works as they make a living, while other countries are "producers"; where people have to work excessively hard to acquire the bare necessities of life. One can conclude that a portion of the world's population feeds off another portion; this is, by definition, cannibalism.

If people in "consumer" countries were aware that their lifestyles are possible at the cost of human lives, they'd find it difficult, and that's putting it mildly, to consider themselves civilized people. Once people are ashamed of their actions, they may then feel tempted to change the scheme of things, which would counter the Monolith's plan.

What the Monolith has done to make sure this doesn't transpire is to distance, disconnect, people from the monstrous crimes they perpetrate; however, whether a person is aware of the harm they've done, or not, does not negate the fact they're responsible for the deeds they've performed; ignorance cannot be used as a defense.

When criminals, and their victims, live in separate countries, the task is simple, but the case today is that due to the Monolith's scheme operating as long as it has; we now live in a "global village", run by an economic system called "globalization", the victims, the people being cannibalized, are in the same countries as those e who cannibalize them; but how can one ignore a crime when the evidence of it is literally, and figuratively, speaking, in your face?

The only way is to make sure those inflicting the damage are not aware they cause it, and that way they can continue to view themselves as, good, even if they are the direct cause of a person's death.

The Monolith does this is by giving people money they can spend, while making them believe they've earned it, which requires work, but if people worked they'd have the chance to wizen up, so the solution to this quagmire was to make up job titles, and make sure that no matter what transpires in the "work place", nobody feels the compulsion to do something better; this explains why grievances are not dealt with today; people don't realize that they are the cause of the grievance.

The Monolith, therefore, must not only give people money to spend, but people must believe they've earned it, and, thus not have the inclination to view themselves as incompetent, and negligent; certainly not to the extent others could possibly be harmed as a result.

People are given the impression they're responsible for their wealth, but people would have to be talented, and intelligent, for this to be possible, which would entail having a strong mind; if that were the case they would be able to question what's going on around them; in order to avoid such a scenario, the Monolith, therefore, decided that a sector of the population in the "consumer" countries would have their rights denied; such as the right to have a home, (a safe, secure, protected, environment to live in); as well as dental, and medical, care; these people are called "marginalized".

The funding that should have gone to providing these services isn't wasted; the I.D. of the marginalized is used for money laundering; social assistance cheques are issued using the stolen I.D.; the cheques are then directed to a town council, for example, and the revenue is then used to make up the pay-checks for the people who have meaningless jobs with impressive sounding, but, senseless, titles. The public is disconnected from the fact that these funds are actually the proceeds of crimes, and they have been solicited in the act of murder.

Initially the Monolith made people believe people are homeless because they're ill, or somehow the situation they're in is their fault; events have evolved to the stage, however, where people are so disconnected from their fellow man that they don't care why people are homeless, and they've not bothered by the suffering they see. The common "consumer" now behaves the same way as the people who worked in the Nazi concentration camps, and inflicted pain, hardship, and suffering, on such a scale it's difficult to conceptualize, but at the same time believed they were good people, and had the capacity to enjoy their lives.

Britain no longer has shelters; there are no places of refuge for people down on their luck; the referral centers that claim to refer people to these non-existent shelters actually steal their I.D., and deprive them of all the services they have the right to receive due to being citizens; medical care; dental care; eye care, etc.. Most die within weeks due to the combined effect of being exposed to the elements; stress; weak immune systems, and consuming food full of toxins given by the referral centers.

In such a society laws, obviously, do not exist, and due to people being the opposite of how they present themselves, one can conclude that the police are the primary orchestrators of this diabolical scheme that eradicates a portion of the population.

People are told it's not fair, or proper, to judge others in a multi-cultural society because each should be able to live his life in accordance with the standards, and values of the cultural/ethnic background he's from; doing so makes a person, good, or so they're old, because one is not discriminating, which is bad thing, however, when one adopts this mind-set one fails to measure the value of anything relative to anything else, which happens to be the natural thing one should do, because it strengthens the mind.

Technology also affords people the luxury of being told what to do no matter where we are, which can only enhance a person's reluctance to think for himself; "mind the gap"; one hears at every train station in Britain; now you don't have to take the time to remember there's a space between the train and the platform; "Surfaces can be slippery when wet"; no kidding; now you don't have to figure that out for on your own; "For safety reasons, don't allow children to run in bus stations"; that's a good idea; why didn't I think of that? Caution; automatic door ahead; don't expect anybody to turn the knob, and open the door themselves.

Here's a warning for all those who fail to open doors for themselves, but instead insist on pressing a button whenever possible; eventually you won't know how to open a door if there isn't a button to press; automatic flush toilets mean that when people become accustomed to using them, a toilet will remain un-flushed when it fails to flush itself; people return to their primitive hunter/gatherer ways when they lived among trees in the forest; they swat, defecate, and go; if they can't move on for some reason, they're content, the same as other primates, gorillas, in particular, to stand; sit, and sleep, in their own faeces. Britain is no less monstrous an island to live on than the one inhabited by Dr. Moreau and the products of his failed medical experiments; others might see people that appear somewhat normal, but due to being acquainted with the Thatcher Illusion, I know that behind every overly solicitous persona; "Hi, where are you from? Where are you from, originally?" The Londoner asks a complete stranger; there lurks a hideous deformed creature; a product of laboratory run by the Monolith.

People have reduced themselves to such a level they view others as disposable objects they can use as they please, and discard whenever they like.

Since arriving in this land, that has produced some of the greatest writers, composers, and scientists, the world has ever seen; I am proud to state that having studied the history of this land, and having worked to instil the best of the culture within myself, that I consider myself as one who represents Britain; if one mistreats me, he also disrespects Britain.

The British Isles are now overrun with foul smelling rodents that leave a path of destruction whenever they travel; they can commit a criminal act in public, without a concern that someone will intercede.

In order to prove the police are behind the genocide taking place within Britain; they are key participants in every phase of its execution, I travelled to the town of Plymouth, where, as the internet indicates, I have an address, and I was on the electoral role in the year 2009.

In entered the town's job center; known as "Job Center Plus", located at; Old Tree Court, 64 Exeter Court, Plymouth, PL4-0FJ, on Monday morning, on, February the 1st, 2016, at approximately 9:30am, and approached the reception desk, and said that I'd once lived in Plymouth, and would like to access my file in order to obtain some information for my records. I was told take a seat, and a floor manager would see me shortly.

I strolled over to where a number of cushioned seats were located, and I was just in the process of making myself comfortable, when a dark haired man, likely in his fifties; first name, "John", according to the card attached to the necklace, approached, and asked how he could help me. I repeated what I said earlier at the front desk, and John replied; "I can only access your file if you have your National Insurance Number"; since a file exists, there must be a N.I. # associated with the file.

"I can provide that", I answered.

I walked over to his desk; the first in a series beside a wall covered with misty windows. I noticed a camera attached to the ceiling not far from the desk; I now knew that all that would next transpire would be recorded, and could not be refuted in the future.

I uttered a few letters, and numbers, and John replied that the series I'd given him didn't contain the right amount of letters or numbers. I then declared that I wasn't certain at that time what my number was, and asked why it was essential I know it in order to have access to the file.

He asked for my I.D., and handed over my passport - according to the internet there are only two Nigel Shindlers in Britain, and only one, Nigel

Edward Swindler, that would be me, and the only one address under my name, and that's in Plymouth, Devon.

"Because you can't remember your N.I. #, I'll have to ask a number of security questions"

"I have no problem with that; fire away", I answered, "You already have my I.D, though", John looked up from his computer; clearly not pleased with my last remark.

"Well; go ahead", I said; which appeared to agitated him even more; his pupils got smaller, and perspiration now appeared on his forehead; drops of sweat were sprinkled close to his hair line.

John continued staring at me, with squinted eyes, before he said;

"I need to know your full address".

"Unfortunately, I'm not certain of the number; I can give you every other detail of the address; my landlord's name, but I can't remember the street number."

"Sorry, I can't help you."

"That's ridiculous; I can't answer that question; give me another; banks have a series of questions they ask to verify your I.D; so should you." The remarkable thing is I'd already shown him my passport.

"What is it you see in my file?" I asked in order to expose a lie; a social worker from Deptford Reach in London was previously in contact with the Plymouth Job Center over the phone in my presence, and she was told that the only way someone at the Job Centre could gain access to the file was by me providing either the full address, or the reference number.

"I see you had a communication with someone in 2009".

"I returned to Canada in 2008; that's not me."

"That is someone I communicated with in this office."

"What's that got to do with me?"

"You applied for your National Insurance Number, September the 15th, 2015."

"That's a lie; you were in contact with Deptford Reach, September 15th.I mailed a form to the Home Office on September 24th, asking for a trace on my National Insurance Number."

"I can't help you."

"Why?"

"I'm not happy."

"I'm miserable today; why should I care whether you're happy or not?"

"Because of the two communications I had, I can't help you."

"What has your communications got to do with me? It's my file; not yours. What else do you see in my file?"

"Your reference number?"

"Fine; I'll take that", I said; giving him the impression I was eager to get out of there.

He tore of a portion from a sheet of paper, and wrote down the reference number; as soon as he handed it over to me, I said;

"Thank you; when you spoke with the social worker at Deptford Reach - which was recorded - you stipulated that one of two things is required to access the file; my National Insurance Number, or the reference number; here's the reference number." I then gently placed the piece of paper with the reference number written on it in front of him.

John promptly snatched up the piece of paper; tore it into shreds, and dumped the pieces into the trash can behind his desk. I could tell from the smirk on his face that he relished what he was doing.

"You have to leave now."

"I have every right to be here."

I then turned around, and spoke to a woman who was watching us a short distance away; "Could you tell me what time it is?"

"Ten to ten", she said.

"Great; I entered this place about ten minutes ago; the time would have been twenty to ten."

John motioned with a hand for security to come over; two men; one had tattoos on both arms, quickly stomped over, and stood immediately behind me; trying to be appear as intimidating as possible.

"I need a reason."

"You need to leave." I was told by one of the security men.

"That's fine; just tell me the reason, and I'll take it to the police." I got a piece of paper, and I had a pen in my hand; prepared to write down what John had to say, when, all of a sudden, both security men came after me; one grabbed the bags out of my hands, and handed them to John, while the other man-handled me, by grasping my arms; I would assume they were making every effort to intimidate me. I let them do whatever they pleased; the only thing that was important to me was that it be recorded, (what is taking place is a murder; my National Insurance Number has been used by the town of Plymouth as a source of revenue since my departure in 2008).

As soon as I was outside the building, I made my way toward the nearest police station. I'd only taken a few steps before I noticed a police community support officer walking toward me; what transpired next proves that there are no laws operating in Britain at this time; the people who are supposed to uphold the law, such as the woman heading in my direction, are wolves in sheep's clothing; ghouls masquerading as human; something, quoting a Mary Shelley's book, Frankenstein; "More abject than a blind mole or a worm."

She was an elderly woman; probably in her sixties, but I should think, due to her profession; younger.

"Excuse me; I'd like to report a crime; I've just been assaulted by two men in the Job Center; it was all caught on C.C.T.V."

The woman took a note pad out of a pocket, and asked for details. I said the culprits were two security guards; both were wearing white shirts; one had tattoos down both his arms. I told her I wasn't given a reason why I should leave before the two guards grabbed me, and physically ejecting me from the building. I told her I wanted to charge both with assault. She said, "O.K.; then you'll need the C.C.T.V. footage".

She went into the building, and began talking to members of the staff. I stood adjacent to the front doors until she came out of building a couple of minutes later.

"A police officer will be coming down", she said, before asking for my personal details; name, and date of birth. She began sharing these details with someone she was speaking with on the phone that was resting on one of her shoulders, when I overheard her give an incorrect spelling of my name; that was when I said; "My last name is spelt without a C; it's, S.H.I.N. D.L.E.R".

"Sorry; I have dyslexia."

"My God, then you're not qualified to do the job you do; your errors can cause a lot of problems." I knew she'd just told a lie; people with dyslexia quite often read letters backwards; a "d" may look like a "b", for example, but there's no reason why a "c" would be added to a word because of dyslexia.

"I find that offensive; we have discrimination laws in this country; you need to be careful about what you say."

"You can sue me for libel; what I said is true; errors such as misspellings of names in your profession can cost people's lives. Nurses are in the same position; if they misspell a prescription, a person can die."

"You are really being offensive." She said in a flat tone of voice that failed to convey that she was upset at all.

The woman I asked for the time while inside the Job Center joined us as I stood beside the community officer directly adjacent to the two front glass doors, and apparently she was offended by something I'd said also; "I have dyslexia as well; what you're saying is discriminatory."

If she did have dyslexia, then she would have known that the community officer had fabricated a lie about the miss-spelling of my name, and she had decided to join in.

"No it isn't; it's a fact of life, and I have the right to offend; that's not a crime". The community support police officer; badge #; 30324, then asked the young woman; "Could I have your name please?"

The young woman with straggly, grimy looking, dark hair, had a grin on her face when she answered, "Certainly!"

"Wow; you intend on charging me"; I said to the community officer.

"That's right", she said with a malicious grin beaming from her face.

"You've got to be joking! That's fine; you try and make up a charge against me; I'll turn it around in court." I asked the young woman who claimed she had dyslexia, and I'd offended her, for her name.

The community support police officer then interjected, "Don't give your name!"

"I have a right to know her name; she's made an allegation about me."

"Say nothing", #30324, commanded. The young woman kept her mouth shut, for the same reason a rat in a cage runs on a metal wheel until a food pellet is dispensed that entices the rat to leave the wheel in order to eat it.

"I'm going."

"No; you can't go until the police officer arrives."

"Go fuck yourself."

I proceeded to stomp up the hill, thoroughly disgusted, and appalled, by the people I'd encountered, but also thankful that I'd hardly slept on the night train coming to Plymouth, and hadn't had the time to eat breakfast, because I may have otherwise decided to defend myself in the Job Center, rather than allow the pasty faced; overweight; smelly, security guards manhandle me; I genuinely felt "dirtied" by their skin touching mine. Evil emanates from those who don't have a soul; it's no mystery to me why I rebuke people who have turned to the dark; I can hear admonition in the harsh tone of my voice. Dark opposes light; this not only makes sense; it's physics.

I continued walking, until I heard a voice behind me say,

"Can I talk with you?"

I turned my head, and saw a police officer a few steps behind me; I said, "No", and continued to walk. Shortly after something happened that should never occur in a civilized society; the officer grabbed my shoulder, and jostled me as if I was a rag doll, then planted his body in front of mine to block my path.

"If you don't want me to put you in lock-up for 24 hours, you're gonna go over there"; he pointed to doorway a few metres away, "and talk to me." It was evident that he wasn't a rational man; someone with an intact faculty of reason, so I walked over to the place he indicated.

"What's your name?"

"Have I committed an offense?" If I hadn't committed an offense, I'm not obliged to give personal details to the police.

"You've offended my colleague", said the officer with number, 13547, on both shoulders; at that point he'd just confessed to a crime, and it was hard for me not smile with satisfaction.

"I want your name."

"I gave it to the community officer". I could tell that my refusal to help him out of his sticky situation wasn't making him happy.

"I want you to give me your name."

"You don't need it; I gave it to community officer."

"You said that she has dyslexia."

"No I didn't; she told me she has dyslexia."

Who should then come trotting up but community officer, 30324, who gave the impression of someone who'd be far more comfortable at home sitting in an easy chair, watching telly, with a hot water bottle nestled next to her tummy to keep her company; hardly the type of person you'd expect, or want, to help out if you were being mugged in a back alley by a couple of thugs.

They both realized, because they're programmed to, that their conduct could get them into trouble, so they were trying to find a way to squirm out of the situation they'd made for themselves; neither showed the slightest bit of interest in serving, and protecting, a member of the public – that would be me - but rather looking after their own interests; defending the guilty.

"You offended me by saying I'm not qualified because I have dyslexia." I was thinking then that quite often when my father encountered a situation as ridiculous as the one I was now facing he'd say; "I had no idea you're a delicate flower."

"I want you to apologize." Officer 13547 demanded.

It was hard for me to believe I was dealing with these dim witted fools; I thought they belonged in some freak show; a Ripley's Believe it or Not, museum; they shouldn't be walking about giving the impression they have something in common with a person who has a brain. All I wanted at that juncture was to get away from them as fast as possible; I felt soiled just being in their presence.

"I am sorry you were offended; that wasn't my intention". I wasn't prepared to take back what I said, but I was genuine about not intending to offend her. If she knew anything about me, and the things I write about, she'd have known that I don't consider dyslexia a disability; some people's minds work differently from others; that doesn't make a person special, in my opinion, or unique; it's just the way they are.

The singular motive driving their actions was to make sure each believes they've done nothing wrong, and, thus, nothing to feel ashamed, or guilty about, and, of course, correspondingly, be held accountable for.

There is no planning involved in their behaviour; they merely response to stimuli; neither is capable of developing any insight into their behaviour, and because they conceive no harm emanating from their actions, they see no reason to indulge in introspection.

Satisfied she's in the right, and that she's fulfilled her duty as a police community support worker, #30324, then toddles along; her rear end sashaying from side to side, reminding me of a hippopotamus, and resumes her daily routine doing nothing while appearing to be doing something; for this she gets paid a handsome wage.

Her job is to support the police, not the community, as her title seems to indicate, and as any well informed person in Britain knows the police make no effort to support any community.

She, #30324 is a "walking contradiction"; she's paid to parade about in a uniform meant to dissuade people from getting involved in criminal behaviour, while she engages in criminal behaviour herself.

Officer, 13547, was no doubt successful in convincing himself he didn't commit a crime once I gave him the impression I was sorry the community officer's feelings were hurt when I said I believed she wasn't qualified to do her job; actually, all I said was that it wasn't my intention to hurt her feelings; that doesn't imply that I regret anything I said, or did.

The police officer's next task was to get the two security officers off the hook; that meant erasing the incriminating video footage.

"Now tell me what happened at the Job Center."

I was quite certain nothing would be done about the behaviour of the guards; if the matter were to go to court, and the case was successful, what would most likely follow would be a class action lawsuit of mammoth proportion that would not only expose the demonic nature of the people who participate in these schemes, but also exhaust Home Office's revenue due the settlement each case would deserve.

I next appealed to his non-existent capacity to empathize; I declared that I wanted to access my file because there were things I needed of deep sentimental value. It was a pathetic sight to see him place a hand on his heart as a way to convince me he was capable of some level of compassion; the people he kills as a member of Britain's police force represent nothing more than a number to him; objects he values as much as the foil used in a candy bar wrapper.

I asked for his name, and number; he, in turn, asked for a piece of paper, and a pen, to write them down. I told him that wasn't necessary; he could just tell me what both were; at that point he suddenly appeared to be deeply troubled. I suspect that what was going through his mind was the suspicion that I had a recording device on me, and I would expose to the public all the hideous nonsense that spews out of the mouths of those involved in the profession of policing in Britain.

"No problem, I can do that right now", he said while running away.

"How do I get them from you?" I yelled out as he bolted in the direction of the Job Center

"Go to the station later on today." He yelled out, while turning his head to look at me.

I never thought there would come a day when being a coward would be a prerequisite to becoming a police officer in Britain.

I went directly to the police station; it was a few hundred metres away. When I got to the front desk I told the morbidly obese woman standing behind the counter that had thick plastic shield in front, about the incident in the Job Center, and I wanted both security officers charged with assault; , as well as the police officer who jumped me from behind..

"Certainly; can you give me the number of the officer at the scene?"

I told her that I asked for his name, and badge number, but he ran off.

I made a guess, and she checked the number I gave her, but it wasn't correct; so she contacted police community service worker #30324, to get the information. I had to wait what seemed about 10 minutes before she returned to the front desk.

"Here's the situation; apparently you were lawfully evicted."

"Could you tell me what the lawful reason was?"

"I have no idea."

"I can't believe I said I was sorry earlier; try this on for size; you're all murderers."

"Please don't say that; there are children present."

Every night in Plymouth two vans stop in front of the down town police station; sandwiches; tea; coffee; soup, and a piece of fruit, are handed to homeless people from the back of each.

Young girls; elderly men, and middle-aged women; people belonging to every age group, help dish out this grub, while they make sure everyone is told they wish them all the best, all the while knowing their mission is to kill them.

They know they've done their job when a homeless person dies; toxins in the food weaken the already weak immune systems of the homeless; due to the lifestyles they are forced to lead; having to depend on Day Centers for toiletries; clothing, meals, and showers, they soon die.

> The Day Centers frequented by homeless people in Britain, resemble the Concentration Camps run by the Nazis During the Holocaust.
>
> Nazism, War, and Genocide, edited by Neil Gregor; includes the description below of what soldiers found when they liberated Jews from these camps.

> "The mass of prison inmates also suffered greatly in the last months of Nazi rule. Cells were filthy, and hopelessly overcrowded, and inmates were starving, and disease ridden."

By Karl Vick/Berlin with Simon Shuster

Fairy tales are where you find them, but any numbers seem to begin in the dark German woods where Angela Merkel spent her childhood.

The girl who would grow up to be called the most powerful woman in the world came of age in a glade dappled by the northern sun and shadowed by tall pines.

Her family's house stood three stories, and the steep rake of its tile roof held an attic window in the shape of a half-open eye. Strangers walked on the paths below, passing residents who often moved at curious gaits. Cries of anguish were sometimes heard. To adults, Waldhof was home to the Lutheran seminary run by Merkel's father, an isolated compound—"forest court" in English—that hosted students and other short-term visitors while also functioning as a home and workplace for mentally disabled adults. But to a child of 3, Angela's age when her family arrived, it was a world unto itself, and would remain so until she went to school in the adjoining town of Templin. There, she came to realize that, like the 17 million other residents of East Germany, she actually was living within the walls of a fortress.

Merkel remained a captive for the first 35 years of her life, biding her time. As an adult, she lived in East Berlin, riding an elevated train beside the barricade whose 1961 construction she recalled as the first political memory of her life. When it fell in 1989, she gathered the qualities cultivated as a necessity in the East—patience, blandness, intellectual rigor and an inconspicuous but ferocious drive—and changed not only her life but the course of history.

The year 2015 marked the start of Merkel's 10th year as Chancellor of a united Germany and the de facto leader of the European Union, the most prosperous joint venture on the planet. By year's end, she had steered the enterprise through not one but two existential crises, either of which could have meant the end of the union that has kept peace on the continent for seven decades. The first was thrust upon her—the slow-rolling crisis over the euro, the currency shared by 19 nations, all of which were endangered by the default of a single member, Greece. Its resolution came at the signature plodding pace that so tries the patience of Germans that they have made it a verb: *Merkeling.*

The second was a thunderclap. In late summer, Merkel's government threw open Germany's doors to a pressing throng of refugees and migrants; a total of 1 million asylum seekers are expected in the country by the end of December. It was an audacious act that, in a single motion, threatened both to redeem Europe and endanger it, testing the resilience of an alliance formed to avoid repeating the kind of violence tearing asunder the Middle East by working together. That arrangement had worked well enough that it raised an existential question of its own, now being asked by the richest country in Europe: What does it mean to live well?

Merkel had her answer: "In many regions war and terror prevail. States disintegrate. For many years we have read about this. We have heard about it. We have seen it on TV. But we had not yet sufficiently understood that what happens in Aleppo and Mosul can affect Essen or Stuttgart. We have to face that now." For her, the refugee decision was a galvanizing moment in a career that was until then defined by caution and avoidance of anything resembling drama. Analysts called it a jarring departure from form. But it may also have been inevitable, given how Angela Merkel feels about walls.

What was not inevitable but merely astounding was that the most generous, openhearted gesture of recent history blossomed from Germany, the country that within living memory (and beyond, as long as there's a History

Channel) blew apart the European continent, and then the world, by taking to gruesome extremes all the forces its Chancellor strives to hold in check: nationalism, nativism, self-righteousness, reversion to arms. No one in Europe has held office longer—or to greater effect—in a world defined by steadily receding barriers. That, after all, is the story of the E.U. and the story of globalization, both terms as colourless as the corridor of a Brussels office building. The worlds Merkel has mastered carry not a hint of the forces that have shaped Europe's history, the primal sort a child senses, listening to a story, safe in bed,

Jesco DenzelPack Leader Merkel, here hosting heads of G-7 nations ahead of a June meeting in southern Germany, has marshalled international consensus on crises in Ukraine and Syria

In some ways, living in East Germany was like living on a stage set. The German Democratic Republic called itself a sovereign nation, but it was Moscow's closest satellite in the Soviet bloc. It's deeply paranoid government put great store on appearances, employing thousands to spy on other citizens. It minted coins that felt strangely light in the palm—they were made of aluminium—and many streets were facades. "I stayed there for six or nine months in 1981. My impression is it was 1947 or '48," says Peer Steinbrück, a Social Democrat who both lost to Merkel and served as her Finance Minister. "Behind Unter den Linden, all these buildings were still destroyed. Bullet scars on the walls."

Erika Benn had the same feeling when she arrived in Templin in 1965 from university at Leipzig to teach Russian: "I said; where have I ended up? My God." The medieval town had a history, with a church that dates to the 14th century. But churches were merely tolerated in the GDR, which was officially atheist.

That made public life delicate at Waldhof. Merkel's father, Horst Kasner, had moved his family there in 1957, after leaving Hamburg, where Angela, the first of three children, was born. Most people were moving in the other direction, to the West. But the Lutheran Church enjoyed a standing in German society that brought a measure of deference even from Marxist-Leninists. Its parishes in the East became refuges for dissidents, something like embassies. That in turn brought anyone associated with them additional scrutiny, though Kasner's situation was tempered by his enthusiasm for socialism—at least as he understood it—and an evident talent for navigating the state apparatus.

It also helped that the pastor embraced a school of theology that steered clear of social activism and instead sought to reconcile the work of modern philosophers like Immanuel Kant with religious belief, according to a former adviser to Merkel. The discussions young Angela grew up amid in the parsonage were erudite and rigorous. Her mother Herlind, trained as an English teacher, was never allowed to teach the language. At school, Angela enrolled in Russian with Frau Benn.

The retired teacher keeps a file folder on her star student. Pulling out a black-and-white group photo, she points out Merkel in the back row, recognizable mostly by her helmet hair. "That's how she was: the girl in the back," says Benn. "She's about almost invisible. It's so typical of her, I can't even tell you."

As an adolescent, Merkel both lived inside her head and exulted in the outdoors. Physically clumsy, she avoided sports but camped with friends, all while excelling at school. As she got older, she explored as much of the world as a citizen of the Soviet bloc was permitted. The system's limits on wanderlust rendered Merkel, waiflike in her youth, with her face pressed up against the glass of a warm shop window.

She journeyed to Bulgaria and stared over the border toward the forbidden hillsides of Greece. She watched, as almost everyone in the GDR did, television stations beamed from West Germany, and dreamed of visiting California. Merkel understood that she would not be permitted to go there until she was 60, the age at which East Germany trusted its citizens to travel to the West. Yet she began to plan for it. Patience was a lesson of life in the East, as was realism.

"You know I grew up in the GDR," Merkel told a security conference in Munich in February, where she was peppered with demands that Russian President Vladimir Putin's incursion into Ukraine be answered with military force rather than the economic pressure Merkel had spearheaded. "As a 7-year-old child, I saw the Wall being erected. No one—although it was a stark violation of international law—believed at the time that one ought to intervene militarily in order to protect citizens of the GDR and whole Eastern bloc of the consequences of that, namely to live in lack of

76

freedom for many, many years. And I don't actually mind. Because I understand this, because it was a realistic assessment that this would not lead to success."

Merkel plays the long game, in other words. For a career, she shrewdly chose a path in the field that communists worshipped instead of God: science. She studied physics at Leipzig University and married another scientist, Ulrich Merkel. She ended the marriage after five years but kept his name, even after marrying her current husband, Joachim Sauer, a quantum chemist, after years spent living together.

THE RATS

The village hadn't been abandoned all that long.
Soon man appeared; he scurried forth through the night.
He appeared human, but, actually, he was a rat.
He could gnaw on anything, as long as it wouldn't bite back.
The streets, stores, and homes,
Were full of the sounds of their scampering feet.
Their claws scrapped along floors;
Against doors; along halls, and down corridors.
Relentlessly, day in, day out, they searched for food, wherever it might be;
Around cupboards; within holes; between cracks, and under sofas;
They never grew tired of getting their fill.
As soon as one meal was finished, the next began.
It didn't take long, however,
Before their gluttony, and greed, ravaged wires,
And damaged the boards that kept homes in one piece.
The rats appeared not to notice;
They were too busy filling their stomachs.

Mindlessly they went about their days,
Happy to eat garbage; spew stench, and sleep in piles of filth;
Until there came a day when the buildings could no longer endure
The habits of these scavenging rodents;
The walls crumbled; the floors cracked open, and ceilings fell;
While the rats scurried for cover in their underground lairs.
Having no other choice,
They fled this village they'd come to call home,
And went in search of another;
That was built by men who could construct towers out of clay.
They were so talented, and strong,
The rats thought they were responsible for
The Sun rising, and setting,
Each day.

The End Result;

A person dies to make room for a fraudulent bank account.

Totalitarian regimes seek to totally rule a population through a "central body"; they dictate how people should speak, act, behave, cloth themselves, and so on and so forth; if final rulings are made by a single individual, the result is the much the same. How was Hitler any worse than Stalin?

When people are allowed to decide for themselves how they wish to live their lives, while no justice system is in place - "inverted totalitarianism" - disorder, anarchy, chaos, and violence, will inevitably result.

Totalitarian regimes, and fascist dictators, create laws to justify their acts; people in Hyper-democratic societies, (First World Countries) lie, deceive, cheat, and steal, to get whatever happens to be their fancy at any particulate time. When people are given a free hand to rule their own affairs, a brute will soon be the result.

Psychological projection, also known as blame shifting, is a theory in psychology in which humans defend themselves against their own unpleasant impulses by denying their existence while attributing them to others. For example, a person who is rude may constantly accuse other people of being rude.

(Wikipedia)

The following documents reveal how Lloyd's bank, (all banks in Britain operate the same way), steals the I.D., and homes of people; practically all the homeless people in Britain have been subjected to the same hideous barbaric crime.

I acquired a home in Sou...
and a month late...
the lock was changed on...
denie...
My I.D., and home addre...
accounts for the purpose...
co-ordination of this sche...

Cannibal.

Steve white
Bank Manager
Horley Branch

Direct Line 0845 3000 000
Fax 01293 785575

Lloyds Bank plc
11 High Street
Horley
RH6 7BJ
www.lloydsbank.com

Horley
Surrey
RH6 7BJ

or **call us on:**	0845 3 000 000
or fax us on:	01293 785575

Letter Date: 25 November 2015

Dear Mr N Shindler,

I would like to confirm the details of your appointment with us:

Interviewer:	STEPHEN WHITE
Date:	27/11/2015
Time:	11:00
Location:	9438 HORLEY
	11 High Street
	Horley
	Surrey
	RH6 7BJ
Reference No:	47701329

It would be helpful if you could bring along the following items:

- DETAILS OF SAVINGS, INVESTMENTS AND PENSIONS
- DETAILS OF INSURANCE POLICIES
- NATIONAL INSURANCE NUMBER
- INCOME AND EXPENDITURE DETAILS

If you want to change or cancel this appointment, please call in to the branch or phone us on the above number. If this becomes necessary, please quote your reference number.

We look forward to seeing you.

Yours sincerely

Shamir Oleemahomed

Shamir Oleemahomed

Stephen White, the manger of the Harley branch, opened an account for me on November 27th, 2016; when I attempted to open another account at the same branch with Naomi Jones, I discovered Lloyd's had stolen both

my home address and name. As soon as I noticed the changes in the paperwork, I was told to leave the bank.

Logon to Image Repair

Username

Password

Navigation

View Customer Home Page

Back

Zoom: 25% | Rotate 90° | Select Page: < | (1) | >

LLOYDS BANK

Customer Name:	Mr N Shindler
Address:	53 ASH STREET
	SOUTHPORT
	MERSEYSIDE
	PR8 6JE
Sort Code:	30-94-38
Account Number:	36013968

Current Accounts and Other Services

Classic Account

Eligibility

Unplanned Overdrafts

Data Protection Act

Checklist

Customer's signature

Previous Page | **Page 1 of 1** | Next Page   Print

http://searchretrieve.service.group:30000/Stores SearchRetrieve/tf/tc?cz=21902&tidx... 08/01/2016

Naomi Jones
Senior Personal Banking Advisor
Horley Branch

Direct Line 0345 3000000
Facsimile 01293 785575

Lloyds Bank plc
11 High Street
Horley
Surrey
www.lloydsbanking.com

LLOYDS BANK

Customer Name: **Mr N Shindler**
Address: **Passage Day Centre**
St Vincent Centre
Carlisle Place
London
SW1P 1NL

Sort Code: **30-94-38**

Account Number: **36250568**

Current Accounts and Other Services

Thank you for opening/ converting your account with Lloyds Bank plc. Your account number and the sort code for your branch are shown at the top of this document.

Our agreement with you is made up of general conditions (contained in the Personal Banking terms and conditions leaflet) and additional conditions. These include the conditions below and the Banking Charges guide which contains our standard fees. If there is any overlap or conflict between the additional conditions and the Personal Banking terms and conditions, the additional conditions apply.

Classic Account

Your Classic account is our standard current account.

Eligibility

To have the account you must be 18 or over.

Unplanned Overdrafts

If you make a payment which means your account goes overdrawn or over your Planned Overdraft limit, if you have one, we will charge the Unplanned Overdraft fees set out in our Banking Charges guide.

The following interest rates will apply:

Classic Account Personal Overdraft Rates		
Account	Interest Paid Monthly	
	% Per Month	% EAR
Classic Account	1.53	19.94

*Equivalent Annual Rate - Variable.

Data Protection Act

It is important that you understand how the personal information you give us will be used and we strongly advise that you read our Privacy Statement which you can find at (www.lloydsbank.com/privacy2.asp) or you can ask us for a copy.

By signing the Application Form, you agree that we can use your information in the ways described.

Checklist

You confirm that you would like to receive a card.

You confirm that you have received the following documents.

Your banking relationship with us (the Personal Banking terms and conditions) ☐

Welcome Pack ☐

Banking Charges Brochure ☐

Financial Services Compensation Scheme Information Sheet ☐

Customer's signature

Date

CUSTOMER NOTES

Mr Nigel Shindler
Passage Day Centre
St Vincent Centre
Carlisle Place
London
SW1P 1NL

DATE	USER	METHOD	SUBJECT	ACTION

30-Nov-2015 9754720 000946 I   ADDRSS   AMD
   Address updated
   for Nigel Shindler change made
    by 9754720 via OSP-COA on 30/11/2015
   at Horley

12-Dec-2015 8957660 001133 I   ADDRSS   AMD
   Address updated
   for Nigel Shindler change made
    by 8957660 via OSP-COA on 12/12/2015
   at London Oxford Street 399

S4712

Financial Services Compensation Scheme

INFORMATION SHEET

Basic information about the protection of your eligible deposits	
Eligible deposits in Lloyds Bank plc are protected by:	The Financial Services Compensation Scheme ("FSCS")[1].
Limit of protection:	£75,000 per depositor per bank[2]. The following trading names are part of your bank: Lloyds Bank, Lloyds Bank Private Banking, C&G Savings and WorldWide Service.
If you have more eligible deposits at the same bank:	All your eligible deposits at the same bank are "aggregated" and the total is subject to the limit of £75,000[2].
If you have a joint account with other person(s):	The limit of £75,000 applies to each depositor separately[3].
Reimbursement period in case of bank's failure:	20 working days[4].
Currency of reimbursement:	Pound sterling (GBP, £) or, for branches of UK banks operating in other EEA Member States, the currency of that State.
To contact Lloyds Bank plc for enquiries relating to your account:	You can visit one of our branches, call us, go online or write to us at the address below: 25 Gresham Street, London, EC2V 7HN.
To contact the FSCS for further information on compensation:	Financial Services Compensation Scheme, 10th Floor Beaufort House, 15 St Botolph Street, London, EC3A 7QU. Tel: **0800 678 1100** or **020 7741 4100** Email: **ICT@fscs.org.uk**
More information:	http://**www.fscs.org.uk**
Acknowledgement of receipt by the depositor:	

LLOYDS BANK

Your Interview Summary Document

The purpose of this document is to provide you with a summary of your interview. I have highlighted products that may meet your needs, these are highlighted to help you make your selection and I have not provided you with advice. If we recommend a Home Solutions product I have provided you with advice for this product only. If you have been unable to provide full details, the recommended product may not be suitable.

If your circumstances change it is important that you check your eligibility and advise us of the change to your circumstances.

Your Needs and suggested outcomes

Mr N Shindler, you told me that you wanted to:
- open an account to pay in your income. We highlighted a Current Account would meet your need and you chose to progress this.
  - Travel Insurance

Products to Meet Your Needs

Mr N Shindler, I highlighted a Classic account to meet your banking needs as this account has no monthly account fee and you would not get value for money from a bank account with insurance benefits.
In addition this product would also give you the following benefits:
- An interest and fee free Planned Overdraft of up to £25 (subject to application and approval)

However this account does not meet some of your insurance needs:
- Travel Insurance

Your product selection

Mr N Shindler, you have chosen

Classic account 309438 36250568 because the banking facilities suit my needs.

This product contains the following benefits:
- An interest and fee free Planned Overdraft of up to £25 (subject to application and approval)

However this account does not provide the following cover to meet your needs:
- Travel Insurance

Important Information
If you are not happy with your choice of account or service, you can cancel it within 30 days of opening the account or taking the service. We will then help you to move to another account we offer or will return your money to you with any interest you have earned on it

The letter Naomi attended to erase, after I insisted that I wanted copies

Signature Mandate Form

LLOYDS BANK

IMPORTANT INFORMATION - PLEASE READ CAREFULLY

On completion, the Specimen Signature slip below must be detached and sent to the SMD Unit daily in the pre-printed SMD Personal envelope.

Please ensure all sections are fully completed using black ink.

Please do not mark or write on the front of the specimen signature slip below except in the designated boxes.

- This form must only be folded **once (in half)** to fit the **A5** SMD personal envelope (525645)
- Please ensure the envelope is SEALED
- Send the SMD Personal Envelope to SMDU.

Branch sort code

3 0 9 4 3 8

Date completed

2 8 | 0 1 | 2 0 | 1 6

Account number(s)

3 6 2 5 0 5 6 8

Signing Instructions:

Either party to sign   Other - see attached

First customer's name

Nigel Schindler

Second customer's name

First customer's signature

Second customer's signature

of the documents, is a c; I have never provided any form of I.D. to Lloyd's bank with the name, "Schindler".

The Passage attempted to open a bank account with my I.D. earlier. When I reported the incident at Belgravia Police Station, I was told that no crime had been committed. Below is the letter authorizing me to use The Passage as a mailing address.

St Vincent's Centre
Carlisle Place
London SW1P 1NL

Tel: 020 7592 1850
Fax: 020 7592 1870

info@passage.org.uk
www.passage.org.uk

Registered charity number
1079764

Founding Patron
Cardinal Basil Hume

Patron
Archbishop Vincent Nichols

To Whom It May Concern

Re: Proof of address Letter

Date: 06/10/2015

This letter is to introduce our client and to confirm the following details:

Client's Name: **Nigel Shindler**

Date of Birth: **16/08/1964**

Current Address: **St Vincent's, Carlisle Place, London, SW1P 1NL**

We can confirm that the client is very well-known to us and has been a regular user of our services. He can use our service as his care of address.

If you require any further information, please do not hesitate to contact us.

Yours faithfully,

Matina Papaioannou

The Passage, St Vincent's Centre, Carlisle Place, London, SW1P 1NL.
Tel: 020 7592 1864

The following four pages comprise the statement I provided to Belgravia Metropolitan police.

PLACE: BELGRAVIA POLICE STATION
DATE: OCTOBER 19, 2015
TIME: 6:46 pm

ON THE MORNING OF SEPTEMBER 26, 2015, AT APPROXIMATELY 7:00 am, I REPORTED VIA OUTSIDE PHONE, AT BELGRAVIA POLICE STATION MY SUSPICION THAT MY FATHER, MAXWELL SHINDLER, HAD BEEN MURDERED, AND REQUESTED, DUE TO MY FATHER BEING A CITIZEN OF BRITAIN, THAT AN INVESTIGATION BE CONDUCTED BY THE AUTHORITIES IN BRITAIN, DUE TO THE TORONTO POLICE SERVICES, REFUSING TO INVESTIGATE, OR WILLING FOR ME TO MAKE OUT A "MISSING PERSONS" REPORT.

I WAS GIVEN THE REFERENCE # CAD 229726 — OCTOBER SEPTEMBER BY POLICE OFFICER CF25528 FIRST NAME, CHRISSIE.

AT APPROXIMATELY 8:00PM, I RETURNED TO BELGRAVIA POLICE STATION, AND BRIEFLY SPOKE WITH TWO OFFICERS, WHO THEN CLAIMED THEY SPOKE TO THEIR SUPERVISOR FOR ADVICE, AND THEY WERE INFORMED THAT I SHOULD FIRST GO TO FOREIGN AFFAIRS, DUE TO MY FATHER LIKELY GOING MISSING IN CANADA, AND THEN THE METROPOLITAN POLICE WOULD WORK IN TANDEM WITH THE FOREIGN AND COMMONWEALTH OFFICE.

I STRONGLY SUSPECTED MY FATHER HAD BEEN MURDERED BY THE PEOPLE HE AND I WERE EXPOSING IN OUR BOOKS AS THE MONOLITH I BELIEVE, AS WELL, THAT ON JUNE 21/22, 2014, THERE WAS AN ATTEMPT ON MY LIFE, RESULTING IN MY BEING

ADMITTED TO OTTAWA GENERAL HOSPITAL BY AN
AMBULANCE FROM GATINEAU, QUEBEC.

ON OCTOBER 6, 2015, I WAS DUE TO
TRAVEL TO HORLEY IN ORDER TO OPEN A BANK
ACCOUNT AT THE HALIFAX BRANCH, AT 3:30pm.
I WAS PROMISED A TICKET TO GET THERE BY
"THE PASSAGE", BY A WORKER NAMED, "JOSH",
BUT THIS WAS NOT TO BE THE CASE.

AT APPROXIMATELY 1:00 pm, ON OCTOBER
6, 2015, I WAS PREPARED TO WALK TO VICTORIA
STATION, AND HAVE "JOSH" BUY ME A TICKET TO
HORLEY. AS WE PASSED THE HALIFAX BANK,
AT 2 WILTON ROAD, LONDON, SW1V 1AN, "JOSH"
ENQUIRED IF THIS BRANCH COULD ALSO OPEN
AN ACCOUNT, I SAID, "Yes: AS LONG AS THEY
HAVE A MAILING ADDRESS.

I WAS TOLD THAT THE RECEPTION DESK
THAT THE ADDRESS I'D PROVIDED TO THE HORLEY
BRANCH WOULD SUFFICE. I SHOWED MY PASSPORT AS
I.D. THEN AN APPOINTMENT WAS ARRANGED FOR
OCTOBER 13, 2015, AT 12:00 am WITH SAGAL
FAISAL ALI.

AS JOSH AND I THEN BEGAN WALKING BACK
TO THE PASSAGE, I TOLD "JOSH" THAT I'D
LIKE TO SEE IF HSBC, BELGRAVIA BRANCH,
333 VAUXHALL BRIDGE ROAD, LONDON, SW1V 1ET,
COULD MAKE AN EARLIER APPOINTMENT.

WE SAT DOWN, AND SPOKE WITH "GORDON",
WHO SAID THAT IF "JOSH" COULD RETURN WITH
AN OFFICER OFFICIAL MAILING ADDRESS FORM,
I WOULD BE ABLE TO OPEN AN ACCOUNT AS
EARLY AS OCTOBER 8, 2015.

WE RETURNED TO THE PASSAGE, AND IT
TOOK "JOSH" CLOSE TO TWO HOURS TO RETURN
WITH A MAILING ADDRESS FORM, BUT IT
WASN'T SIGNED. HE WALKED AWAY, AND
RETURNED SECONDS LATER WITH A SIGNATURE
ADDED.

I RETURNED TO HSBC, AND SAW RAQUEL
LOPES, I HANDED HER THE FORM "JOSH"
HANDED ME, AND SHE SAID, "THE MANAGER
NEEDS TO SEE THIS". I WAS INTERROGATED
BY RAMESH PINDORIA, AND HIS CONCLUSION
WAS THE MAILING ADDRESS FORM WAS NOT
VALID, AND HE ASSURED ME HE WOULD BE
WILLING TO MAKE A STATEMENT TO THE POLICE.

I ATTENDED MY APPOINTMENT ON OCTOBER
13, 2015, AT NOON, AT THE HALIFAX BRANCH AT
2 WILTON ROAD. I WAS TOLD THEY WANTED TO
PHOTOCOPY I.D., AND MY PASSPORT, AND SEND THAT
TO HEAD OFFICE, AND THEY WOULD DECIDE
WHETHER I WAS ELIGIBLE FOR AN ACCOUNT.

I SAID THAT I HAD ALREADY BEEN GIVEN
ASSURANCE THAT I HAD THE REQUIRED I.D, AND
I WOULD LIKE TO SPEAK TO HEAD OFFICE; MY
REQUEST WAS DENIED. I ASKED TO SEE THEIR
TERMS AND CONDITIONS, HOWEVER, ANISH VARSANI,
BUSINESS MANAGER, SAID THAT I DIDN'T NEED
THEM, AND HE WAS REFUSING TO PROVIDE THEM.

I SAID, "THEN YOU NEVER INTENDED TO OPEN
AN ACCOUNT, WHY THEN DO YOU NEED TO MAKE
PHOTOCOPIES OF MY DOCUMENTS, AND, I.D.?
NO EXPLANATION PROVIDED.

I ASKED THE PRIEST AT THE PASSAGE TO VIEW THE MAILING FORM. HE STATED IT LOOKED INVALID. HE LATER TOLD ME MATINA PAPAIOANNOU WAS NOT AT THE PASSAGE ON OCTOBER 6, 2015.

I HAVE SINCE ASKED STAFF AT THE PASSAGE FOR COPIES OF FORMS I'VE SIGNED, BUT MY REPEATED REQUESTS HAVE BEEN IGNORED. I HAVE SINCE BEEN DENIED FOOD, TOILETRIES, AND LAUNDRY SERVICES AT THE PASSAGE.

THE PEOPLE RESPONSIBLE FOR MY FATHER GOING MISSING, AND THE ATTEMPTS ON MY LIFE, HAVE BLOCKED MY HAVING ACCESS TO THE ROYALTIES OF MY BOOKS, AND BY DEPOSING, (KILLING BOTH MY FATHER AND I) STOP OUR BOOKS BEING PUBLISHED.

ONE OF THE MOST POWERFUL FAMILIES RUNNING THE UNITED JEWISH CONGRESS IS THE BRONFMAN FAMILY. SIMONE SHINDLER, AND NICOLA SHINDLER, (MY SISTERS) CHOSE AS HUSBANDS PEOPLE CONNECTED TO THE BRONFMAN FAMILY, WHO HAVE TIES TO BOTH THE WHITE HOUSE, AND THE U.S. CONGRESS.

IN THE BOOK "THE GREAT AMERICAN LIE, WORLD DESTRUCTION", I SHOW HOW THOSE RUNNING ISRAEL, AND AMERICA, (MENTIONED ABOVE), ARE COMMITTING ATROCITIES ON A GLOBAL SCALE WITH THE SOLE AIM TO ENRICH THEMSELVES. I, Nigel Shindler, swear that this statement is true, to the best of my knowledge, so help me God.

After my I.D., and address in Southport, was used by Lloyd's bank to open fraudulent bank accounts, I asked the staff at The Passage for a letter authorizing me to use The Passage as a mailing address; I was first handed a copy of a "Consent Form"; when I said that wouldn't suffice,

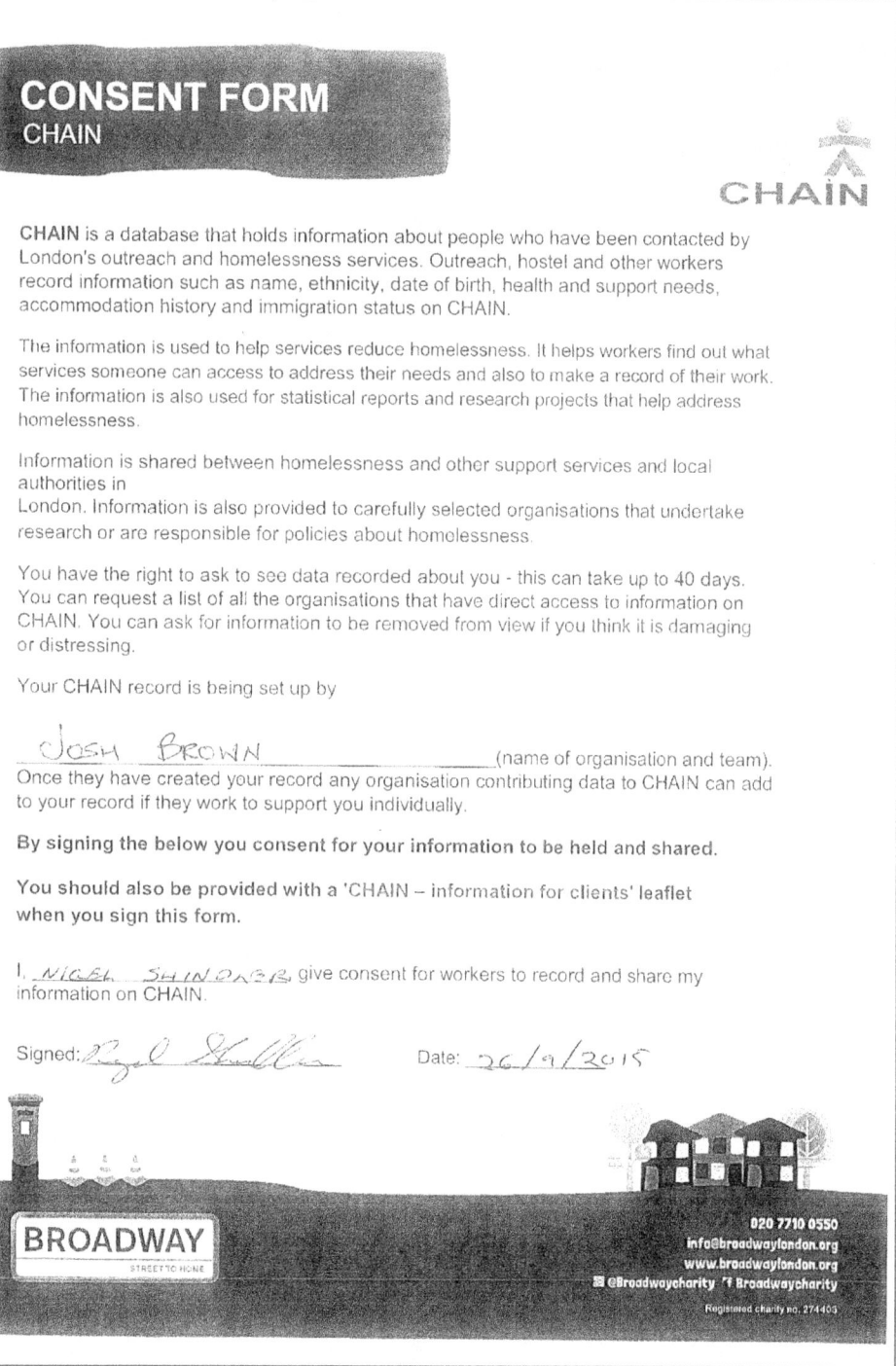

I was handed a document that did not include The Passage's postal code.

Postal Contract Agreement:

Terms and conditions:

• Only *rough sleepers* and *clients of no fixed abode* are entitled to use the Passage Day centre as a postal address. Clients must inform the Day centre if their accommodation status changes at any time

• Post can only be collected at these times: *9.30am -11.30am* and *1.30pm -5.45pm, Monday - Friday*

• Post will be kept for a maximum of *4 weeks*. After this time, post will be *returned to the sender or disposed of.*

• Client use of the Passage Day centre as a postal address will be reviewed on a monthly basis.

• At any time, The Passage Day centre reserves the right to withdraw a client's use of its postal address if it is felt or there is evidence that it is being used *inappropriately* e.g., as a *business. residential* or *bail* address

Non-compliance with these terms and conditions may result in the immediate termination of use of the postal service.

Please speak with a member of staff if you have any queries about the above terms and conditions.

I agree to the above terms and conditions.

CLIENT: SIGN & PRINT: ....

DATE OF BIRTH: ___/___/___

STAFF: SIGN & PRINT: ......Paul Fleming.................

DATE: 11 / 2 / 16

EXPIRY DATE (12 weeks): 11 / 5 / 16

The same day I made the statement at Belgravia Police Station; I was abducted by members of Belgravia Police Station; locked in the rear of a police van, and transported to Guilford. Upon arrival I was dumped on the outskirts of the town; they assumed I didn't have access to funds, and

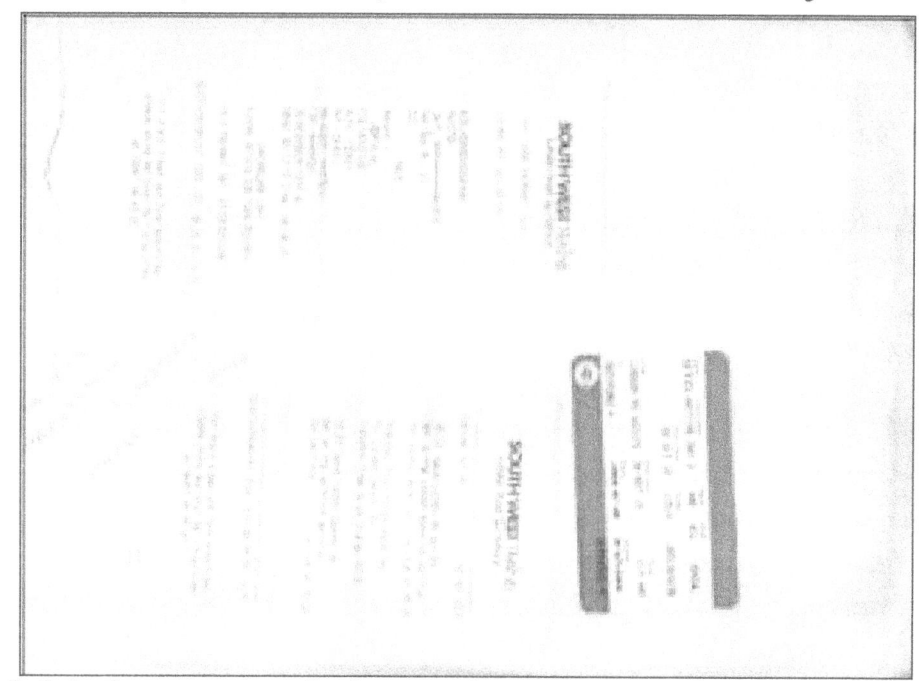

didn't have proper clothing to keep warm, and would soon die.

People like to behave as a herd; behaving the same as everybody else gives them a sense of security, and belonging; the most unfortunately result of this mentality is that practically everybody in Britain today is a cannibal due to participating in this diabolical scheme involving, I.D. theft, and money laundering. The form of reasoning used to justify their deeds is;

"If everybody else is doing something,
then it's alright for them to do the same"

Below is a quote from the book, "How to Stop Your Children Fighting", which illustrates just how ludicrous this mentality is; it was written by Adele Faber & Elaine Mazlish, and published in 1987, by Sidgwick & Jackson Limited, Great Britain.

The book fits the genre, "pop psychology"; it's the sort of material that popular in schools these days; it contains a load of wishy-washy nonsense that happens to sound quite impressive; "Newspeak", in book form.

"From the time they opened their eyes in the morning, (reference is to one of the author's children), till the time they closed them at night, they seemed committed to a single purpose – making each other miserable.

It baffled me. I had no way to account for the intensity, savagery, and never-ending-ness of the fighting between them.

Was there something wrong with them?

Was there something wrong with me?

Not until I shared my fears with other members of Dr Ginott's parent-guidance group did I begin to relax. It was pure happiness to discover that my misery had lots of company. Mine was not the only day punctuated by name-calling, tattling, punches, pinches, shrieks and bitter tears. I wasn't the only one walking around with a heavy heart, jangled nerves, and feelings of inadequacy. **(Misery loves company — I would say, on the other hand, that her kids have serious psychological problems).**

You would think, having been young siblings ourselves once that we would all have known what to expect. Yet most of the parents in the group were as unprepared as I for the antagonism between their children. Even now, years later, as I sit here leading my first workshop on sibling rivalry, I realize how little has changed. People can't wait to express their dismay at the disparity between their rosy expectations, and rude reality. **(There is definitely something wrong with the author if she thinks she's qualified to deal with sibling rivalries among other people's children.)**

Two wrongs don't make a right; the whole world can think something's right, and it can still be wrong. What exists among those who reside in Britain today is "mass psychogenic illness", everybody has it; the exceptions are very few, and far between.)

Projection

Description

When a person has uncomfortable thoughts or feelings, they may project these onto other people, assigning the thoughts or feelings that they need to repress to a convenient alternative target.

Projection may also happen to obliterate attributes of other people with which we are uncomfortable. We assume that they are like us, and in doing so we allow ourselves to ignore those attributes they have with which we are uncomfortable.

- *Neurotic projection* is perceiving others as operating in ways one unconsciously finds objectionable in yourself.
- *Complementary projection* is assuming that others do, think and feel in the same way as you.
- *Complimentary projection* is assuming that others can do things as well as you.

Projection also appears where we see our own traits in other people, as in the false consensus effect. Thus we see our friends as being more like us than they really are.

Example

I do not like another person. But I have a value that says I should like everyone. So I project onto them that they do not like me. This allows me to avoid them and also to handle my own feelings of dislike.

An unfaithful husband suspects his wife of infidelity.

A woman who is attracted to a fellow worker accuses the person of sexual advances.

Discussion

Projecting thoughts or emotions onto others allows the person to consider them and how dysfunctional they are, but without feeling the attendant discomfort of knowing that these thoughts and emotions are their own. We can thus criticize the other person, distancing ourselves from our own dysfunction.

One explanation is that the ego perceives dysfunction from 'somewhere' and then seeks to locate that somewhere. The super ego warns of punishment if that somewhere is internal, so the ego places it in a more acceptable external place - often in convenient other people.

Projection turns neurotic or moral anxiety into reality anxiety, which is easier to deal with.

Projection is a common attribute of paranoia, where people project dislike of themselves onto others such that they believe that most other people dislike them.

Projection helps justify unacceptable behaviour, for example where a person claims that they are sticking up for themselves amongst a group of aggressive other people.

Empathy, where a person experiences the perceived emotions of others, may be considered as a 'reverse' form of projection, where a person projects other people onto themselves. Identification may also be a form of reverse projection.

Projection is one of Anna Freud's original defense mechanisms.

So what?

To work authentically with other people, avoid projecting your woes onto them. When you see others in a negative light, think: are you projecting? Also understand that when others criticizing you; they may well be criticizing a projection of themselves.

When others are using projection, you can hold up a mirror to show them what they are doing. As usual, this may well be met with other forms of resistance.

Psychological Projection

Psychological projection, also known as **blame shifting**, is a theory in psychology in which humans defend themselves against their own unpleasant impulses by denying their existence while attributing them to others. For example, a person who is rude may constantly accuse other people of being rude.

According to some research, the projection of one's negative qualities onto others is a common process in everyday life.

Historical Precursors

A prominent precursor in the formulation of the projection principle was Giambattista Vico, and an early formulation of it is found in ancient Greek writer Xenophanes, which observed that "the gods of Ethiopians were inevitably black with flat noses while those of the Thracians were blond with blue eyes." In 1841, Ludwig Feuerbach was the first to employ this concept as the basis for a systematic critique of religion.

Psychoanalytic Developments

Projection (German: *Projektion*) was conceptualized by Freud in his letters to Wilhelm Fliess, and further refined by Karl Abraham and Anna Freud. Freud considered that in projection thoughts, motivations, desires, and feelings that cannot be accepted as one's own are dealt with by being placed in the outside world and attributed to someone else. What the ego repudiates is split off and placed in another.

Freud would later come to believe that projection did not take place arbitrarily, but rather seized on and exaggerated an element that already existed on a small scale in the other person. (The related defence of projective identification differs from projection in that there the other person is expected to become identified with the impulse or desire projected outside, so that the self maintains a connection with what is projected, in contrast to the total repudiation of projection proper.)

Melanie Klein saw the projection of good parts of the self as leading potentially to over-idealization of the object. Equally, it may be one's conscience that is projected, in an attempt to escape its control: a more benign version of this allows one to come to terms with outside authority.

Theoretical Examples

Projection tends to come to the fore in normal people at times of crisis, personal or political but is more commonly found in the neurotic or psychotic in personalities functioning at a primitive level as in narcissistic personality disorder or borderline.

Carl Jung considered that the unacceptable parts of the personality represented by the Shadow archetype were particularly likely to give rise to projection, both small-scale and on a national/international basis. Marie-Louise Von Franz extended her view of projection, stating that "wherever known reality stops, where we touch the unknown, there we project an archetypal image".

Psychological projection is one of the medical explanations of bewitchment used to explain the behavior of the afflicted children at Salem in 1692. The historian John Demos asserts that the symptoms of bewitchment experienced by the afflicted girls were due to the girls undergoing psychological projection of repressed aggression.

Victim blaming: The victim of someone else's accident or bad luck may be offered criticism, the theory being that the victim may be at fault for having attracted the other person's hostility.

Projection of marital guilt: Thoughts of infidelity to a partner may be unconsciously projected in self-defence on to the partner in question, so that the guilt attached to the thoughts can be repudiated or turned to blame instead, in a process linked to denial.

Bullying: A bully may project his/her own feelings of vulnerability onto the target(s) of the bullying activity. Despite the fact that a bully's typically denigrating activities are aimed at the bully's targets, the true source of such negativity is ultimately almost always found in the bully's own sense of personal insecurity and/or vulnerability. Such aggressive projections of displaced negative emotions can occur anywhere from the micro-level of interpersonal relationships, all the way up through to the macro-level of international politics, or even international armed conflict.

Projection of general guilt: Projection of a severe conscience is another form of defence, one which may be linked to the making of false accusations, personal or political.

Projection of hope: Also, in a more positive light, a patient may sometimes project his or her feelings of hope onto the therapist.

Counter-Projection

Jung wrote, "All projections provoke counter-projection when the object is unconscious of the quality projected upon it by the subject." Thus, what is unconscious in the recipient will be projected back onto the projector, precipitating a form of mutual acting out.

In a rather different usage, Harry Stack Sullivan saw counter-projection in the therapeutic context as a way of warding off the compulsive re-enactment of a psychological trauma, by emphasizing the difference between the current situation and the projected obsession with the perceived perpetrator of the original trauma.

Clinical Approaches

Drawing on Gordon Allport's idea of the expression of self onto activities and objects, projective techniques have been devised to aid personality assessment, including the Rorschach ink-blots and the Thematic Apperception Test (TAT).

Projection may help a fragile ego reduce anxiety, but at the cost of certain dissociation, as in dissociative identity disorder. In extreme cases, an individual's personality may end up becoming critically depleted.

In such cases, therapy may be required which would include the slow rebuilding of the personality through the "taking back" of such projections.

ROOM 101
FACING YOUR GREATEST FEARS

Mass Psychogenic Illness

Mass psychogenic illness (MPI), also called **mass sociogenic illness** or just **sociogenic illness**, is "the rapid spread of illness signs and symptoms affecting members of a cohesive group, originating from a nervous system disturbance involving excitation, loss or alteration of function, whereby physical complaints that are exhibited unconsciously have no corresponding organic aetiology." MPI is distinct from other collective delusions, also included under the blanket terms of mass, in that MPI causes symptoms of disease, though there is no organic cause.

There is a clear preponderance of female victims. The DSM-IV-TR does not have specific diagnosis for this condition but the text describing conversion disorder states that "In 'epidemic hysteria', shared symptoms develop in a circumscribed group of people following 'exposure' to a common precipitant."

Current State of Research

According to Balaratnasingam and Janca, "mass hysteria is to date a poorly understood condition. Little certainty exists regarding its etiology."

Besides the difficulties common to all research involving the social sciences, including a lack of opportunity for controlled experiments, mass sociogenic illness presents special difficulties to researchers in this field. Balaratnasingam and Janca report that the methods for "diagnosis of mass hysteria remains contentious. According to Timothy Jones of the Tennessee Department of Public Health, MPI "can be difficult to differentiate from bioterrorism, rapidly spreading infection or acute toxic exposure."

These troubles result from the residual diagnosis of MPI. Singer, of the Uniformed Schools of Medicine, puts the problems with such a diagnosis thus: "[y]ou find a group of people getting sick, you investigate, you measure everything you can measure . . . and when you still can't find any physical reason, you say 'well, there's nothing else here, so let's call it a case of MPI.'" There is a lack of logic in an argument that proceeds: "There isn't anything, so it must be MPI." It precludes the notion that an organic factor could have been overlooked. Nevertheless, running an extensive number of tests extends the probability of false positives.

British psychiatrist Simon Wesseley distinguishes between two forms of MPI:

1. Mass anxiety hysteria "consists of episodes of acute anxiety, occurring mainly in schoolchildren. Prior tension is absent and the rapid spread is by visual contact."
2. Mass motor hysteria "consists of abnormalities in motor behavior. It occurs in any age group and prior tension is present. Initial cases can be identified and the spread is gradual. . . . [T]he outbreak may be prolonged."

While his definition is sometimes adhered to, others such as Ali-Gombe et al. of the University of Maiduguri, Nigeria contest Wesseley's definition and describe outbreaks with qualities of both mass motor hysteria and mass anxiety hysteria.

An evolutionary psychology explanation for this disorder, as well as for conversion disorder more generally, is that the symptom may have been evolutionary advantageous during warfare. A non-combatant with these symptoms signals non-verbally, possibly to someone speaking a different language, that she or he is not dangerous as a combatant and also may be carrying some form of dangerous infectious disease. This can explain that conversion disorder may develop following a threatening situation, that there may be a group effect with many people simultaneously developing similar symptoms, and the gender difference in prevalence.

Commonalities in Outbreaks

Qualities of MPI outbreaks often include;

- symptoms that have no plausible organic basis;
- symptoms that are transient and benign;
- symptoms with rapid onset and recovery;
- occurrence in a segregated group;
- the presence of extraordinary anxiety;
- symptoms that are spread via sight, sound or oral communication;
- a spread that moves down the age scale, beginning with older or higher-status people;
- a preponderance of female participants

Also, the illness may recur after the initial outbreak.

Common Symptoms

Jones compiles the following symptoms based on their commonality in outbreaks occurring in 1980–1990:

Symptom	Percent reporting
Headache	67
Dizziness or light-headedness	46
Nausea	41
Abdominal cramps or pain	39
Cough	31
Fatigue, drowsiness or weakness	31
Sore or burning throat	30
Hyperventilation or difficulty breathing	19
Watery or irritated eyes	13
Chest tightness/chest pain	12
Inability to concentrate/trouble thinking	11
Vomiting	10
Tingling, numbness or paralysis	10
Anxiety or nervousness	8
Diarrhea	7
Trouble with vision	7
Rash	4
Loss of consciousness/syncope	4

Predisposition for Psychogenic Illness

The hypothesis that those prone to extroversion or neuroticism, or those with low IQ scores, are more likely to be affected in an outbreak of hysterical epidemic has not been consistently supported by research. Bartholomew and Wesseley state that it "seems clear that there is no particular predisposition to mass sociogenic illness and it is a behavioural reaction that anyone can show in the right circumstances."

Females are affected with mass psychogenic illness at greater rates than males. Adolescents and children are frequently affected in cases of MPI.

History and Examples

Middle Ages

The earliest studied cases linked with epidemic hysteria are the dancing manias of the Middle-Ages, including St. John's Dance and tarantism. These were supposed to be associated with spirit possession or the bite of the tarantula. Those afflicted with dancing mania would dance in large groups, sometimes for weeks at a time. The dancing was sometimes accompanied by stripping, howling, the making of obscene gestures, or even (purportedly) laughing or crying to the point of death. Dancing mania was widespread over Europe.

Between the 15th and 19th centuries, instances of motor hysteria were common in nunneries. The young ladies that made up these convents were typically forced there by family. Once accepted, they took vows of chastity and poverty. Their lives were highly regimented and often marked by strict disciplinary action. The nuns would exhibit a variety of behaviors, usually attributed to demonic possession. They would often use crude language and exhibit suggestive behaviors. One convent's nuns would regularly mew like cats. Priests were often called in to exorcise demons.

18th to 21st centuries

In factories

MPI outbreaks occurred in factories following the industrial revolution in England, France, Germany, Italy and Russia as well as the United States and Singapore.

W. H. Phoon, Ministry of Labor in Singapore gives a case study of six outbreaks of MPI in Singapore factories between 1973 and 1978. They were characterized by (1) hysterical seizures of screaming and general violence, wherein tranquilizers were ineffective (2) trance states, where a worker would claim to be speaking under the influence of a spirit or *jinn* (or genie) and (3) frightened spells: some workers complained of unprecedented fear, or of being cold, numb, or dizzy. Outbreaks would subside in about a week. Often a *bomoh* (medicine man) would be called in to do a ritual exorcism. This technique was not effective and sometimes seemed to exacerbate the MPI outbreak. Females and Malays were affected disproportionately.

Especially notable is the "June Bug" outbreak: In June 1962, a peak month in factory production, sixty two workers at the Montana Mills dressmaking factory experienced symptoms including severe nausea and breaking out on the skin. Most outbreaks occurred during the first shift, where four fifths of the workers were female. Of 62 total outbreaks, 59 were women. Entomologists and others were called in to discover the pathogen, but none was found. Kerchoff coordinated the interview of affected and unaffected workers at the factory and summarizes his findings:

1. Strain – those affected were more likely to work overtime frequently and provide the majority of the family income. Many were married with children.
2. Affected persons tended to deny their difficulties. Kerchoff postulates that such were "less likely to cope successfully under conditions of strain."
3. Results seemed consistent with a model of social contagion. Groups of affected persons tended to have strong social ties.

Kerchoff also links the rapid rate of contagion with the apparent reasonableness of the bug and the credence given to it in accompanying news stories.

Stahl and Lebedundescribe an outbreak of mass sociogenic illness in the data center of a mid-western university town. Ten of thirty-nine workers smelling an unconfirmed "mystery gas" were rushed to a hospital with symptoms of dizziness, fainting, nausea and vomiting. They report that most workers were young women either putting their husbands through school or supplementing the family income. Those affected were found to have high levels of job dissatisfaction. Those with strong social ties tended to have similar reactions to the supposed gas, which only one unaffected woman reported smelling. No gas was detected in subsequent tests of the data center.

In schools

Thousands were affected by the spread of a supposed illness in a Serbian province of Kosovo, exclusively affecting ethnic Albanians, most of which were young adolescents. A wide variety of symptoms were manifested, including headache, dizziness, impeded respiration, weakness/adynamia, burning sensations, cramps, retrosternal/chest pain, dry mouth and nausea. After the illness had subsided, a bipartisan Federal Commission released a document, offering the explanation of psychogenic illness. Radovanovic of the Department of Community Medicine and Behavioural Sciences Faculty of Medicine in Safat, Kuwait reports:

This document did not satisfy either of the two ethnic groups. Many Albanian doctors believed that what they had witnessed was an unusual epidemic of poisoning. The majority of their Serbian colleagues also ignored any explanation in terms of psychopathology. They suggested that the incident was faked with the intention of showing Serbs in a bad light but that it failed due to poor organization.

Rodovanovic expects that this reported instance of mass sociogenic illness was precipitated by the demonstrated volatile and culturally tense situation in the province.

The Tanganyika laughter epidemic of 1962 was an outbreak of laughing attacks rumored to have occurred in or near the village of Kanshasa on the western coast of Lake Victoria in the modern nation of Tanzania, eventually affecting 14 different schools and over 1000 people.

On the morning of Thursday 7 October 1965, at a girls' school in Blackburn in England, several girls complained of dizziness. Some fainted. Within a couple of hours, 85 girls from the school were rushed by ambulance to a nearby hospital after fainting. Symptoms included swooning, moaning, chattering of teeth, hyperpnea, and tetany. Moss and McEvedy published their analysis of the event about one year later. Their conclusions follow. Note that their conclusion about the above-average extroversion and neuroticism of those affected is not necessarily typical of MPI.

- Clinical and laboratory findings were essentially negative.
- Investigations by the public health authorities did not uncover any evidence of pollution of food or air.
- The epidemiology of the outbreak was investigated by means of questionnaires administered to the whole school population. It was established that the outbreaks began among the 14-year-olds, but that the heaviest incidence moved to the youngest age groups.
- By using the Eysenck Personality Inventory it was established that in all age groups the mean E [extroversion] and N [neuroticism] scores of the affected were higher than those of the unaffected.
- The younger girls proved more susceptible, but disturbance was more severe and lasted longer in the older girls.
- It was considered that the epidemic was hysterical, that a previous polio epidemic had rendered the population emotionally vulnerable,

and that a three-hour parade, producing 20 faints on the day before the first outbreak, had been the specific trigger.

- The data collected were thought to be incompatible with organic theories and with the compromise theory of an organic nucleus.

Another possible case occurred in Belgium in June 1999 when people, mainly schoolchildren, became ill after drinking Coca-Cola. In the end, scientists were divided over the scale of the outbreak, whether it fully explains the many different symptoms and the scale to which sociogenic illness affected those involved.

A possible outbreak of mass psychogenic illness occurred at Le Roy Junior-Senior High School in upstate New York, United States, in which multiple students began suffering symptoms similar to Tourette syndrome. Various health professionals like Dr. Jennifer McVige, Dr. Laszlo Mechtler and personnel from the New York Department of Health had ruled out such factors as Gardasil, drinking water contamination, illegal drugs, carbon monoxide poisoning and various other potential environmental or infectious causes, before diagnosing the students with a conversion disorder and mass psychogenic illness.

Starting around 2009, a spate of apparent poisonings at girls' schools across Afghanistan began to be reported, symptoms included dizziness, fainting and vomiting. The United Nations, World Health Organization and NATO's International Security Assistance Force carried out investigations of the incidents over multiple years, but never found any evidence of toxins or poisoning in the hundreds of blood, urine and water samples they tested. The conclusion of the investigators was that the girls were suffering from mass psychogenic illness.

Terrorism and biological warfare

Bartholomew and Wessely anticipate the "concern that after a chemical, biological or nuclear attack, public health facilities may be rapidly overwhelmed by the anxious and not just the medical and psychological casualties." Additionally, early symptoms of those affected by MPI are difficult to differentiate from those actually exposed to the dangerous agent.

The first Iraqi missile hitting Israel during the Persian Gulf War was believed to contain chemical or biological weapons. Though this was not the case, 40% of those in the vicinity of the blast reported breathing problems.

Right after the 2001 anthrax attacks in the first two weeks of October 2001, there were over 2300 false anthrax alarms in the United States. Some reported physical symptoms of what they believed to be anthrax.

Also in 2001, a man sprayed what was later found to be a window cleaner into a subway station in Maryland. 35 people were treated for nausea, headaches and sore throats.

Response to Outbreaks

Timothy F. Jones, of the Tennessee Department of Health recommends the following action be taken in the case of an outbreak;

- Attempt to separate persons with illness associated with the outbreak.
- Promptly perform physical examination and basic laboratory testing sufficient to exclude serious acute illness.
- Monitor and provide oxygen as necessary for hyperventilation.
- Minimize unnecessary exposure to medical procedures, emergency personnel, media or other potential anxiety-stimulating situations.
- Notify public health authorities of apparent outbreak.
- Openly communicate with physicians caring for other patients.
- Promptly communicate results of laboratory and environmental testing to patients.
- While maintaining confidentiality, explain that other people are experiencing similar symptoms and improving without complications.
- Remind patients that rumors and reports of "suspected causes" are not equivalent to confirmed results.
- Acknowledge that symptoms experienced by the patient are real.
- Explain potential contribution of anxiety to the patient's symptoms.
- Reassure patient that long-term sequelae from current illness are not expected.
- As appropriate, reassure patient that thorough clinical, epidemiologic and environmental investigations have identified no toxic cause for the outbreak or reason for further concern.

Some responses by authorities to MPI are not appropriate. Intense media coverage seems to exacerbate outbreaks. Once it is determined that the illness is psychogenic, it should not be given credence by authorities. For example, in the Singapore factory case study, calling in a medicine man to perform an exorcism seemed to perpetuate the outbreak. **(Wikipedia)**

False Memory **is a novel by the best-selling author Dean Koontz,
released in 1999.**

Editions

The main idea of the story is the creation of false memories or a memory
that never occurred.

False Memory was first released by Cemetery Dance Publications as a
limited edition hardcover (ISBN 1-881475-85-9) that came in two different
versions:

- A limited edition of 698 signed, numbered, and slipcased copies
  (signed by Dean Koontz and Phil Parks who did the illustrations for the
  Cemetery Dance versions).
- A lettered edition of 52 signed, lettered, and traycased copies (also
  signed by Dean Koontz and Phil Parks)

Plot Summary

Martie Rhodes helps her friend Susan Jagger who suffers
from agoraphobia; attend visits to psychologist Dr. Ahriman. Martie's
husband, Dusty Rhodes, tries to help his brother Skeet, by providing
employment in his roofing business. Skeet was in rehab for drug use, and
when he first appears in the story, he is high and tries to commit suicide by
jumping off a roof. Dusty decides to take him back to rehab due to drug
overdose.

Martie suddenly develops a mysterious case of autophobia, fear of
oneself, and returns home to find herself frightened by her own reflection.
Later, her condition worsens, and soon she becomes afraid of pointed
objects, although she is actually afraid of the harm she might cause with
them.

When Dusty leaves Skeet at the rehab center, he notices a shadow
lurking in his brother's room window. From this point on, strange things
begin happening to both Dusty and Martie, involving Skeet, Martie's
autophobia, and hypnotism.

The couple eventually discovers that they've both been progressively
brainwashed and programmed to obey Dr. Ahriman, a sexual psychopath
who drugs and indoctrinates his patients, then either repeatedly rapes them
or orders them to commit murders or suicide for his amusement. Dr.
Ahriman orders Susan to commit suicide by slitting her wrists after
discovering that she videotaped him having sex with her. The doctor has
also programmed Skeet, which explains his inability to fully recover from
drug use and distorted thinking. Dr. Ahriman establishes control, sending
patients almost instantly into a detached state of consciousness by stating a

name and then reading them a short haiku. He tries to justify this by stating that, by ordering certain patients to commit horrific crimes—mass murders, bombings, random shootings—he can force legislation in order to make the world a "better place."

Dr. Ahriman is eventually killed by another patient, who had a fear of Keanu Reeves, based on his character in *The Matrix*. The woman believes that Dr. Ahriman is one of the Machine agents trying to control her. Dusty and Martie, receiving a substantial inheritance from Martie's friend Susan, slowly begin to restore their shattered live

False Memory Syndrome

False memory syndrome (**FMS**) describes a condition in which a person's identity and relationships are affected by memories that are factually incorrect but that they strongly believe. Peter J. Freyd originated the term, which the False Memory Syndrome Foundation (FMSF) subsequently popularized. The term is not recognized as a mental disorder in any of the medical manuals, such as the ICD-10 or the DSM-5; however, the principle that memories can be altered by outside influences is overwhelmingly accepted by scientists.

False memories may be the result of recovered memory therapy, a term also defined by the FMSF in the early 1990s, which describes a range of therapy methods that are prone to creating confabulations. Some of the influential figures in the genesis of the theory are forensic psychologist Ralph Underwager, psychologist Elizabeth Loftus and sociologist Richard Ofshe.

Definition

False memory syndrome is a condition in which a person's identity and interpersonal relationships center on a memory of a traumatic experience that is objectively false but that the person strongly *believes*. Note that the syndrome is not characterized by false memories as such. We all have inaccurate memories. Rather, the syndrome is diagnosed when the memory is so deeply ingrained that it *orients* the individual's entire personality and lifestyle—disrupting other adaptive behavior. False memory syndrome is destructive because the person assiduously avoids confronting evidence that challenges the memory. Thus it takes on a life of its own; the memory becomes encapsulated and resistant to correction. Subjects may focus so strongly on the memory that it effectively distracts them from coping with real problems in their life.

The FMS concept is controversial, and the Diagnostic and Statistical Manual of Mental Disorders does not include it. Paul R. McHugh, member of the FMSF, stated that the term was not adopted into the fourth version of the manual due to the pertinent committee being headed by believers in recovered memory.

Recovered Memory Therapy

Recovered memory therapy is used to describe the therapeutic processes and methods that are believed to create false memories and false memory syndrome. These methods include hypnosis, sedatives and probing questions where the therapist believes repressed memories of traumatic events are the cause of their client's problems. The term is not listed in DSM-IV or used by any mainstream formal psychotherapy modality.

Memory consolidation becomes a critical element of false memory and recovered memory syndromes. Once stored in the hippocampus, the memory may last for years or even for life, regardless that the memorized event never actually took place. Obsession to a particular false memory, planted memory, or indoctrinated memory can shape a person's actions or even result in delusional disorder.

Mainstream psychiatric and psychological professional associations now harbor strong skepticism towards the notion of recovered memories of trauma. They argue that self-help books, and recovered memory therapists can influence adults to develop false memories. According to this theory, psychologists and psychiatrists may accidentally implant these false memories. The American Psychiatric Association and American Medical Association condemn such practices, whether they are formally called "Recovered Memory Therapy" or simply a collection of techniques that fit the description. In 1998, the Royal College of Psychiatrists Working Group on Reported Recovered Memories of Sexual Abuse wrote:

No evidence exists for the repression and recovery of verified, severely traumatic events, and their role in symptom formation has yet to be proved. There is also striking absence in the literature of well-corroborated cases of such repressed memories recovered through psychotherapy. Given the prevalence of childhood sexual abuse, even if only a small proportion are repressed and only some of them are subsequently recovered, there should be a significant number of corroborated cases. In fact there are none.

That such techniques have been used in the past is undeniable. Their continued use is cause for malpractice litigation worldwide. An Australian psychologist was de-registered for engaging in them.

Human memory is created and highly suggestible, and can create a wide variety of innocuous, embarrassing, and frightening memories through different techniques—including guided imagery, hypnosis, and suggestion by others. Though not all individuals exposed to these techniques develop memories, experiments suggest a significant number of people do, and will actively defend the existence of the events, even if told they were false and deliberately implanted. Questions about the possibility of false memories created an explosion of interest in suggestibility of human memory and resulted in an enormous increase in the knowledge about how memories are encoded, stored and recalled, producing pioneering experiments such as the lost in the mall technique. In Roediger and McDermott's (1995) experiment, subjects were presented with a list of related items (such as candy, sugar, honey) to study. When asked to recall the list, participants were just as, if not more, likely to recall semantically related words (such as sweet) than items that were actually studied, thus creating false memories. This experiment, though widely replicated, remains controversial due to debate considering that people may store semantically related items from a word list conceptually rather than as language, which could account for errors in recollection of words without the creation of false memories. Susan Clancy discovered that people claiming to have been victims of alien abductions are more likely to recall semantically related words than a control group in such an experiment.

The lost in the mall technique is a research method designed to implant a false memory of being lost in a shopping mall as a child to test whether discussing a false event could produce a "memory" of an event that did not happen. In her initial study, Elizabeth Loftus found that 25% of subjects came to develop a "memory" for the event which had never actually taken place. Extensions and variations of the lost in the mall technique found that an average of one third of experimental subjects could become convinced that they experienced things in childhood that had never really occurred—even highly traumatic, and impossible events.

Experimental researchers have demonstrated that memory cells in the hippocampus of mice can be modified to artificially create false memories.

Court Cases

Sexual abuse cases

The question of the accuracy and dependability of a repressed memory that someone has later recalled has contributed to some investigations and court cases, including cases of alleged sexual abuse or child sexual abuse (CSA). while others have been deemed confabulations or "false memories" that were not legally admissible. The research of Elizabeth Loftus has been used to counter claims of recovered memory in court and it has resulted in stricter requirements for the use of recovered memories being used in trials, as well as a greater requirement for corroborating evidence. In addition, some states no longer allow prosecution based on recovered memory testimony. Insurance companies have become reluctant to insure therapists against malpractice suits relating to recovered memories.

Supporters of recovered memories believe that there is "overwhelming evidence that the mind is capable of repressing traumatic memories of child sexual abuse." Whitfield states that the "false memory" defense is "seemingly sophisticated, but mostly contrived and often erroneous." He states that this defense has been created by "accused, convicted and self-confessed child molesters and their advocates" to try to "negate their abusive, criminal behavior." Brown states that when pro-false memory expert witnesses and attorneys state there is no causal connection between CSA and adult psychopathology, that CSA doesn't cause specific trauma-related problems like borderline and dissociative identity disorder, that other variables than CSA can explain the variance of adult psychopathology and that the long-term effects of CSA are non-specific and general, that this testimony is inaccurate and has the potential of misleading juries.

Malpractice cases

During the late 1990s, there were multiple lawsuits in the United States in which psychiatrists and psychologists were successfully sued, or settled out of court, on the charge of propagating iatrogenic memories of abuse, incest and satanic ritual abuse.

Some of these suits were brought by individuals who later declare that their recovered memories of incest or satanic ritual abuse had been false. The False Memory Syndrome Foundation uses the term *retractors* to describe these individuals, and have shared their stories publicly. There is debate regarding the total number of retractions as compared to the total number of allegations, and the reasons for retractions. **(Wikipedia)**

"Pictures In The Dark"

Follow the light that glows
Through your bedroom window.
Tonight, tonight, the fading twilight.
There's a hollow deep in the woods
Where you know you're crazy
To go, to go, not even meant to know
There are...

Pictures in the dark, I see all around.
Voices calling underground
And I'm watching the stars since the
World was found.

One, two, three.

[Chorus:]
Pictures in the dark, I see
Morpheus comes to me.
Pictures in the dark, I see.
Aurora sets you free.

And in the deepest dark
You come to a maze in.
The night, the night, the fading twilight.
And you shiver the glistening path
Where you know you're crazy
To go, to go, not even meant to know.
There are...

Pictures in the dark, I see all around.
Voices calling underground
And I'm watching the stars since the
World was found.

One, two, three.

Lost in my dreams.

This night will never end.
You can only fly in your dreams.
Midnight will be your friend.
Drift away on starlight beams.
Clocks are ticking the night away.
You can only fly until dawn ascends.

[Chorus]

Pictures in the dark I see,
Aurora sets you free

Holding a fantasy that changes you a way through the door,
The door (and) through the day is born
As the sunlight shines through the window,
will you remember the night,
The night, that crazy starlight
There are...
Pictures in the dark,
see all around

[Chorus]

The moon shines starlight beams
And you'll be flying in your dreams.

[Chorus]

Mike Oldfield

Words inspired by the life of Robert F. Kennedy

At no other time in history did man reveal his capacity to destroy, and cause suffering, than during the Second World War. During this time bonfires were fuelled with books written by the greatest minds ever, while the public watched, and treated it as a spectacle.

These act was not pardoned; forgiven, or forgotten; history has shown that man have has taken himself as a consequence to the brink of self-annihilation. Surely, by now man has learned his lesson, and The Lord will provide us all His saving grace;

The Jews.

Chapter Two

Killing Machines

Plymouth has a port; it's one of the most famous in the world, primarily due to the Mayflower full of pilgrims, being launched from here in 1620. The Port is still in use today; it holds back the waters from the Ocean that stream into the English Channel, so the harbour is a safe place for ships to dock.

Plymouth appears on the surface to be as welcoming to people, as the harbour is for boats, tragically, however, this is a hideous illusion, despite one being able to reach the city center by walking a short distance from the train station down a shallow till until one greets a pedestrian market area full of stores. A little further, on the other side of a street that runs through the heart of Plymouth is the Barbican District full of cobbled streets, and ornamented lampposts, that remind one of the time Charles Dickens lived in; history is brought to life for people to experience, and enjoy.

What a first time visitor to this town will find peculiar, however, is that there aren't that many, relatively speaking, tourists to be found; why should this be so?

I found out on February the 2nd, 2016, when I went to the train station to purchase a train ticket to a nearby town called Dartmouth. When I arrived I couldn't find one teller that was hospitable in the slightest; I mentioned my destination to a teller, and she responded; "What?" Not, "Pardon me?"; "Excuse me?"; "I didn't catch what you said; would you mind repeating it, please?" Her hair was dishevelled; when she spoke I had to lean forward to hear what she was saying. She was hard to look at, and also listen too.

The repeated where I wanted to travel too, "Dartmouth!" She belched, "You can get there by train, and bus, but I don't think the train goes there this time of year."

"I did some research before coming here; it's awfully confusing on the internet; there appears to be a number of different ways one can get there by public transportation."

"There is; it's winter now, so you're limited in how you can get there. I'll get a map. That should make it easier to me to find out how you get to Dartmouth."

Does what I've described sound like a region of Britain that is structured to have a thriving tourist industry? The answer, quite obviously, is; no? It could take some while on the internet to figure out how to get there due to various sites having conflicting information, and because it takes, no matter the time of year, and the mode of public transportation one uses, a lengthy period of time to travel the short distance to the coastal town, Dartmouth.

Tourists, generally speaking, like to see as much as possible in a day, but due to the way Britain is now, it could take several days to explore just a short stretch of the county of Devonshire's coast. One could conclude due to the factors mentioned that the people who live in this part of Britain do not want tourists to come here, and deliberately make an effort to hamper a person from having as easy passage through this region for this reason; why would Britain do such a thing? Does She not want to acquire more money from the tourist industry?

The answer is that more money is not the prize Britain is seeking these days; the coffee shops; malls; theatres, and movie houses, in Plymouth are already a full capacity with the town's residents; Britain is a society full of consumers that spend a large portion of the day buying products; in order for this to be possible they steal people's I.D. Britain, therefore, contrary to how travel books make it appear; is a jungle; the people live by the laws of the jungle; as soon as a tourist gets trapped in heavy foliage, figuratively speaking, cannibals soon arrive at the scene, and shortly after no body will see that tourist again; the person will go missing. The only proof remaining that the person existed is his I.D. that no one claim is their own.

Britain has radically changed from how it used to be three decades ago; it was quite easy back then to contact a person. People registered to vote so their names could be found on a list anyone could access; which included a person's profession, and their home address; most people had their name; address, and home phone number, in the telephone directory, as well.

Practically everybody who lives in Britain now, in contrast, is reluctant, that being an understatement, to hand out such information, and they will fail to provide a logical reason why they don't. The most common answer is; "I'm not required to give you my name". They rarely say, yes, or no; even if a person is required to do so, they rarely provide their name.

The simple; plain, and obvious, truth of the matter is that if a person has done nothing wrong; if he obeys the law; he's got nothing to hide. People steal the I.D. of other people, and use their I.D. to open bank accounts; and use other people's names to acquire club memberships; library cards; all sorts of things. Their intention is to hide the money they spend because it has been acquired illegally; they do not want it traced back to themselves; that way they can avoid judicial penalty for committing these crimes; "thou shall not covet another person's property"; is a commandment that doesn't registers in the minds of Britain's brainwashed public today.

I bought tickets to Plymouth from Victoria Station, and was provided a timetable that makes absolutely no sense.

Timetable and Fares Information

Ticket :	ADVANCE	From :	Three Bridges	To :	Plymouth
Type :	SINGLE	Route :	APREADING	RCard :	None.
Total :	£23.00		1 Adult @ £23.00		
Restrictions :	GWRADVANCE				

Outward journey

Date :	Sun 31/01/2016	Changes :	2		
From :	Three Bridges	Depart :	21:35		
To :	Plymouth	Arrive :	05:35	Duration :	8 hr 0 min

Type	Dep.	From	Arr.	At	Dur.	Service
Train	21:35	Three Bridges	23:02	London Victoria	1 hr 27 min	
Underground		London Victoria		London Paddingtn	0 hr 17 min	Zone 1
	Line	Victoria	Change	Oxford Circus	00:04	
	Line	Bakerloo	Exit	Paddington Underground Station	00:08	
Train	23:50	London Paddingtn	05:35	Plymouth	5 hr 45 min	GW117600

Super off-peak tickets are valid on this journey.

Reservations and Catering

Three Bridges - London Victoria (SOUTHERN)

Reservations: Not possible   Catering   No catering services available

London Paddingtn - Plymouth (FIRST GREAT WESTERN)

Reservations: Advised   Catering   Cold Buffet

30-Jan-2016 18:38:15 (RS07025)   Information is only valid at date and time of printing. Fares shown are subject to availability.

Upon arrival at Plymouth I was told they don't accept Travel cards.

=== CARD PAYMENT ===

North Road
Plymouth
PL4 6AB
MASTERCARD
Card: ***********9756
SWIPED
Amount :
    GBP12.80
Ref: 031240
Merchant: ***28351
TID: ****1040
Date: 03/02/16 Time: 11:49:51

CARDHOLDER COPY
PLEASE RETAIN FOR YOUR RECORDS

DECLINED

Card Payment Ref: 242040001YTS

3580T01W30M2040 7382 6597 11:47 03-02-16

Thank you and have a good journey
www.GWR.com
for times, tickets, news and offers

Southern
DEBIT/CREDIT CARD SALES VOUCHER
CARDHOLDER'S COPY

SWIPED   MASTERCARD
Card No.(PAN): **** **** **** 9756

Sequence Num.   09
Merchant ID: 54463311
Auth Code: 14348

Total Value: £59·90

Description: 3 RAIL TICKETS
Issuing Office: VICTORIA

Issuing NLC: 5426
Terminal ID: 27620072
Machine Number: 5153
Window Number: 87
Transaction No.:21060
        30 JNR 16 18:44

*****NOT VALID FOR TRAVEL*****
Please retain for your records

Please debit my Account

Printed 18:44 on 30.JAN.16

VALID ONLY WITH TRAVEL TICKET

Class	Ticket Type	Date of issue	Price
STD	SEAT	30·JNR·16	£0·00X

From
LONDON PADDINGTN   ONE   57601   5153542687

To
PLYMOUTH *   Valid at
                   23:50 HOURS ON 31·JNR·16
Coach   Seats
A   07F

SEAT RESERVATION

Printed 18:44 on 30-JAN-16

⊖ **Day Travelcard**

Class	Ticket Type	Start Date	Price
STD	SUP OFFPK DAY	31·JNR·16	£15·90X

Valid until
31·JNR·16ᴬ   5153542687

Between
THREE BRIDGES * & LONDON ZONES 1-6   57599
Route
NOT GATWICK EXP

Valid within zone(s) indicated. Not for resale
Valid as advertised

Printed 18:44 on 30-JAN-16

I called every customer service number I could think of, only to discover after close to three hours on the phone at Drake's reception desk, that the customer service numbers are fraudulent; the people who service customers in Britain are situated in Mumbai, and they have no contact with any

---

Path — *Supporting people in housing need*
Rough Sleeper Team
Client Information

Path ID No:

Date:	2|2|16	Referred to Safe Sleep by:	N/A

BASIC INFORMATION

Client's Name: NIGEL SHINDLER.

NINO: Awaiting trace letter response   D.O.B: 16/8/64.

UK Citizen: Yes [✓]   No [ ]   Nationality / Status:

Tel Number or point of contact: N/A

Current Address: NFA

[ ] Evidenced RS   [✓] Not Evidenced RS   [ ] Other (ie SWEP)

Family/ Next of Kin /Others Information

Names (In full): NICOLA SHINDLER

Relationship: SISTER

Address: CANADA (82 CASTLE ROCK ROAD, RICHMOND HILL, TORONTO)

Contact Number: TBC

Family Information (relationships, contact etc): N/A

HOUSING HISTORY

LOCAL Connection to Plymouth? Yes [ ] No [ ]   If 'No' Where is connection? PCC TO ADVISE — SEE HOUSING INFO

If Local connection is not Plymouth, please confirm what assistance can be offered with regard to accommodation with the Local Authority where connection is held.

Address immediately prior to becoming homeless	TRAVELLING UK — ADDRESS IN MERSEYSIDE
Tenure Type:	Landlord:
Date moved in	Date moved out
Reason you left:	

1|

---

train station in Britain; I was then without shelter.

Housing History for the last 5 years (include reason for leaving and details of any arrears owed)

Sept - Nov	53 Ash St, Flat 5, ▪ Southport, Merseyside,	Private rent
Sept 2015	Returned from Canada (renting) PR8 6JE	Couldn't Family home
		Sheltie
	Ended Nov 27, Informed police (confirmed	fraudulent ⟶
	involvement with scheme).	unable to return.
	– Attended Open House in Crawley.	
	– Mainly around UK ⟶ then to South West.	
NOTE:		
Aug 2008	Grenville Rd, Plymouth	Father able to return.

Employment History (Include dates, type, employer and length of employment)

Ex Armed Forces (tick): ☐

Artist, writer, from Dec 2012

⟶ presently with publishing company

Current Income

Is the client in receipt of benefits? Yes ☐ No ☑	Benefit(s) & Amount (s):	(He previously claimed) JSA .
Date of next benefit payment?	N/A	
Any Debts including previous rent arrears and outstanding social loans? Yes ☑ No ☐	Details: Credit card – use for transport / food Credit Bank of Material	

Does the client have any of the following ID?

☑ Passport ☐ Driving licence ☐ Benefits Ltr ☑ Birth Cert ☑ Other

2|

National ID card

Current Medical Information

GP & Surgery Details:	No GP
Please give details of any of the following;	
General Health Issues?	N/A – full physical health check Ok.
Physical Disabilities?	
Mental Health issues?	N/A .
Alcohol issues?	N/A
Drug issues?	N/A .
Any medication prescribed?	No meds required

Offending History (Include dates where possible)

N/A .

Is the client currently working with Probation? Yes ☐ No ☑	*Ensure Offender manager details are added below
Is the client currently on bail? Yes ☐ No ☑	Offence / Date to answer bail :

Local Authority /Other Agencies / individuals offering support (add name and contact numbers)

N/A

3 |

Risk Assessment

Risk	If the answer is 'yes' to any of the questions, please expand below
Do you have a history of damaging property? No ☑	
Have you ever been violent or aggressive towards others? No ☑	
Do you have convictions for violence? No ☑	
Do you have any convictions for Sex Offences? No ☑	
Do you have any convictions for Arson? No ☑	
Do you have any other criminal convictions? No ☑	
Have you ever experienced Depression or anxiety? No ☑	
Do you have a history of self harm or attempted suicide? No ☑	
Have you ever been admitted to a psychiatric hospital? No ☑	
Do you have a mental health diagnoses? No ☑	
Do you have a history of substance misuse? No ☑	
MAPPA? No ☑	
S/O Register? No ☑	

Any additional info / Risk Management suggestions :

N/A

I confirm that the information given in this risk assessment is correct to the best of my knowledge.

Client Signature: ....

Print Name: .... Nicola .... Date: 1/2/16

4|

Gender	Male	☑	Female	☐	Transgender	☐	Age	51

Do you consider yourself to have a disability?		Yes ☐	No ☑

If yes, what sort of disability?	Sight impairment	☐	Physical disability	☐
Mobility	☐	Hearing impairment	☐	Learning disability	☐
Progressive	☐	Mental health disability	☐	Prefer not to say	☐
Wheelchair dependant	☐	Occasional Wheel chair user		☐
Difficulty turning/gripping with hands	☐		

Marriage/Civil Partnership

Married	☐	Widowed	☐	Divorced	☐	Unknown/refused	☐
In Civil Partnership		☐	Separated but legally still in civil partnership		☐
Separated but legally married	☐	**Single**	☑	
Formerly in Civil partnership now dissolved	☐	Surviving partner from a civil partnership		☐

Which group best describes your ethnicity?

White	British	☐	Irish	☐	Other	☐
Black or Black British	Caribbean	☐	African	☐	Other	☐
Asian or Asian British	Indian	☐	Pakistani	☐	Bangladeshi	☐
	Chinese	☐	Japanese	☐	Other	☐
Mixed	White and black Caribbean	☐	White and black African	☐
	White and Asian	☐	Other	☐
Gypsies and travellers	Gypsy	☐	Romany	☐	Irish traveller	☐
	Other	☐	Prefer not to say		☐

Sexuality	Heterosexual ☐	Gay man ☐	Lesbian ☐	Bisexual ☐

Religion	

Pregnancy	Currently pregnant ☐	On Maternity benefit ☐	On paternity benefit ☐

Refuse to answer this section	☑

EEA National		Non EEA National	

51

- I can confirm that the information given in this form and in the attached risk assessment is correct, true and to the best of my knowledge.

- I agree to allow any information disclosed to PATH to be shared with <u>any</u> other organisation that may be able to assist in providing suitable accommodation including:

Housing providers (PCC Housing Options, BCHA George House, Salvation Army Devonport House, Westcountry, DCHA, Stonham etc), Devon & Cornwall Police, Probation, Shekinah Day Service, Benefits Agencies DWP & Job centre, GP and Health Professionals, Mental Health Professionals, and Drug & alcohol workers.

- I understand that each organisation my information could be shared with will have their own confidentiality policy, and in normal circumstances will not disclose my information further. However, if there is a serious risk of harm to others, or yourself, section 115 of the Crime and Disorder Act 1998 provides for my information to be discussed as is considered necessary.

Client
Signature: *(signature)*

Print name: NIGEL EDWARD STRANGER

Date: 2/1/16

✱ FORM COMPLETED BY CRAIG HARRIS (PATH)
CH (1/1/16)

6|

The Shekinah referred me to Town Council for help in acquiring shelter, I saw Jason Cross early the second morning after arriving in Plymouth, and he made out two referrals; one for the Salvation Army; the other for George House, and told me to return at 2pm to see if there was a vacancy at either; below is his official appointment card.

I received no help in getting shelter, and stayed at Plymouth University's library throughout most of the time I was in Plymouth.

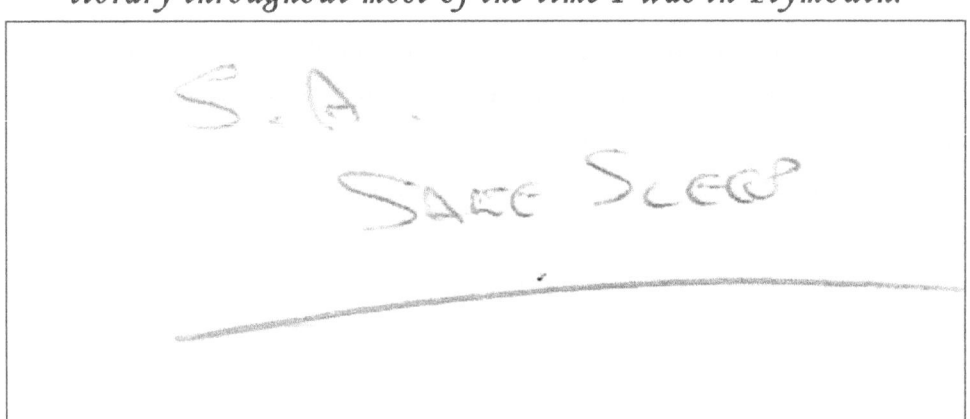

The first day I was at The Shekinah Mission I had a every pawnable item stolen from my luggage; the objective is make sure I had no alternative source of funds I could use escape.

Plymouth University Library Visitor Pass   Number 128394

Please familiarise yourself with the notes as your signature indicates that these have been read and understood

- We can only provide a limited enquiry service for visitors
- In the event of emergency please leave by the nearest exit & report to an assembly point
- This is a no smoking building
- All visitors are subject to the Health & Safety at Work Act 1974 & University Regulations
- All children under the age of 14 must be accompanied by a responsible adult

Your visitor pass **must** be kept with you at all times.

For health and safety reasons, please return to Reception on leaving and record your time of departure.

Soup Run Times

7 Days a week
Shekinah mission – 01752 203480

A. Charles Cross Police Station: 9:00 – 9:15pm

B. The Wedding Cake: 9:30 – 9:45pm

C. Devils Point: 10:00 – 10:15pm

D. Kings Road (Stonehouse Creek): 10:30 – 10:45pm

If you don't get the "runs" after eating this food, you're not far from an early grave, because that means your body failed to evacuate the toxin deposited in the mud like soup; the spoilt sandwiches, and bruised, worm infested, fruit.
I was so ill on the two occasions I eat this food, I felt like I'd swallowed a barrel of rat poison.

The person who is not given to the opportunity to travel is, thus, confined without consent; this is called "kidnapping". Due to not having "connections", the person can be killed; declared "missing"; I.D. used for money laundering; and who's the wiser? Definitely not the rats that participate in this scheme.

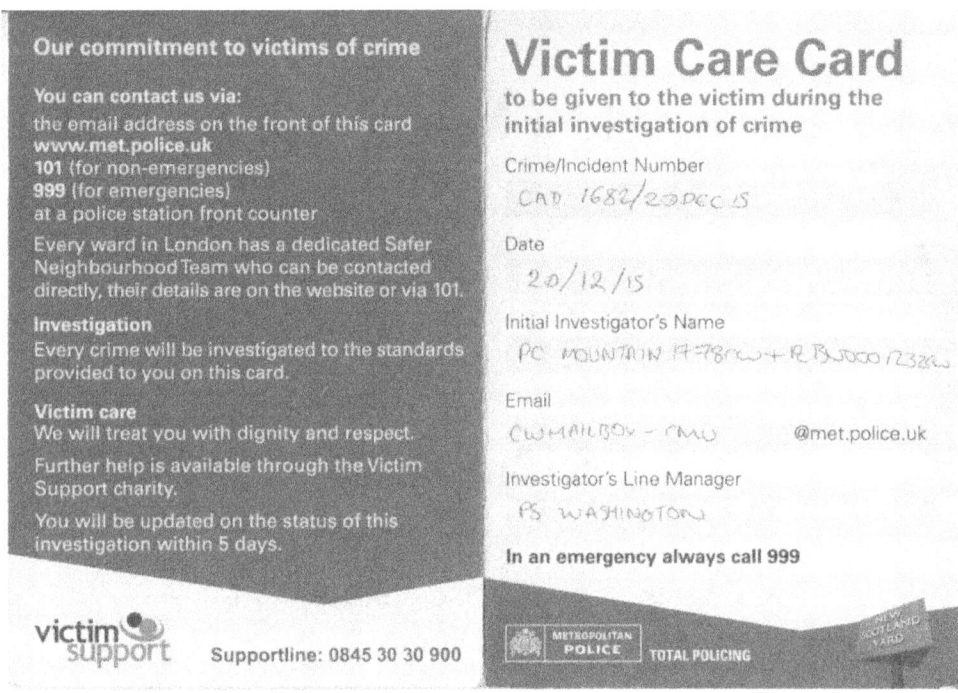

Victim Care Card
to be given to the victim during the
initial investigation of crime

Crime/Incident Number

CAD 1682/20 DEC 15

Date

20/12/15

Initial Investigator's Name

PC MOUNTAIN F778CW + R BJ000 7232CW

Email

CWMAILBOX - CMU   @met.police.uk

Investigator's Line Manager

PS WASHINGTON

In an emergency always call 999

METROPOLITAN POLICE   TOTAL POLICING

A security guard at Victoria Coach Station attempted to steal my belongings while I was seated in the waiting area; he then threatened to get violent with me if I didn't eave the station. I called the police from a phone booth immediately outside the station, two female officers arrived shortly after, and I was told they would investigate. I haven't heard a word from the police about the matter since.

The officer should have discussed with you:

- His/her role as the initial investigator of your crime.
- Any particular vulnerability that you feel you may have and offered referral to Victim Support.
- Whether you have been a victim of crime in the past.
- The details of the crime, any potential evidence such as forensics, CCTV or house-to-house enquiries and your expectations of the police response.
- Crime prevention advice.
- Your contact details including email address if applicable
- The next steps including what happens to your crime report and decisions about further investigation.
- Your understanding of the information given

The Next Steps

Initial investigation — Current stage

↓

Crime recorded on police system — Before the end of the officer's shift

↓

Crime reviewed and crime number issued — Within 36 hours of initial investigation

↓

Crime transferred to investigating officer for further investigation

Investigation closed unless further evidence comes to light

Rate your PC

Ensuring victims receive the best possible service, improving public confidence and victim satisfaction are key to what we do in the Metropolitan Police Service.

Please follow the link below in order to rate the service you received from us:

http://www.smartsurvey.co.uk/s/westminster

BUD00 1232cw

Name of PC: MOUNTAIN   **Number:** ff8cw

Sergeants name: PS WASHINGTON

Team and cluster: C-SOUTH

CAD/Crime No: 1682/20DEC15

Date: 20/12/15

CALL YOUR
LOCAL
POLICE
101
IN AN EMERGENCY
ALWAYS CALL 999

www.met.police.uk

METROPOLITAN POLICE   TOTAL POLICING   Westminster

I.D. theft; fraudulent bank accounts, and money laundering, enable Britain's consumers to spend money without having to earn it. Most of Britain's residents have impressive sounding job titles, but do very little of anything throughout the day. Mrs. Kate Knight, for example, "works "in the same building as Plymouth's Citizen's Advice Borough;; what does she do? Her job, as she put it to me; entails "signposting", which means, according to her description, that she hands people pamphlets; they aren't posted anywhere so people have to make an appointment to see her in her office; which contains a desk, and a telephone. I told her she's a cannibal because people lose their homes, and die, in order to sustain her lifestyle. She didn't disagree, but I was told to leave the office because she had another appointment.

Shelter

Kate Knight
Advice Assistant

Ernest English House
Buckwell Street
Plymouth PL1 2DA
t   0344 515 2345
e   kate_knight@shelter.org.uk

shelter.org.uk
Registered charity in England and Wales (263710)
and in Scotland (SC002327)

I spoke to the woman who manages the staff at Victoria Station's ticket booth, and was told to fill out a "customer comment" form. I returned later that day with the form completed, and when I handed it to her she told me she would give me a refund. She walked away with the form, and when she returned she handed me a photocopy of the form, and the tickets I'd bought; there wasn't a refund; there was no apology; and I haven't heard anything from National Rail on the matter since; why would I? The employees are doing what they're told to do, and they continue doing what they do, because they enjoy destroying people's lives because they are **Evil.**

Moisten along this edge

Your details

Contact us:
Southern: 0345 127 29 20
Gatwick Express: 0345 850 1530

Personal details

Title Mr ☐ Mrs ☐ Ms ☐ Other [D][R]
Initials [N][B] Suurname [S][H][I][N][D][L][E][R]
Address [C][J] [A][S][H] [S][T] [A][L][A][T] [S]
[M][E][R][S][E][Y][S][I][D][E] [S][O][U][T][H][P][O][R][T]
Postcode [P][R][8] [G][J][E] Best contact number
Email [S][J][N][a][S][F][R][O][M][E][A][R][T][H][@][G][M][A][I][L][.][C][O][M]

Your journey

From ... THREE BRIDGES ...
To ... PLYMOUTH ...
PADDINGTON
Departure time (24 hrs) [2][3][5][0] Arrival time (24 hrs) [5][4][5][0]
Date of travel (DD/MM/YY) [3][0][/][0][1][/][1][6]

Your comments

We value your opinion about our services, good or bad. Let us know what you think by completing this form and sending it to our Customer Services team. If you prefer, you can hand this form to a member of staff, or speak to them directly. You can also call us, or send us your comments via our website. We will always get back to you, unless you tell us you'd rather we didn't. We aim to respond to your comments fully within ten working days. If this is not possible we will write to you to let you know. We will then keep in touch regularly until we can give you a full reply.

Ticket details, or attach ticket here

If you hold a daily or weekly ticket, please attach it here. If you hold a Season ticket, please attach a copy or complete the details below (Please note that in order to consider compensation for a delay, we will need a copy of the ticket)

from Season ticket holders: Single [X] Return ☐ ticket type)
Season ticket holders: Monthly ☐ Annual ☐ Weekly ☐ Other ☐
Photocard no. ☐☐☐☐☐☐☐☐☐☐☐☐
(season ticket holders only)

Ticket price £ ☐☐[6][9][.][0][0]

ON 30/01/16, I WOULD TICKETS AT VICTORIA STATION, LONDON, TO TRAVEL FROM THREE BRIDGES, ZONE 1-6 LONDON, TO PLYMOUTH.
UPON ARRIVAL AT PLYMOUTH I WAS TOLD THEY DO NOT ACCEPT TRAVEL CARD, I WAS THEN STRANDED IN PLYMOUTH UPON 11/02/16, WHEN A CURATE AT ST ANDREW'S CHURCH BOUGHT A COACH TICKET TO RETURN TO LONDON.
FOR 10 DAYS I WAS WITHOUT PERMA-NANT SHELTER, I DID NOT HAVE PROPER FOOD, I WAS ASSAULTED ON SEVERAL OCCASIONS, REPORTED EACH TO THE POLICE, AND HAD ALL MY VALUABLES STOLE A VALUE $500 ≈ £250

Signature *Nigel Shindle* Date 11/02/2016
(email.)

Do you require a reply? Yes [X] No ☐   For office use only ☐☐☐☐☐☐☐☐☐☐☐☐

The information provided will be used in accordance with our Privacy Policy. Visit southernrailway.com/privacypolicy for more information

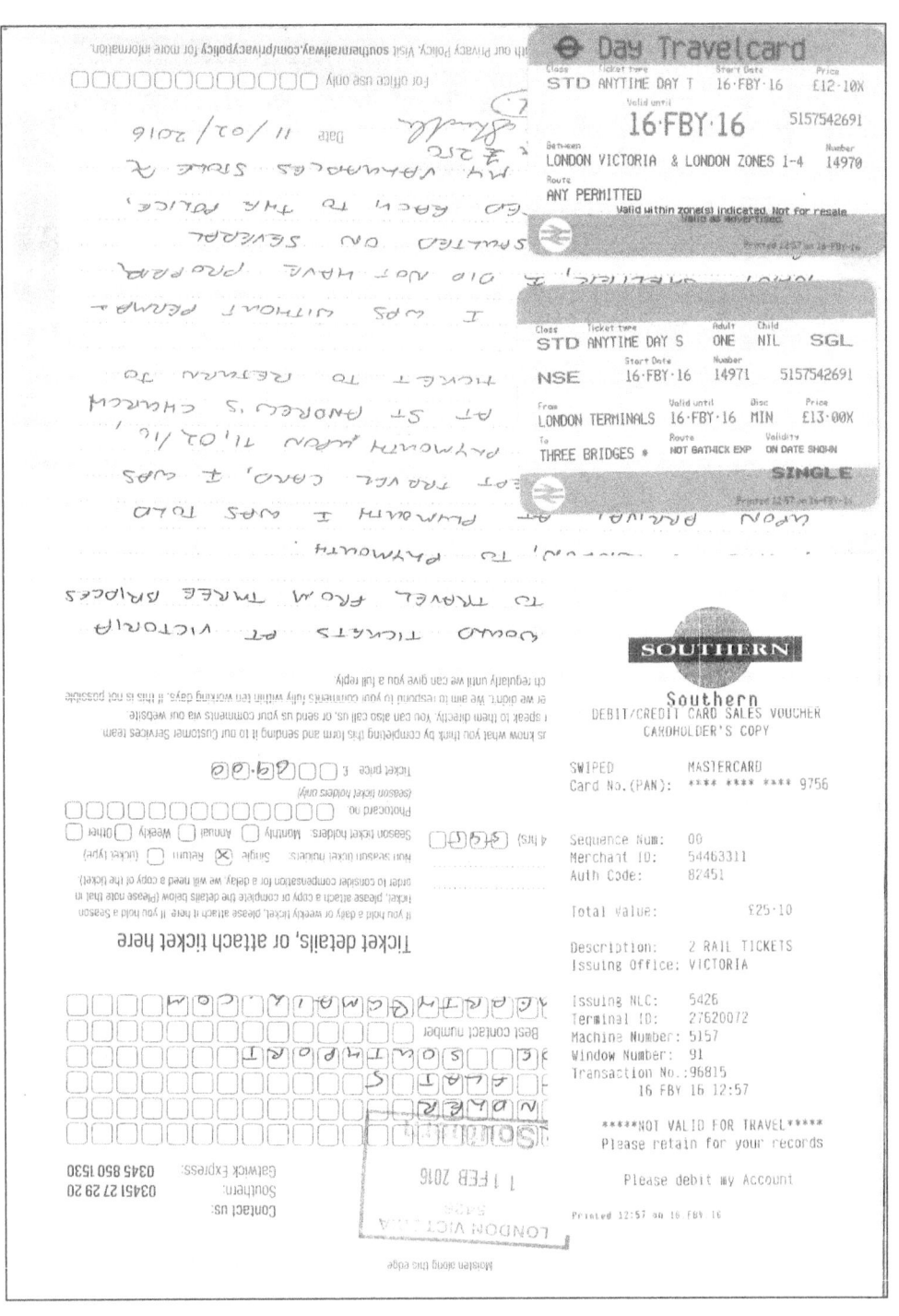

Day Travelcard

Class	Ticket type		Start Date	Price
STD	ANYTIME DAY T		16·FBY·16	£12·10X

Valid until
16·FBY·16   5157542691

Between		Number
LONDON VICTORIA & LONDON ZONES 1-4		14970

Route
ANY PERMITTED

Valid within zone(s) indicated. Not for resale
Valid as advertised.

Printed 1257 on 16·FBY·16

Class	Ticket type	Adult	Child	
STD	ANYTIME DAY S	ONE	NIL	SGL

	Start Date	Number	
NSE	16·FBY·16	14971	5157542691

From	Valid until	Disc	Price
LONDON TERMINALS	16·FBY·16	MIN	£13·00X

To	Route	Validity
THREE BRIDGES *	NOT BATH1CK EXP	ON DATE SHOWN

SINGLE

Printed 1257 on 16·FBY·16

SOUTHERN

Southern
DEBIT/CREDIT CARD SALES VOUCHER
CARDHOLDER'S COPY

SWIPED   MASTERCARD
Card No.(PAN): **** **** **** 9756

Sequence Num: 00
Merchant ID: 54463311
Auth Code: 82451

Total Value: £25-10

Description: 2 RAIL TICKETS
Issuing Office: VICTORIA

Issuing NLC: 5426
Terminal ID: 27620072
Machine Number: 5157
Window Number: 91
Transaction No.:96815
              16 FBY 16 12:57

*****NOT VALID FOR TRAVEL*****
Please retain for your records

Please debit my Account

Printed 12:57 on 16 FBY 16

The following handwritten form content appears upside-down on the page:

 with our Privacy Policy, visit southernrailway.com/privacypolicy for more information.

For office use only ☐☐☐☐☐☐☐☐☐☐☐☐☐

Date 11 /02/ 2016

... my grievances since I ... ... to each to the police, ... on several ... I did not have proper ... I was without permit... ... ticket to return to ... at St Andrew's Church ... Plymouth upon 11/02/16, ... travel card, I was ... Plymouth at ... ... to Plymouth.

Oamo tickets at Victoria to travel from three bridges

... regularly until we can give you a full reply.

... know what you think by completing this form and sending it to our Customer Services team ... speak to them directly. You can also call us, or send us your comments via our website ... er we didn't. We aim to respond to your comments fully within ten working days. If this is not possible

Ticket price £ ☐☐.☐☐

(Season ticket holders only)
Photocard no ☐☐☐☐☐☐☐☐☐☐☐☐☐

Season ticket holders: Monthly ☐ Annual ☐ Weekly ☐ Other ☐
Non season ticket holders: Single ☐ Return ☒ (ticket type)   4 (hrs) ☐☐☐

order to consider compensation for a delay, we will need a copy of the ticket)
ticket, please attach a copy or complete the details below (Please note that in
If you hold a daily or weekly ticket, please attach it here. If you hold a season

Ticket details, or attach ticket here

☐☐☐☐☐☐☐☐☐ ... @gmail.com
Best contact number ☐☐☐☐☐☐☐☐☐☐☐

Contact us:
Southern: 03451 27 29 20
Gatwick Express: 0345 850 1530

1 FEB 2016
LONDON VICT...
5426

Moisten along this edge

My commentary was so valued by the rail service that a few days later a middle aged man, (according to his label his name is, John), refused to sell me a ticket at Three Bridges; I was, however, able to later purchase a ticket at Victoria Station, proving that the staff at Plymouth Train Station had deliberately blocked me from leaving, and, as I have proven,

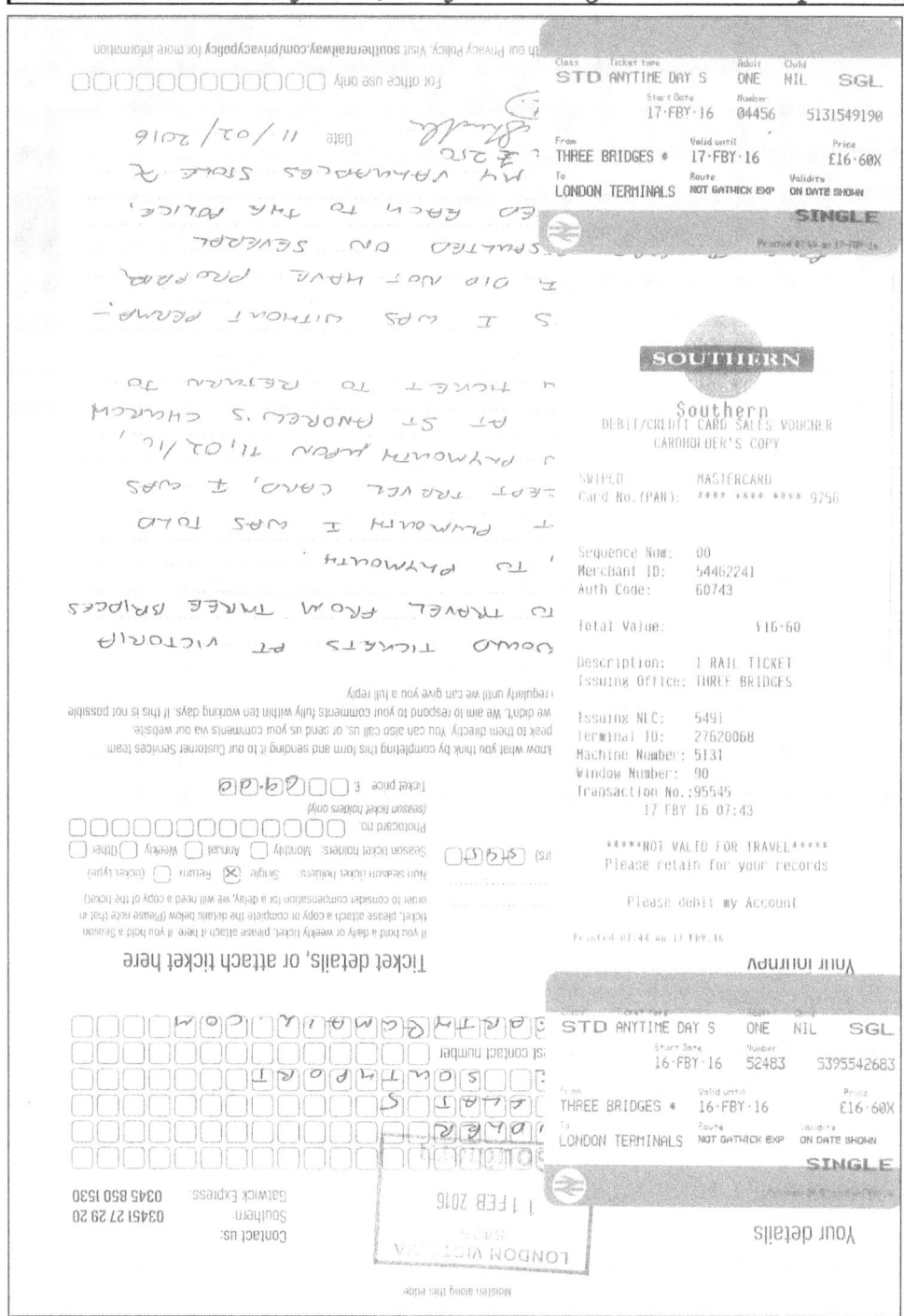

THE DARK

the intent was murder, financial gain; there is also a sexual component, which compels them to repeat the same act.

To whom it may concern

I am writing in regard to an incident that occurred at Horley Station, early evening, 26/01/2016. I informed the young man, who appeared to be in his late teens, early twenties, at the ticket booth, (he was only willing to provide his first name, Mark; he informed me I don't need his full name because he's not required to provide it); he's bald, and speaks with a cockney accent, that I would like a refund for a ticket I've bought, and also buy a ticket for Victoria Station in London, and a Zone 1-6 Day Travelcard for the underground.

He told me he couldn't provide a refund for the ticket I'd bought. I told him he could, and he simply didn't know how to do it. He was adamant that he was correct, and was certain also that he was wiser than myself, (we had a brief exchange of words after I discovered he couldn't do his job properly, and as is so often the case, it became childish, and silly, due to level of the mentality of the person I'm dealing with with).

He then stated that if I wanted a refund, there was a ten pound administration fee. I knew that was ridiculous, and reiterated that he didn't know what he was doing. He printed off a number of pages that he stated prove that the fee is ten pounds; I have since read these pages, which, actually, confirm there's no charge. He then said that if I didn't like that, I could go to another station. It was raining quite heavily at the time, and I certainly wasn't prepared to compromise my health, and well-being, due to the ignorance, and incompetence, of a cocky young man who appeared to have no respect for his elders.

"I'll pay at Victoria"; I said, and went through the gate by using the ticket I'd requested a refund for. I spoke briefly with the man at the gate who was much older than the immature young lad at the ticket booth about what had just transpired, but he appeared indifferent. I then asked to speak with the supervisor; he told me he was the supervisor.

I then walked to the appropriate platform for the train to Victoria, and moments later, to my utmost surprise, I discovered the supervisor had followed me, and appeared to have another piece of useless paper he wanted to hand to me. I declined, and said the ten I've got is more than enough. He then asked WHERE I'M FROM. I failed to fathom the relevance, and said; "I'm asked that bizarre question all the time; people want to know where I'm from, and also, where I'm from originally - meaning, where I'm born; meanwhile they don't know the names of the people who live across the street from where they live; I thought my answer was frank, and honest. He then said, "I'M AMERICAN, AND THAT'S WHY I'M RUDE". I stated that his words were due to the defense mechanism, "projection", and, actually, he was the one who had shown dim-wittedness one usually associates with the sort of crude; uncultivated; savage like, behaviour, he'd shown.

He returned to the ticket area, and I followed shortly after, and showed him proof that both he, and the bald, Goth looking, young man, in the ticket booth were incorrect; the refund on a two way trip between Horley and Three Bridges is 90p. He cleared his throat, and nothing more. I saw him later on the platform, and made it known to him that I thought I deserved an apology... I was ignored.

Thank you in anticipation of your response.

I would like to be compensated for the inconvenience caused by the two people mentioned that night, and I would like to know how Southern intends on improving it's service; a lot of people are reliant on Southern to get to and from work every day, thus, this matter is not just a personal matter, and one that adversely affects a great number of people other than myself.

        Kind regards

        Sincerely

Mail Delivery Subsystem <mailer-daemon@googlemail.com> 14:48 (1 hour ago)
to me

This is an automatically generated Delivery Status Notification

THIS IS A WARNING MESSAGE ONLY.

YOU DO NOT NEED TO RESEND YOUR MESSAGE.

Delivery to the following recipient has been delayed:

  comments@souternrailway.com

Message will be retried for 2 more day(s)

Technical details of temporary failure:
The recipient server did not accept our requests to connect. Learn more
athttps://support.google.com/mail/answer/7720
[souternrailway.com 141.8.225.226: socket error]

----- Original message -----

DKIM-Signature: v=1; a=rsa-sha256; c=relaxed/relaxed;
    d=gmail.com; s=20120113;
    h=mime-version:date:message-id:subject:from:to:content-type;
    bh=IM1g/gL4aoYSYqXaHyNhHR8/hvLptv5vOIo+Uirakxk=;

b=LNrFzgaQE+wO4oAtKB7Wq20aLPT1MWZ8kzrzl7/Bim3NrrQO1m3T
m8cKZChgQKapn8

 p7QbTVFtOgukZpSAcgPeFA5T/gs38Q3sFrHVRqW1efaMnCljwVXpO/x
rJLM5pHOuBP5G

 MP1l436YU5SxBhzXMZiGdXNs88QW3yBuvXR/asT1dQuddfPITF3YT
U2Zl+xChMJ5Juj4

 tdb9u2P2vxeC3DR4fD9kGBGxvBHCSNLFH7R2dhgmXMt7ELRU7RR
XVRbpsjGS2AaSQTFS

 1HW4RV/RE5qPq11ldApWe2o9XG8/jvDeM8NLNoRocExP+MTCDRY
n8Djnmn3ewWCRtyry
    dBtQ==
X-Google-DKIM-Signature: v=1; a=rsa-sha256; c=relaxed/relaxed;
    d=1e100.net; s=20130820;
    h=x-gm-message-state:mime-version:date:message-id:subject:from:to
     :content-type;
    bh=IM1g/gL4aoYSYqXaHyNhHR8/hvLptv5vOIo+Uirakxk=;

b=fwFlZigsIvp7h/LsrxvUV8YgI8PK/9N8DSy/b9fZbcl0MUvjoQeB9NhhU

136

LOPSdX5ue

j/1gYpkGLBsOKdhBGiqf8IREoPEajw2KvkagrMGSQBAFuEDGecsTaUi
a9ZIf9kwF8omb

x1+Fhevn3dvZJbeBo+a2OLsNE4k7eLCEp16SSJKHWqyysF9EPVoz8kJt
hZdtOH6ZIvwM

KOJoKO83BGkWuCa4AOXBEyVAZdhFb3jFUNyx/pFVGuRZBNiyDeX
sS+kZWnjl+fjlS0P4

zZNHE5NtDAVMoX83YNF/mTgbnGBOwJ4EsCLnlGlh3x5qb7NQUbQ
w2GbJcqc7MlVcuo6+
        Nukg==
X-Gm-Message-State:
AG10YOTn3XB6we8JJSesUVzP4uHGmI+kTUfkUpK4cx/gbzMXd06fm
YVcbOtmZqH9lhYrPb9DgAgbfoufenZ5AQ==
MIME-Version: 1.0
X-Received: by 10.194.9.42 with SMTP id
w10mr28700500wja.159.1453903694585;
 Wed, 27 Jan 2016 06:08:14 -0800 (PST)
Received: by 10.28.18.67 with HTTP; Wed, 27 Jan 2016 06:08:14 -0800
(PST)
Date: Wed, 27 Jan 2016 14:08:14 +0000
Message-ID:
<CACD2JikwYoCyLdoadj=g7fBR3Fi5v92oYzpNbUCv8r7M49ppmQ@m
ail.gmail.com>
Subject: DENIAL OF RIGHT TO TRAVEL
From: Nigel Shindler <songsfromearth@gmail.com>
To: comments@souternrailway.com, Nigel Shindler
<trinitymanifesto@gmail.com>
Content-Type: multipart/alternative;
boundary=047d7b45049640e426052a515592
----- Message truncated -----

On 27/01/2016, at approximately 4:15 pm, I attempted to buy a rail ticket from London Victoria Rail Station to Glasgow Scotland, but I was told the purchase was declined, furthermore, I was told due to the way the "system" is designed, I would be unable to obtain a receipt as proof of such being the case, and I was told I should take the attendant's word this was the case.

Southern

Go Ahead House
26-28
Addiscombe
Road
Croydon
CR9 5GA

machine Number: 5158
Window Number: 92
Transaction No.:3972C
    26 JNR 16 19:40

*****NOT VALID FOR TRAVEL*****
Please retain for your records

Please credit account-holder

Printed 19:41 on 26.JNR.16

What sort of world does Virgin Rail inhabit that it believes it is proper to treat customers this way? My day to day dealing with people indicate that most are accustomed to lying most of the time, while others are comfortable lying practically all the time; case in point; earlier the same day as I was purchasing a Day Travel Pass at the same rail station, the computer crashed right in front of my eyes, and I was then told I would have to go to the next wicket to make my purchase; having done so the attendant at that wicket disclosed that the computers crash all the time; the primary reason being, in her opinion, is that they use various pieces of technology that are made by different companies. I've experienced the same problem with my home computer on several occasions.

When my ticket for Glasgow was declined I offered what I have just detailed as a plausible explanation; the attendant refused to accept such a premise. I told him what the attendant earlier that day had noted, and at that point he informed me that it was WRONG of her to say such a thing, and wanted to know who she was, in fact, he was insistent that somehow he'd find out who the culprit was, no matter what.

I would assume on the basis of these two incidents that your employees are told to lie to your customers; would you care to comment, and also explain why my travel ticket was denied?

Thank you in anticipation of your response.

I would not have been able to escape Plymouth, and tell this story, if it weren't for the kindness of one person; a curate at Plymouth's St. Andrew's Church.

Hi Nigel, **Thank** you for your email and the book. I'll look at in a couple of days when I have some more time. Below is the Coach ticket for tomorrow afternoon (12.10). I'll print out a copy and leave it at the church office for you. I hope you have a safe journey back to London. Do pop in to St Andrew's again if you're back in Plymouth.

Every blessing,
Lawrence
From: tickets@nationalexpress.com [mailto: tickets@nationalexpress.com]
Sent: Tuesday, February 9, 2016 2:46 PM
To: Lawbraschi@Hotmail.com
Subject: National Express confirmation email

Confirmation

Journey Ref Outbound: **NLCG-01-3E8AF**

	Thanks, your booking is confirmed.
For ticket validation	Please print your e-Ticket. By clicking the 'Printer Friendly Ticket' button above, you will be provided with your ticket to print. **Please show your ticket to the driver when boarding your coach.**
	We hope you enjoy your journey.

Prepaid Travel Card
(MasterCard)

5280 6100 0284 9756

Exp. 01/17

NIGEL SHINDLER.

Payment Card not recognised
- declined as a result.
System does not produce receipt. MIKE ADAMS

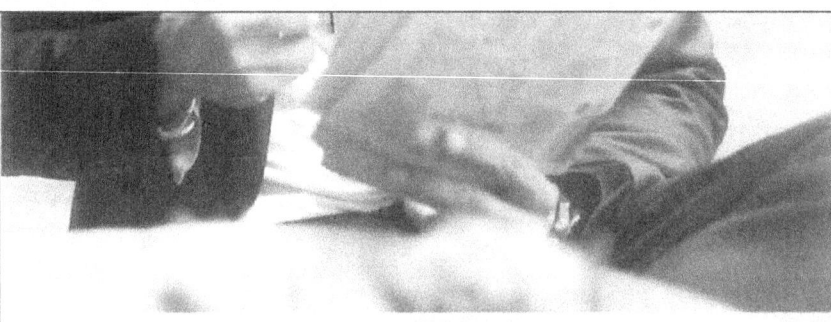

Make time for
the paperwork
that matters

Premier Inn
Business Account

Britain's money laundering system supplies Britain's consumers pay checks, and funds terrorist organizations; the I.D. that is stolen in order to open fraudulent bank accounts hides the fact that Britain's population has been brainwashed. The people who've lost their I.D. stolen are unable to have thieir voices heard because they're dead, and all evidence they once existed is eradicated.

The victims of this genocide can be placed in two categories; those who are stigmatized in Britain, (the mentally challenged; physically disabled; homosexuals, and Jews), and tourists.

The stigmatized are dropped off by the police in areas where they will be unable to acquire the basic necessities of life; tourists unable to access funds succumb to the same fate.

Death occurs far sooner if direct force is used to kill; such happened to me on enight in Horsham; following is the sequence of events.

Late one evening I fell asleep on a train, and wound up in Horsham. All train service had stopped, and there wouldn't be another train till early morning. The town has two hotels;, (I was told); the Premier Inn, and the Travel Lodge, which I was unable to find, although I asked half a dozen people for directions; everyone gave me different instructions.

My travel card wasn't accepted at the Premier Inn, so I then asked if I could sit in the single chair available in the lobby to keep warm till the train station reopened. The receptionist, Mike Adams, said no; the reason being, he said, was previously he allowed a person to sit in the lobby, and his boss got upset with him, so he didn't want to risk losing his job.

I left the lobby, and searched for the other hotel; the following emails detail what happened next.

Dear Mr Shindler,

I have searched through the CCTV system from last night to see if any of the incident was captured. Unfortunately the Barclay's Bank camera is pointing in the opposite direction at the time of the assault. The van can be seen parked up as you described, but the camera pans away just at the wrong time which is very frustrating. I will make enquiries with the window cleaning company and be in touch with you.

Regards
Henry Child
Police Constable

Sussex Police
Divisional Response Team
West Sussex Division - Tel: 101

Dear Mr Child

  I am most eager to hear the results of your on going investigation; did you pick up the book I left at the place I was sitting when I saw a man washing windows in the rain? Horley library has just informed me it's due.

    Thank you in anticipation of your response.

    Nigel Shindler

Mr Shindler,

**I am waiting to hear back from Nationwide Window Cleaning Ltd.
Once I hear back from them I will let you know. I am unable to find
any cctv of the incident itself, therefore if I am able to interview the
driver it will be your word against his I am afraid. I did not go looking
for your book as you had not asked me to, hopefully the library wont
charge you too much for it.**

I will be touch

**Regards
Henry Child
Police Constable**

**Sussex Police
Divisional Response Team
West Sussex Division - Tel: 101**

Mr Child (what is your number?)

You mentioned that there are other cameras you can check in the square; have you done so?

I am quite familiar with how banks protect themselves from thieves; a camera may rotate 90 degrees, or 180 degree, and there is a time span in which each must make a "sweep".

I know how long the incident took, and the range of space it took place within; he not only came at with such force I stumbled backward several yards, he continued to come after me, and I moved further away. It was only when I stood my ground that he retreated to the van. I then walked across to the other side of the square in order to write down his license plate - go get the picture, Mr Child.

There are cameras all along the avenue in which the shopping mall entrance is located where I was seated. I repeated on several occasions while you were writing your statement, (I should be the one writing it, not you), that I left a book at the exact spot where I was seated when I saw the man. I also told you that after he left, I walked over to where he was standing, and found his line of vision was directed exactly toward me. I mentioned that I saw him taking pictures of the Vodofone store a little further up the avenue, and I stated that I saw him washing a window across the street from that store

You failed to write all these details down.

You mentioned that Horsham is a quiet town, and your shift started at 11;00pm, and mine was the fist call you got that night; I can therefore, surmise, that you've had ample time to look into these matters, and have failed to do so. I hadn't caught the man breaking the law, but he did seem enthusiastic about physically harming me?

Why have you not done your utmost to serve, and protect, a British citizen?!

Sincerely

Nigel Shindler

Mr Shindler,

I am not sure why you believe I am not doing all I can to identify the person responsible. If you had a problem about what was 'written down' in your statement then why did you not say so before signing it? I wrote it down in front of you and read it back through. You agreed with what was written and you signed it. I even asked you if there was anything else you wanted to add or anything I had missed out and you said 'No'. As for the CCTV, I am not saying there is no footage of the

man in van throughout the night. What I am saying is that the assault itself is not captured which is what we need to prove the offence. I have worked the town for 7 years and know where all the Police CCTV is. My colleague and I spent some time after I saw you trawling through the town cameras. I am looking into this further but unfortunately Mr Shindler these things take time. Over the coming days I will continue to make enquires. However I am an emergency responder on a small team and therefore I will only get chance to follow this up when time allows. Whilst it was quiet on the night in question I have been very busy since. I am sorry if you feel I have not 'served' and 'protected' you but if I recall I came straight out to you when you called and spent considerable time with you taking your statement and giving you the time you deserve. If that is not service then I am not sure what the issue is?

As for the cameras rotating you are of course correct. However you were assaulted by being pushed and that was a very quick action. If that very action is not covered by a camera due to it 'rotating' then that can't be helped and is just unfortunate. The window cleaning company are based in up north in Leeds so I have made contact with them via email. I will keep you updated. My warrant number is CC216.

Regards

Henry Child
Police Constable

Sussex Police
Divisional Response Team
West Sussex Division - Tel: 101

Just answer the questions?
   The statement does not concern the manner in which you should investigate; it details the nature of the crime.
   The physical evidence proves the crime took place; not, as you put in your previous email; it's your word against his; why did you write that?
      Trawling cameras?
            Pointing in the opposite direction at the time of the assault.
            Pans away just at the wrong time
               Assaulted by being pushed, and that <u>was</u> a very quick action.
               You saw it CC216.

144

> Every action has a reaction; our altercation continued 2, 3 minutes.
>
> What do you have in common with the man that assaulted me.
>
> Nigel Shindler

I am not sure why you believe I am not doing all I can to identify the person responsible. If you had a problem about what was 'written down' in your statement then why did you not say so before signing it? I wrote it down in front of you and read it back through. You agreed with what was written and you signed it. I even asked you if there was anything else you wanted to add or anything I had missed out and you said 'No'. As for the CCTV, I am not saying there is no footage of the man in van throughout the night. What I am saying is that the assault itself is not captured which is what we need to prove the offence. I have worked the town for 7 years and know where all the Police CCTV is. My colleague and I spent some time after I saw you trawling through the town cameras. I am looking into this further but unfortunately Mr Shindler these things take time. Over the coming days I will continue to make enquires. However I am an emergency responder on a small team and therefore I will only get chance to follow this up when time allows. Whilst it was quiet on the night in question I have been very busy since. I am sorry if you feel I have not 'served' and 'protected' you but if I recall I came straight out to you when you called and spent considerable time with you taking your statement and giving you the time you deserve. If that is not service then I am not sure what the issue is?

As for the cameras rotating you are of course correct. However you were assaulted by being pushed and that was a very quick action. If that very action is not covered by a camera due to it 'rotating' then that can't be helped and is just unfortunate. The window cleaning company are based in up north in Leeds so I have made contact with them via email. I will keep you updated. My warrant number is CC216.

Regards

Henry Child
Police Constable

Sussex Police
Divisional Response Team

CC216; I would like answers to the questions I've given you; as well as; are you related to the man who assaulted me; or are you simply like minded men who came across one other?

Mr Shindler,

I find the tone of your emails most inappropriate and rude. I will answer your questions but I have to say that I will not be emailing you again unless I have a further update for you.

The statement does not concern the manner in which you should investigate, it details the nature of the crime.- A 'statement' is there to detail the nature of the crime, which is what your statement does.

   The physical evidence proves the crime took place; not, as you put in your previous email; it's your word against his; why did you write that? I do not question whether the crime took place, and the offending party may well admit to the assault. The point I make is that without an admission from him we have to have other evidence to be able to charge him. These are rules/guidelines set out by our criminal justice system that are out of my control.

   Trawling cameras?
               Pointing in the opposite direction at the time of the assault.
               Pans away just at the wrong time – By trawling cameras I refer to playing back recorded footage of all the Police cameras covering the very area in which the assault took place. As explained to you at the time they pan automatically and I cannot control them.

            Assaulted by being pushed, and that **was** a very quick action.

               You saw it CC216. – not sure what you mean by this…I did not see it, you described it to me.
               Every action, has a reaction; our altercation continued 2, 3 minutes.- The cameras stays on one shot for several minutes before they move to another position.
               What do you have in common with this man? - I am not sure what you mean here.

Mr Shindler, as for your Library book, you are responsible for leaving the book behind. You made no request of me to retrieve the book on your behalf.

regards

Henry Child
Police Constable

Sussex Police
Divisional Response Team
West Sussex Division - Tel: 101

The Terms and Condition for Network Rail Card

2. You must be aged 16 years or over to purchase a Network Railcard.
3. Tickets for your journey should be purchased before boarding the train and when buying tickets you must show the Railcard.
4. You must carry your Railcard with you on your journey and when asked by rail staff, you must show a valid ticket and valid Railcard. If you fail to do so, you and, where applicable, each member of your group will be required to pay the full price Standard Single fare for your journey as if no ticket was purchased before starting your journey and in some cases a Penalty Fare. This does not apply if there was no ticket office at the station at which you began your journey or if the ticket office was closed and there was no ticket machine from which you could buy a discounted ticket.
5. You will be asked to pay the difference between the price of your discounted ticket and the full price Standard fare (or the Penalty Fare if travelling in the Penalty Fares area) if:
   a) you travel beyond the station for which your ticket is issued;
   b) you travel to a destination beyond the area shown on the map in this leaflet, without having first obtained the correct ticket for your journey;
   c) you travel on a route for which a higher fare applies or at a time when reduced fares do not apply.

Totalitarian regimes rule a population through a "central body"; they dictate how people should speak, act, behave, cloth themselves, and so on and so forth; if final rulings are made by a single individual, the result is the same.

When people are allowed to decide for themselves how they should live, while no justice system is in place - "Inverted Totalitarianism" occurs, and disorder, anarchy, chaos, and violence, will inevitably result.

Totalitarian regimes, and fascist dictators, create laws to justify their acts; people in Hyper-democratic societies, (Inverted Totalitarian), lie, deceive, cheat, and steal, to get whatever happens to be their fancy at any particulate time.

When people are given a free hand to rule their own affairs, a brute will soon result.

Inside the Mind of Mass Murderer Adolf Eichmann

www.timesofisrael.com › Jewish Times

German historian Bettina Stangneth's 'Eichmann Before Jerusalem' explores the Final Solution architect's personal and professional life before his Jerusalem trial

By JP O' Malley

LONDON — The Holocaust tends to come up so frequently nowadays as a reference point for the barbaric hatred that gripped Central and Eastern Europe during the 1940s, that it's very easy to forget just how long it took after World War II to speak about it in terms of evidence and legal prosecutions.

The first major trial addressing issues relating to the Holocaust took place at Nuremburg before the International Military Tribunal (IMT) in November 1945 and October 1946.

The focus of these trials, however, did not specifically relate to the genocide of Jewish people. The main conviction was for "crimes against the peace" as defined in the Nuremburg Charter. The initial charge here was against waging an aggressive war that defied international treaties, and not one against exterminating a specific ethnic group, and anyway, getting a proper admission of guilt from senior Nazi sources was complicated, primarily because Adolf Hitler, Heinrich Himmler and Reinhard Heydrich were all dead.

Meanwhile, the only man left alive to have played a central role within the senior hierarchy that oversaw the administrative duties of the Nazi gas chambers was Adolf Eichmann. And at that stage, he was still residing in an American prisoner of war camp in Bavaria, Germany. He soon escaped.

By 1950 Eichmann fled to Argentina where he would live in exile for over a decade. But in May 1960 he was eventually kidnapped by Mossad forces and brought to Israel.

The Eichmann trial in Jerusalem during 1961-62 saw him convicted and hanged for his crimes. It is seen as a major sea change in how the world perceived the Holocaust.

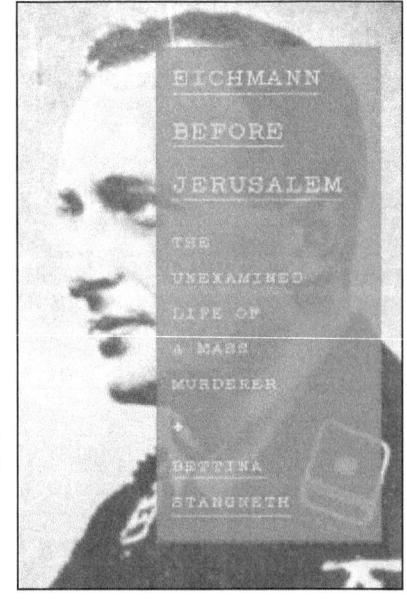

'Eichmann Before Jerusalem' by Bettina Stangneth (courtesy)

Although this trial was grounded on the same legal principles that came out of the first Nuremburg trials, the actual charges put to Eichmann during that case were not just crimes against humanity. Instead, the more symbolic charge of "crimes against the Jewish people" was fittingly put against him.

German philosopher and historian Bettina Stangneth recently published an English translation of "Eichmann Before Jerusalem," a book that attempts to explore Adolf Eichmann's personal and professional life before he was put on trial.

The narrative seeks to get inside the mind of a mass murderer and asks questions like: who really was Adolf Eichmann? What were the ideological beliefs that compelled him to orchestrate the greatest genocide ever witnessed in human history?

Through a solid body of research, the book is intent on deliberately avoiding falling into the trap of believing the numerous mythologies that Eichmann himself helped created during his infamous trial in Jerusalem.

Stangneth begins our conversation by stressing just how important it is to remember how the Holocaust has been studied, talked about, analyzed and placed in historical discourse since Eichmann's trial.

"Before 1961 we didn't have that many big archives and documents with knowledge about the Holocaust," she explains.

'Eichmann was the first Nazi who admitted to the public that the Holocaust really happened'

"Eichmann was the first Nazi who admitted to the public that the Holocaust really happened. He explained that he guessed there was probably between five and six million victims. This kind of admission from a Nazi was huge at that time."

"To hear Eichmann say something new about this crime really changed our knowledge about the Holocaust. Nobody after 1961 could say that this crime did not happen. This is the reason why Nazis, even today, hate Eichmann."

Stangneth's book is a fascinating — albeit extremely disturbing — account of Eichmann's rise through the ranks of the Nazi party, where he eventually became the administrative face of the Holocaust. In August 1939, under Eichmann's leadership, the Central Office for Jewish Emigration was set up in Prague. The following January, Eichmann was then put in charge of coordinating plans to relocate Jews to the east.

By the winter of 1941-42 Eichmann had even bragged to fellow Nazi colleagues to have coined the term "Final Solution," which by that stage had taken on a very definite meaning: the extermination of the Jewish race.

"Eichmann was a kind of spider in the Nazi network," Stangneth, explains. "And most of the information around the killing of Jews found its way onto his desk.

'Eichmann was a kind of spider in the Nazi network'

"After the Wannsee conference in January 1942 Eichmann was given responsibility with reports from both the Jewish ghettos and from the concentration camps. So he had a lot of knowledge about the numbers of victims there were. This obviously makes him so important today for a historian looking at the Holocaust."

Stangneth's narrative then documents Eichmann's activities in the post war period. We read about his time living alone in a hut near a forest in Northern Germany, then running a chicken farm in Bergen, located just two miles from a former concentration camp, before finally immigrating to Argentina.

While in Buenos Aires, Eichmann made the acquaintance of William Sassen, a Dutch journalist and Nazi sympathizer, who had written propaganda disguised as news for the Germans during the war.

Adolf Eichmann, on trial in 1961 (photo credit: Wikimedia Commons)

Sassen and Eichmann during these years developed a deep friendship. They bonded around the deluded idea of rebuilding the Third Reich from an expatriate community in Argentina, where many Nazis had found refuge after the war.

Sassen also hosted at his home during 1956 and 1957 what became known as the Dürer circle: a group of Nazi sympathizers, including Eichmann, who regularly held meetings, debates, discussed Nazi literature, and eventually recorded conversations about the extermination of the Jews.

This evidence, which is often referred to as the Sassen transcripts, is still vitally important, even today, in confirming Eichmann's guilt as a committed Nazi ideologue.

"Those transcripts are extremely helpful if you want to find out something more about Nazi thinking," says Stangneth. "Because contained with them is Eichmann explaining his methods for manipulating and lying against the Jews."

"The kind of nonsense Eichmann would say to Jews [to get them to go to concentration camps} was along the lines of: you will get a new ghetto, you'll have to do some short-term labour in the east, or that it's only for the time during the war and after that you will be able to come back."

"But what makes the Sassen transcripts so interesting is that Eichmann explained his own methods of deception. Later in Jerusalem, at his trial, he used this method of deception again."

The most famous book that exists on this subject is "Eichmann in Jerusalem: A Report on the Banality of Evil." It was written in 1963 by Hannah Arendt, a German Jew who covered the trial for the New Yorker. In it, Arendt presents the following argument: that Eichmann was merely obeying orders to move up the career ladder within the Nazi party. We now know that this opinion was slightly misinformed.

'Eichmann and Sassen knew they had lost the war. But they did not want to lose the war over the interpretation of history'

The counterargument that Stangneth puts forward is what makes her book such a compelling piece of historical research.

For example, Eichmann and Sassen, we now understand, were well aware of how the Holocaust literature that emerged in Europe and the US during the 1950s began to challenge the version of history they sought to own.

"These guys had a lot of experience with propaganda under Hitler," says Stangneth. "And they knew what one can do by holding power over the interpretation of events. Eichmann and Sassen knew they had lost the war. But they did not want to lose the war over the interpretation of history. "
Over a decade ago, when Stangneth began working on this project, she was fully convinced of Hannah Arendt's thesis that Eichmann was merely a pen pusher who put absolute faith in his own personal benefit from obeying a hierarchical system. And she believed that the Holocaust's chief perpetrator was devoid of any proper intellectual engagement or ideological obsessions about the Third Reich.

But it's a testament to her dedication as a researcher that she changed her mind with each new document she dissected in scrupulous detail.

In fact, Eichmann was such a dedicated Nazi ideologue, that according to Stangneth, he didn't believe that the German philosopher Immanuel Kant could be reconciled within the warped racial biological struggle that the Third Reich was intent on fighting. Other Nazis did.

"The Nazis understood there was a huge connection between the German philosopher and the Jews," says Stangneth. "So for a man like Eichmann, Kant's philosophy and Jewish thinking was very closely linked. The Nazis believed that philosophy was dangerous because it was the weapon of the Jews. Eichmann had to learn to fight against this idea. So he had to know the weapon of the enemy, which was the philosophy that Jews were reading."

'Eichmann [during his trial] tried to look like a Jew before the Jews in Israel. He tried to give himself the image of a philosopher as a method for fighting'

Strangely, Eichmann actually began quoting Kant during his trial in Jerusalem, says Stangneth. Again though, we must remember how this ties in with his powers of deception, she warns.

"Eichmann [during his trial] tried to look like a Jew before the Jews in Israel. He tried to give himself the image of a philosopher as a method for fighting."

"This idea contrasts with Hannah Arendt's concept of the banality of evil. The point is that Eichmann and the Nazis knew about the voice of reason and the power of thinking. But they were convinced that this was dangerous. So there is a difference between the ability of thinking, and then the mistrust of thinking itself."

"For Nazis, blood, and not brain, and the relationship to the country, was where the power lay. They knew well about the power of thinking, but they thought there was a much bigger power: which they believed was the German race. And that is the difference between the banality of evil, and what I would call a kind of academic evil."

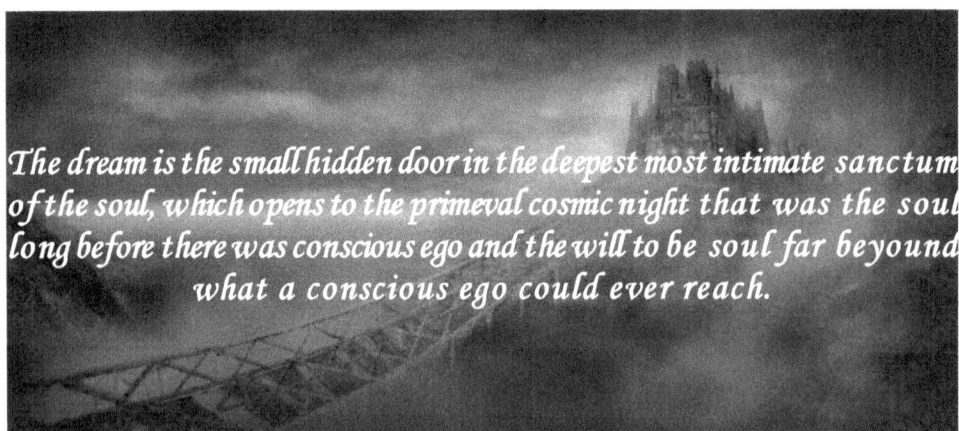

The dream is the small hidden door in the deepest most intimate sanctum of the soul, which opens to the primeval cosmic night that was the soul long before there was conscious ego and the will to be soul far beyound what a conscious ego could ever reach.

Carl Jung

Jung was kind.
Not wanting us to be blind,
He taught us to be kind to the young.

During youth we develop as a person.
Be careful with those during this time,
Otherwise with a sentence they will be dealt to prison.

There are many parts to man, Jung knew;
Also parts unseen.
Some, if you are careful, you will gradually know,
While others are mysterious, and appear merely as a show.

To explore this stage that lies beneath the floor,
We need first to find some door to get there.
We need to follow each stair;
Be careful, though, it can be scary down there.

What may lurk in the shadows can cause a scare,
But, throughout it all, we shouldn't have a care;
It is not as if there lies a bear down there.
Though, at first, there is darkness,
Our eyes acclimatize,
Then we see it all, we are just plain kindness
Deeper we go; closer to the surface we arrive.
Then progressively more, an ever opening flower
Will reveal all we are meant to feel.

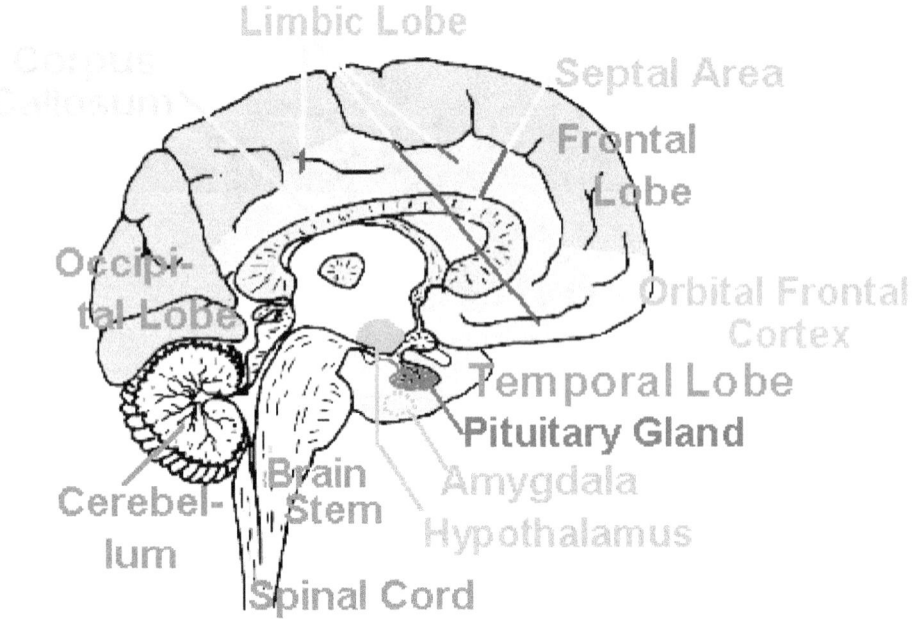

Limbic Lobe
Septal Area
Frontal Lobe
Occipi-tal Lobe
Orbital Frontal Cortex
Temporal Lobe
Pituitary Gland
Cerebel-lum
Brain Stem
Amygdala
Hypothalamus
Spinal Cord

THE HYPOTHALAMUS

The **Hypothalamus** is the main output of the limbic system. It is located on lower part of the top end of the brain stem.
The limbic system is basic to motivation, emotion, and reward processes.
It controls emotional expression through the hypothalamus, which has control over the body's emotional responses systems:

- The autonomic nervous system, which controls internal organs (e.g., gut, heart) and the release of adrenalin (epinephrine) by the adrenal medulla.

- Pituitary gland of the **endocrine system** just below it. The pituitary is often called "master gland of the body," because it regulates the action of many other glands, including the adrenal cortex (vital for response to stress and control of salt balance in the body), the gonads (sex glands: ovaries in females and the testicles in males), the thyroid gland (controls metabolism), etc.
- skeletal muscle system, which shows emotion through facial expression and body posture

The hypothalamus is particularly important for maintaining the proper balance of the body's internal environment (e.g., body temperature, blood sugar, water and salt concentration, oxygen level, etc.). The hypothalamus helps maintain homeostasis by its control over the autonomic nervous system and the endocrine system.

Stimulating the hypothalamus can activate many functions related to motivation, emotion, and reward, whereas damage disrupts such functions. For example, electrodes implanted in the lateral hypothalamus can activate very strong reward. Rats and other animals with such electrodes will make thousands of responses per hour to turn on 0.5-second bursts of stimulation (Olds, 1969). Such stimulation can also elicit (trigger) vigorous eating in animals that have just finished a meal (Valenstein, 1976)

Repeated stimulation several times a day for several weeks will make animals eat so much that they put on a lot of fat. (Steinbaum & Miller, 1965). Damage in this same area depresses eating (Teitelbaum & Epstein, 1962). Stimulation in the hypothalamus can also trigger aggression indicating anger and/or fear, and many other functions related to motivation and emotion.

Sexual Masochism

Definition

The essential feature of sexual masochism is the feeling of sexual arousal or excitement resulting from receiving pain, suffering, or humiliation. The pain, suffering, or humiliation is real and not imagined and can be physical or psychological in nature. A person with a **diagnosis** of sexual masochism is sometimes called a masochist.

The *Diagnostic and Statistical Manual of Mental Disorders,* also known as the *DSM,* is used by mental health professionals to diagnose specific mental disorders. In the 2000 edition of this manual (the Fourth Edition Text Revision also known as *DSM-IV-TR)* sexual masochism is one of several **paraphilia.** Paraphilias are intense and recurrent sexually arousing urges, fantasies, or behaviours.

Description

In addition to the sexual pleasure or excitement derived from receiving pain and humiliation, an individual with sexual masochism often experiences significant impairment or distress in functioning due to masochistic behaviours or fantasies.

With regard to actual masochistic behaviour, the person may be receiving the pain, suffering, or humiliation at the hands of another person. This partner may have a diagnosis of **sexual sadism** but this is not necessarily the case. Such behaviour involving a partner is sometimes referred to as sadomasochism.

Masochistic acts include being physically restrained through the use of handcuffs, cages, chains, and ropes. Other acts and fantasies related to sexual masochism include receiving punishment or pain by means of paddling, spanking, whipping, burning, beating, electrical shocks, cutting, rape, and mutilation. Psychological humiliation and degradation can also be involved.

Masochistic behaviour can also occur in the context of a role-playing fantasy. For example, a sadist can play the role of teacher or master and a masochist can play the role of student or slave.

The person with sexual masochism may also be inflicting the pain or suffering on himself or herself. This can be done through self-mutilation, cutting, or burning.

The masochistic acts experienced or fantasized by the person sometimes reflect a sexual or psychological submission on the part of the masochist. These acts can range from relatively safe behaviours to very physically and psychologically dangerous behaviour.

The *DSM* lists one particularly dangerous and deadly form of sexual masochism called hypoxyphilia. People with hypoxyphilia experience sexual arousal by being deprived of oxygen. The deprivation can be caused by chest compression, noose plastic bag, mask, or other means and can be administered by another person or be self-inflicted.

Causes and Symptoms

Causes

There is no universally accepted cause or theory explaining the origin of sexual masochism, or sadomasochism in general. However, there are some theories that attempt to explain the presence of sexual paraphilias in general. One theory is based on learning theory that paraphilias originate because inappropriate sexual fantasies are suppressed. Because they are not acted upon initially, the urge to carry out the fantasies increases and when they are finally acted upon, a person is in a state of considerable distress and/or arousal. In the case of sexual masochism, masochistic behaviour becomes associated with and inextricably linked to sexual behaviour.

There is also a belief that masochistic individuals truly want to be in the dominating role. This causes them to become conflicted and thus submissive to others.

Another theory suggests that people seek out sadomasochistic behaviour as a means of escape. They get to act out fantasies and become new and different people.

Symptoms

Individuals with sexual masochism experience sexual excitement from physically or psychologically receiving pain, suffering, and/or humiliation. They may be receiving the pain, suffering, or humiliation at the hands of another person, who may or may not be a sadist, or they may be administering the pain, suffering, or humiliation themselves.

They experience distressed or impaired functioning because of the masochistic behaviours, urges, and fantasies. This distress or impairment can impact functioning in social, occupational, or other contexts.

Demographics

Although masochistic sexual fantasies often begin in childhood, the onset of sexual masochism typically occurs during early adulthood. When actual masochistic behaviour begins, it will often continue on a chronic course for people with this disorder, especially when no treatment is sought.

Sadomasochism involving consenting partners is not considered rare or unusual in the United States. It often occurs outside of the realm of a mental disorder. More people consider themselves masochistic than sadistic.

Sexual masochism is slightly more prevalent in males than in females.

Death due to hypoxyphilia is a relatively rare phenomenon. Data indicate that less than two people per million in the United States and other countries die from hypoxyphilia.

Diagnosis

The *DSM* criteria for sexual masochism include recurrent intense sexual fantasies, urges, or behaviours involving real acts in which the individual with the disorder is receiving psychological or physical suffering, pain, and humiliation. The suffering, pain, and humiliation cause the person with sexual masochism to be sexually aroused. The fantasies, urges, or behaviours must be present for at least six months.

The diagnostic criteria also require that the person has experienced significant distress or impairment because of these behaviours, urges, or fantasies. The distress and impairment can be present in social, occupational, or other functioning.

Sexual masochism must be differentiated from normal sexual arousal, behaviour, and experimentation. It should also be differentiated from sadomasochistic behaviour involving mild pain and/or the simulation of more dangerous pain. When this is the case, a diagnosis of sexual masochism is not necessarily warranted.

Sexual masochism must also be differentiated from self-defeating or self-mutilating behaviour that is performed for reasons other than sexual arousal.

Individuals with sexual masochism often have other sexual disorders or paraphilias. Some individuals, especially males, have diagnoses of both sexual sadism and sexual masochism.

Treatments

Behaviour therapy is often used to treat paraphilias. This can include management and conditioning of arousal patterns and masturbation. Therapies involving cognitive restructuring and **social skills training** are also utilized.

Medication is also used to reduce fantasies and behaviour relating to paraphilias. This is especially true of people who exhibit severely dangerous masochistic behaviours.

Treatment can also be complicated by health problems relating to sexual behaviour. Sexually transmitted diseases and other medical problems, especially when the sadomasochistic behaviour involves the release of blood, can be present. Also, people participating in hypoxyphilia and other dangerous behaviours can suffer extreme pain and even death.

Prognosis

Because of the chronic course of sexual masochism and the uncertainty of its causes, treatment is often difficult. The fact that many masochistic fantasies are socially unacceptable or unusual leads some people who may have the disorder not to seek or continue treatment.

Treating a paraphilia is often a sensitive subject for many mental health professionals. Severe or difficult cases of sexual masochism should be referred to professionals who have experience treating such cases.

Prevention

Because it is sometimes unclear whether sadomasochistic behaviour is within the realm of normal experimentation or indicative of a diagnosis of sexual masochism, prevention is a tricky issue. Often, prevention refers to managing sadomasochistic behaviour so it primarily involves only the simulation of severe pain and it always involves consenting partners familiar with each other's limitations.

Also, because fantasies and urges originating in childhood or adolescence may form the basis for sadomasochistic behaviour in adulthood, prevention is made difficult. People may be very unwilling to divulge their urges and discuss their sadistic fantasies as part of treatment.

Mr Nigel Shindler,

The court case involving the suspect in the theft of your luggage ref – 0515213433, is Friday 15th January at South Sefton Magistrates court. I have no further information regarding this, but information can be sourced from the court itself.

I have no knowledge of any other crimes you may have reported. I advise you to collate any reference numbers and contact telephone number 101 to enquire who is dealing with them incidents.

Regards

Constable ROTHWELL 6363
Planned Demand Team
Maghull Police Station
Tel - 101

Nigel Shindler <trinitymanifesto@gmail.com>   **Jan 15**

Dear Mr Rothwell

I was obstructed from attending court proceedings; in due time the case will be handled by another court; the matters are linked; and to answer the question you wanted to speak to me about in person; the answer is Yes! The theft of my property is connected to the "missing person's" report I requested concerning my father, Maxwell Shindler; D: O: B August 7, 1931, Liverpool, U.K.

See you in court
Sincerely
Nigel Shindler D: O: B

P.S. - I must advise you that it is your best interests to obtain legal representation.

Nigel Shindler <trinitymanifesto@gmail.com>   Jan 15

to ME.CourtSupport

Crown Attorney's office;

I was unable to attend court proceeding today; please send all court documents pertaining to this case.

Kind regards

Nigel Shindler
D.O.B.; 16/08/1964

Good Afternoon

The court does not provide any paperwork relating to any case, for an outcome you can ring the Court 0151 285 6241 and we will give you the result or witness care may be able to provide you with information.

Regards

From: Nigel Shindler [mailto:trinitymanifesto@gmail.com]
Sent: 15 January 2016 12:19
To: ME-CustomerSupport
Subject: Nigel Shindler vs., Martin Magill

This email was scanned by the Government Secure Intranet anti-virus service supplied by Vodafone in partnership with Symantec. (CCTM Certificate Number 2009/09/0052.) In case of problems, please call your organisations IT Helpdesk.
Communications via the GSi may be automatically logged, monitored and/or recorded for legal purposes.
This e-mail (and any attachment) is intended only for the attention of the addressee(s). Its unauthorised use, disclosure, storage or copying is not permitted. If you are not the intended recipient, please destroy all copies and inform the sender by return e-mail.

Internet e-mail is not a secure medium. Any reply to this message could be intercepted and read by someone else. Please bear that in mind when deciding whether to send material in response to this message by e-mail.

This e-mail (whether you are the sender or the recipient) may be monitored, recorded and retained by the Ministry of Justice. E-mail monitoring / blocking software may be used, and e-mail content may be read at any time. You have a responsibility to ensure laws are not broken when composing or forwarding e-mails and their contents.

The original of this email was scanned for viruses by the Government Secure Intranet virus scanning service supplied by Vodafone in partnership with Symantec. (CCTM Certificate Number 2009/09/0052.) This email has been certified virus free.

Nigel Shindler <trinitymanifesto@gmail.com>   Jan 26 (9 days ago)

to Bowen

Contempt of Court Act 1981

1981 CHAPTER 49

An Act to amend the law relating to contempt of court and related matters.

[27th July 1981]

C1By Criminal Justice Act 1991 (c. 53, SIF 39:1), s. 101(1), Sch. 12 para. 23; S.I. 1991/2208, art. 2(1), Sch. 1 it is provided (14.10.1991) that in relation to any time before the commencement of s. 70 of that 1991 Act (which came into force on 1.10.1992 by S.I. 1992/333, art. 2(2), Sch. 2) references in any enactment amended by that 1991 Act, to youth courts shall be construed as references to juvenile courts.

Commencement Information

I1Act not in force at Royal Assent. Act partly in force at 27.8.1981 see s.21(2)(3).

I request that my case against Martin Magill be handled by another court, as I informed your office once before, i am the prosecutor in this case.
Please notify me at your earliest convenience when another court is chosen.

Thank you in anticipation of your response.

Nigel Shindler

Nigel Shindler <trinitymanifesto@gmail.com>   Jan 27 (8 days ago)

to Bowen

You have 48 more hours to respond.

Mr Schindler:

I am responding to your emails regarding the case of Regina vs. Martin Magill. I work for the Magistrates Court. As a prosecuting witness in

this case, your interests are represented to the court by the Crown Prosecution Service (CPS). The CPS maintain a Witness Liaison Service who will communicate with you and keep you abreast of developments regarding the case you are involved in. I have found the following information on how to contact them and you may find it helpful:

The telephone number for the Witness Liaison Service in Liverpool is: (0151) 239-6439. (Main Switchboard number is: 0151 239-6400)

The email address is: **Victimliaison.Merseycheshire@cps.gsi.gov.uk**
The contact address is CPS Merseyside Cheshire
  2nd Floor
  Walker House
  Exchange Flags
  Liverpool, L2 3YL

Bowen Heather <heather.bowen@hmcts.gsi.gov.uk>   Jan 29 (6 days ago)

As regards your request for information contained in the court file, it is correct that information contained in the file cannot be disclosed unless for specific permitted reasons under law. However, considering your relationship to this case as a prosecution witness, the most appropriate contact for you is the CPS as I have indicated above. They may be in a position to give the information that you require. In any case they can inform you of the conduct of this case.

Sincerely,

Robert Coleman

Legal Advisor
Sefton & St Helens Magistrates Courts
Merton Road
Bootle
L20 3XX

Nigel Shindler <trinitymanifesto@gmail.com>   Jan 29 (6 days ago)

to Bowen

Mr Schindler:
 Who is this person?

Nigel Shindler <trinitymanifesto@gmail.com>   Jan 29 (6
                                                            days ago)

to Bowen

Mr Shindler is the prosecutor.

Nigel Shindler <trinitymanifesto@gmail.com>   Feb 2 (2 days
                                                                    ago)

to Bowen

6Prosecutions instituted and conducted otherwise than by the Service.

(1)Subject to subsection (2) below, nothing in this Part shall preclude any person from instituting any criminal proceedings or conducting any criminal proceedings to which the Director's duty to take over the conduct of proceedings does not apply.

(2)Where criminal proceedings are instituted in circumstances in which the Director is not under a duty to take over their conduct, he may nevertheless do so at any stage.

Nigel Shindler <trinitymanifesto@gmail.com>   Feb 2 (2 days
                                                                    ago)

to Bowen

<u>**Private citizen wins right to prosecute Met police worker ...**</u>
www.independent.co.uk › News › UK › Crime

Mr Nigel Shindler,

The court case involving the suspect in the theft of your luggage ref – 0515213433, is Friday 15th January at South Sefton Magistrates court. I have no further information regarding this, but information can be sourced from the court itself.

I have no knowledge of any other crimes you may have reported. I advise you to collate any reference numbers and contact telephone number 101 to enquire who is dealing with them incidents.

Regards

Constable ROTHWELL 6363

Planned Demand Team
Maghull Police Station
Tel - 101

Nigel Shindler <trinitymanifesto@gmail.com>   **Jan 15**

to Rothwell
Dear Mr Rothwell

I was obstructed from attending court proceedings; in due time the case will be handled by another court; the matters are linked; and to answer the question you wanted to speak to me about in person; the answer is Yes ! The theft of my property is connected to the "missing person's" report I requested concerning my father, Maxwell Shindler; D:O: B August 7, 1931, Liverpool, U.K.

See you in court
Sincerely
Nigel Shindler D:O:B

P.S. - I must advise you that it is your best interests to obtain legal representation.

I, Nigel Edward Shindler   ACCUSE MARTIN MAGILL to be AN ACCESSORY FOR
CONSTABLE ROTHWELL 6363

AMOUNT   £4.99
THANK YOU
13:24 09/01/16
xxxxxxxxxxxxxxxxxxxxxxx

*   NOT AUTHORISED   *
xxxxxxxxxxxxxxxxxxxxxxxx

Become a local hero!

If you have a few hours to give, please support your community by volunteering in our shop. See inside for details...

Biophilia – A Synopsis of the Concept

As presented in Erich Fromm's

I found a first edition of this book online for a good price and bought it. That was a couple of years ago. I finally managed to read it and it was an interesting journey back in time, both in zeitgeist (written and published during and shortly after the Pig's Bay crisis and the threat of nuclear war) and psychodynamic theory.

Fromm is searching for the essence of mankind, the characteristic that defines humans. His take on this is that the basic position of man is to stand apart from nature due to his ability to be aware of himself and his consequential ability to be reflexive.

These abilities separate man from nature and make him stand alone. Fromm refers to this separation as
"a contradiction inherent in human existence" (Fromm, 1964, p. 116). This contradiction is evident in two ways.

1) Albeit being an animal, man's survival instincts are incomplete or not sufficient to survive anymore (they have become blunt). Man relies on speech and tools to survive and that makes him special among all other living beings (although this might not be quite true

anymore today, as we discovered some animals using tools and know more about their communication strategies).

2) We are aware of ourselves and of the fact that we are mortal. In this sense, we transcend nature because we are aware of life itself (the animal is not, which makes it a part of nature).

His quest about how we deal with this contradiction in our existence leads him to the question of whether our action are based on free will or whether they are determined by nature and/or nurture. He brings this conflict and contradiction to the point as follows:

"Man is confronted with the frightening conflict of being a prisoner of nature, yet being free in his thoughts; being part of nature, and yet to be as it were a freak of nature; being neither here nor there. Human self-awareness has made man a stranger in the world, separate, lonely, and frightened"
(Fromm, 1964, p. 117).

As a result, we strive towards overcoming our sense of separateness and to become one again with nature. Our attempts at achieving a sense of belonging, we either regress or progress. Regression leads us back to nature (i.e. **Rousseau**, becoming childlike or childish, the womb), to animal life (rule of strength, violence, etc.) and to our ancestors (religions, laws, etc.). Progression means to develop to become fully human and to regain the lost harmony with nature and to lose the terror of separateness.

Fromm explores humans' 'Genius for Good and Evil' and our regressive and progressive paths by investigating the dimensions of narcissism (benign-malignant), necrophilia-biophilia and incestuous ties (absent – incestuous symbiosis). In their malignant or destructive expressions, he calls these three concepts the syndrome of decay. This syndrome encompasses all tendencies directed against life and finds its expression in necrophilia, narcissism, and incest. I have always been particularly interested in his concept of biophilia. Hence, I summarised the key aspects of biophilia, as well as its opposite necrophilia, below.

Necrophilia or the love of the dead shows itself in sexual perversion or the 'morbid desire to be in the presence of a dead body' (Fromm, 1964, p. 39). However, it is more than that. A person with necrophilous tendencies is drawn to everything that is dead or not alive, including corpses, decay, feces, dirt. They prefer to talk about sickness, funerals, death, destruction, the past; they are 'cold, distant, devotees of law and order' (p. 40) and like the use of force. Necrophiles like everything that does not grow but which is mechanical. 'The necrophilous person is driven by the desire to transform the organic into the inorganic, to approach life mechanially, as if all living persons were things. All living processes, feelings, and thoughts are transformed into things' (p.41). He continues to provide example in a similar vein but I think the picture he draws is emerging.

The opposite of necrophilia is biophilia, the love of life, the attraction to everything that lives and grows. Preserving life and preventing death is one form of biophilia. Biophilous tendencies can be much more varied and tend to integrate and unite, to fuse with different and opposite entities (this starts on a molecular level but also includes sexual union). This productive orientation expresses itself in curiosity, preference of the new over the old and a functional rather than mechanical approach to life. For biophilia to emerge or be sustained, certain societal conditions need to be in place. Chief among them are the absence of injustice and the presence of freedom to create and innovate.

Interestingly, Fromm also had something to say about knowledge management:

'Briefly then, intellectualization, quantification, abstractification, bureaucratization, and reification – the very characteristics of modern industrial society, when applied to people rather than to things, are not the principles of life but those of mechanics. People living in such systems become indifferent to life and even attracted to death'
(Fromm, 1964, p. 59).

The concept of biophilia encompasses people searching for self-awareness, aspirations, and growth. Given the current emphasis on mindfulness in psychological therapies and beyond, it was interesting to rediscover that in the 60s, when this book was published, From was already repeatedly referring to Buddhism and the **eightfold path** leading to awareness to the good in man by discovering him/herself. Moreover, Fromm's approach fits with the psychological, health, and economic theories for which I have the greatest affinity:

Frankl's Logotherapy and Existential Analysis, Antonovsky's Salutogenesis and **Amartya Sen's Capability Approach.**

To Have, or to Be?

To Have or to Be? is a 1976 book by psychoanalyst Erich Fromm, in which he differentiates between having and being.

Fromm mentions how modern society has become materialistic and prefers "having" to "being". He mentions the great promise of unlimited happiness, freedom, material abundance, and domination of nature. These hopes reached their highs when the industrial age began. One could feel that there would be unlimited production and hence unlimited consumption. Human beings aspired to be Gods of earth, but this wasn't really the case. The great promise failed due to the unachievable aims of life, i.e. maximum pleasure and fulfillment of every desire (radical hedonism), and the egotism, selfishness and greed of people. In the industrial age, the development of this economic system was no longer determined by the question of what is good for man, but rather of what is good for the growth of the system. So, the economic system of society served people in such a way in which only their personal interests were intended to impart. The people having unlimited needs and desires like the Roman emperors, the English and Society nowadays has completely deviated from its actual path. The materialistic nature of people of "having" has been more developed than "being". Modern industrialization has made great promises, but all these promises are developed to fulfill their interests and increase their possessions. In every mode of life, people should ponder more on "being" nature and not towards the "having" nature. This is the truth which people deny and thus people of the world have completely lost their inner selves. The point of being is more important as everyone is mortal, and thus having of possessions will become useless after their death, because the possessions which are transferred to the life after death will be what the person actually was inside.

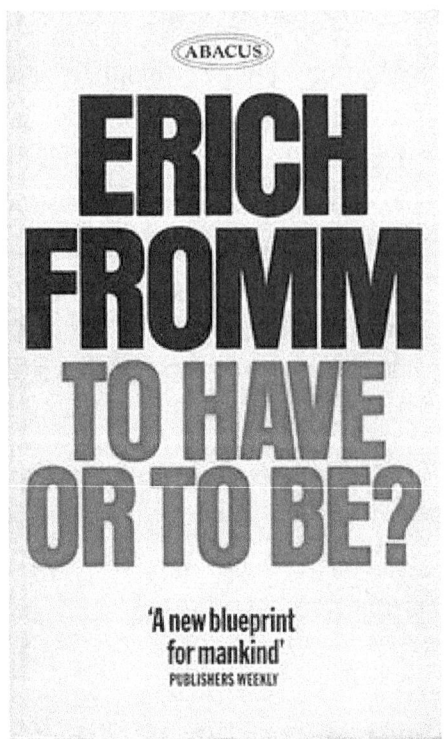

A man's capacity to survive is largely dependent on the degree to which his mind operates proficiently; the longer he is able to live, greater in the amount of time he has to enrich his mind.

The world, in a manner of speaking, has been dying, falling into an ever worsening state of decay since the conclusion of the Second World War.

Man, once given the capacity to be the noblest of creatures, has sunk so low as to now be a cretin, in the vast preponderance of cases; a person unable to successfully execute the simplest of tasks.

When a man continually behaves in a moral, truthful, manner, his thoughts, actions, and deeds, are in sympathy; they synchronize. The collected body of energy, to make a comparison, flows cohesively in the same direction; if one were to make the analogy of it being a body of water, it has the capacity to displace objects; picture a feather travelling of the surface of a swiftly flowing stream.

When a man lies, deceives, steals, or in any way consistently behaves in an immoral manner, the current of energy is at cross purposes; picture streams flowing from different directions clashing as they bombard one another. The capacity to move an object in a definite direction, along a certain course, is no longer a possibility; a feather will most likely be sucked beneath the surface, and become trapped within the currents churning below.

A person who chooses to behave in an immoral manner will inevitably compromise his capacity to reason, and eventually acquire dementia, due to, using a medical term, nerves; "wires" in the brain cross one another, and short circuit; unless new synaptic connections are made, and due to a weakened recognition of the cause behind the trouble, the condition progressively becomes worse.

The Buddha, Siddhartha Gautama, spoke of Four Noble Truths, and an Eightfold Noble Path, which include "right livelihood"; he is referring to encouragement of moral behaviour.

Personally I would never be a member of any organization that would have me as a member – and that's no joke!

I have explained the nature of reality; the history of the British Empire, and also the present condition of the psychology of the masses.

I TRAVELED TO THE FUTURE TO FIND OUT WHY I DIED IN 1984. PEOPLE WENT MAD.

THE CAMELOT PROJECT BEGAN THE YEAR I WAS BORN, 1964.

The brain takes 23 years to develop.

By 2007 man had become primal due to an infantilism of mind. The cerebral cortex is undeveloped, and cannot restrain the aggressive instinct from expressing itself in an instant.

Very minor impulses, and suggestions, are making people do diabolical acts.

Man does not have the capacity for insight, or introspection; he is not aware of this, and kills because he kids himself that he has any attachment to others, as well as to himself.

"I feel this.. I feel that"... People don't say, "I think this...I thought that"; they are fuelled by raw instinctual energy.

Islam is in Britain because the illness feeds itself; it wants to survive; the dark is not merely the absence of light; when it is not controlled by light, and is albe to do as it pleases, it destroys recklessly, and indiscriminately.

*People continually fidget not because they're nervous, unsettled, or excited, but because, but not in all cases, the **aggression emotion** "overflows" into the rest of the body; right down to the soles of people's feet.*

People should minimize the amount of caffeine they consume; in fact all the chemicals that enable the aggressive instinct to surge upward. Coffee, tea, red bull, chocolate bars – these are not social habits any more, but addictions. Pornography has the same affect as caffeine; it stimulants the brain, so the aggressive instinct surfaces.

*Psychiatric drugs are dangerous because they inhibit the flow of the aggression emotion, serotonin reuptake inhibitors, and this is why people become suicidal, and violent, while on these drugs. Aggression that is directed outward is called, sadism, and aggression directed toward oneself is called masochism, Because the quantity of this emotion is so great, **neurological damage** occurs.*

The primary problem, and also the solution, is the development of the mind.

The cause of all illnesses is psychological; prolonged, excessive, stress, for example, creates a hormonal in-balance, and this is what causes both heart-disease, and cancer.

Chapter Three

I Simply Want Some Food

In order for the Monolith to brainwash the masses they have to make sure they are dumbed down; an essential component of this process is making them unaware of their incompetence, and ignorance.

People who work for corporations, for example, defend the actions of their colleagues, irrespective of the cost incurred by others, no matter how many they might be.

Most believe they behave this way because they don't care; this implies that they chose not to care, which means taking an interest in the implications of their actions; the truth is that they are unable to care; their minds haven't developed enough to learn affectively.

A civilized society cannot exist unless people are educated to behave morally; the education system t today, on the contrary, trains people to behave in an immoral fashion; this is easy to accomplish when people fail to question what they are told, and this is why so many work for businesses, and corporations, that are thoroughly corrupt; the neo-psychoanalyst, Erich Fromm, described the masses as, "Eichmanns"; he was referring to the Nazi mass murderer, Adolf Eichmann.

The masses have been brainwashed due to society being de-stabilized; this has been due to the influx of Muslims into Western nations; they come from countries that dissuade progress and open-mindedness - previously Communism, and the threat of nuclear war, were used to create the same affect.

Islam is a fundamentalist belief system, (I call it a death cult because it promotes violent behaviour), which means that its members are dissuaded from questioning what they are told, and this is why they are ill equipped to interpret the meaning of "holy scriptures"; Muslims customarily memorize large portions of the Koran, if not the entire book, but this does not mean they understand what they read in this book.

Due to their lack of education they deny the validity of the science that proves the Earth is approximately four and a half billions of years old; they also deny the science that enables us to understand, for example, evolution, consequently; how can they be educated so they have something to offer society? They can't because they deliberately keep themselves ignorant of any information that doesn't concur with how they perceive things; so why do Muslims immigrate to Britain which continually modernizes, not just for its own betterment, but for the benefit of countries all around the world?! There is the sense in inviting people who are unable to make a contribution; they can only take what doesn't belong to them!! The best, and brightest, who represent the Churches in Britain, consider this a "humanitarian act". What they are actually doing is promoting criminal behaviour.

"Beware of men on pulpits", was one of the warnings Jesus gave!

What history has proven is that when a leader wants to modernize a Muslim country he makes it secular, and this entails eliminating radicals. When the "uprisings" began in the Arab World, December 17, 2010, it was the leaders who had successfully modernized their respective countries who were deposed; Hosni Mubarak, and Muammar Gadhafi. Both countries had stable economies for many years due to the policies implemented by these leaders.

Muslims, as history has proven, are reluctant to exert themselves; which explains why the Shah of Iran, who turned Iran into a modern, First World Country, faced strong opposition, not just in his own country, but the West also; or, more precisely, the Monolith. While the Shah was abroad seeking medical treatment, the Ayatollah Khomeini returned to Iran and became leader; Iranians, it soon became apparent, were more than willing to stop modernization, and were praised by the Ayatollah for doing so.

The Muslims who reside in Britain are like most others; they like to live comfortable lifestyles, however, this is impossible for them to do legally due to being Muslim, hence, how are they able to do it?

They use "referral centres" to conduct their illegal activities, and they're not bothered by the sight of people dying on the streets so they can enrich themselves, because, according to the psychotic, deluded, world they've created for themselves, the people who die in the gutters of Britain's streets are infidels, and the body count serves as proof that they are succeeding in spreading Islam.

How is it that such psychosis could manage to imbed itself in society; the answer; gangsters don't rat on their friends, and they always keep their mouths shut!! A conspiracy of silence; and a propaganda campaign intended to persuade people to believe the vast majority of Muslims are peace loving people; if they were, they wouldn't be Muslim, according to the Koran.

They are nothing more than bullies who insist on always having things their way; in order to live harmoniously with them, one must always agree with their point of view. Muslims are accustomed to committing violent acts because they religiously read about violent acts; violent thoughts lead to violent acts. Teenage girls are encouraged to read violent literature for the same reason; the most popular books on the teenage market these days are full of violent/sexual, behaviour; which means acquiring a form of sexual gratification from harming someone.

Is it a then a mystery why so many children are reluctant to go to school these days in Britain, and why they're afraid to mention the names of those who bully them?

Diseases of the mind proliferate within society due to people avoiding being "rude"; it is now "politically incorrect" to point out the indiscretions of others; this is considered "anti-social". It is now considered proper to look the other way, and ignore the harm being done to another person. The truth is that people like to make other people mad; they do this because they enjoy hurting others. Malignant aggression spreads, and destroys, in the same manner malignant tumours spread cancer; the only way to stop it is to extract it from the body it is infesting; correspondingly, the sick person who displays this type of behaviour needs to be extricated from the public.

Following is an incident that occurred between me and the staff at the Marks & Spencer store within Brighton's train station; it is typical of what occurs in Britain today, not just in the "work place", but anywhere people congregate.

I was staying in Crawley at the time, and discovered there wasn't an M & S Simply Food store in town; I found out that one is located in Brighton, and bought a train ticket to get there.

I didn't intend on going back and forth between Brighton and Crawley anymore than necessary, so my intention was to buy as many groceries as I could; due to haste, however, I neglected to keep my expenditure under my spending limit; following is what happened next, which was recorded on video surveillance camera.

Once the items were rung in, I said that I wished to pay by credit card; swipe. The man behind the counter, whom I presumed was an Arab by the colour of his skin, and his facial features; he also had a goatee, which is something a lot of Muslim men like to have; gave me the impression he'd set up the card machine appropriately, so I then proceeded to swipe my card. The teller then told me my purchase had been "declined".

asked if he knew why? He said he didn't know why. I said;

"There must be a mistake, because I know my bank has authorized my card for this franchise. Are you certain you set up the machine properly?"

He appeared to check things before replying; "Yes".

"Are you sure?" I asked again, "I can only use my card three times at one location."

He repeated that everything was set up fine, and then said; "Why don't you try again; it should be fine."

I swiped the card, and got the same result. What happened next alarmed me, and awakened me to the fact that daily I am surrounded by people who failed to acquire a proper upbringing, and, consequently, aren't mindful of the things I consider common-sense; which is a gentlemanly way of saying that I think most people are morons; the best description, in my opinion, of a well educated person in Britain today is, "a stranger in a strange land."

The teller then reached over the counter; snatched the bank card from my hand, and swiped it; making that the third occasion it'd been swiped, which made sure I was "struck-out"; I wouldn't be able to take food back to Crawley; the umpire is the bank that issues the card, that decides you only get three tries.

"Why did you do that?" I exclaimed; I should have said something else because he had no right, or reason, to do what he'd just done. "You've now made it impossible for me to buy groceries!"

"I was only trying to help", he replied.

"Help? How? I can't use the card again because of what you've done!"

He didn't answer, which isn't surprising, because there isn't an answer.

What happened next amazed me even more than what had just happened; a female teller, picked up the bag of groceries, and proceeded to the end of the counter, and while doing so, said, "I'll do it again."

"What's the point; he's just swiped it a third time", I said in a voice loud enough for practically anyone in the store to hear.

"I'm just helping", she replied.

"How are you helping? How can you possibly help? I've just told you the card's already been swiped three times; that's it!"

Nobody, it was apparent, was listening to me; one teller, then another, had just done whatever he/she pleased.

I then approached the security guard standing at the store's entrance, and I told him about what had just happened. I was quite surprised when he apologized on behalf of the staff, but I told him he couldn't do that; he wasn't one of the people who'd offended me. I told him I wanted receipts as proof of what had happened, and I wanted to speak to the store manager.

A moment later, the female teller approached, (the one who claimed she could help by doing the impossible, and using my bank card a fourth time), I assume she overheard what I'd said to the security guard, and thrust the receipts at me; her hand jammed into my chest, exclaiming that everybody was just trying to help me.

I couldn't fathom how denying a person food in a grocery store could be considered help; from the tone of her voice, however, I had the impression she saw **me** as the problem, because I was unable to appreciate the fact that someone was doing me a good deed; obviously, I'd have to be a buffoon, if that were the case, and most unlike the rocket scientists working in the M & S store. If I'd wanted to swipe the card again I could have done so; I'd already proven I could do that.

I then said, "This level of ignorance should not exist in Britain", which sums up the problem in a nut shell; ignorance makes the Monolith rich; the cost is a depreciation in the value of people's souls.

I asked the young female teller, why she thought everything transpired the way it did.

She said that she thought everything was quite normal, and there wasn't an issue that needed to be explained. I tried explaining that I thought otherwise. I also told her I had a Ph.D. in psychology. She then declared that no matter the level of my education I'd obtained, my opinion couldn't be of greater value to her than anybody else's, and furthermore, according to her perception of things, which she expressed as an absolute; not a matter for debate; there is no hierarchy in society, thus, any degree I had, was only worth the equivalent of the piece of paper.it was written on.

She then proudly declared that she had a Ph.D., in sociology. I replied, "Really; I have a Ph.D. is sociology. Could you tell me what your thesis was on? I'll look up the title."

She became standoffish at that point, and said she wasn't going to tell me; despite not having any regard for the education system, she, according to what she was telling me, had a Ph.D., but was not interested in sharing the subject of her thesis. People, on a regular basis, use "double speak"; they say two things that contradict each other, so they're not actually saying anything. This is a method the Monolith uses to make sure that when people occupy their time, they are not using time wisely; people are kept busy; not productive.

I next spoke to the manager about the conduct of his employees. He said that he would look into the matter, and get in touch with me; I, on the other hand, saw no reason why the matter shouldn't be dealt with right away, so I asked if the man who took the card from my hand, and swiped it without my authorization; could join us, and share his version of what took place.

The manager asked which teller I first dealt with; I pointed to the man, in his mid twenties; wearing a goatee, which had left, I'd noticed, the store a bit earlier with two employees at the same time I began making an issue of his ignorance.

The manager signalled from where we were both standing; close to the rear of the store, that he should leave his station and join him.

What I found unsettling is that although the teller had known for some while that he is the cause behind my not have groceries, he did not appear disturbed; quite the contrary, he appeared proud of how composed he was, and appeared to have a smirk on his face, which made my stomach churn.

He confessed, right away, to swiping the card without my authorization; not once, but twice. I turned to the manager, and said, "Your investigation is now over; he's just admitted everything; now what are you going to do?"

His answer made me think I was listening to a broken record player; he gave the same answer as both the tellers who'd caused me so much grief, "He was only trying to help you."

I repeated the same question I'd already asked; "How can his actions be construed as helping? He took my card; that's theft."

The manager appeared bemused by what I was saying, and retorted, that because I now had the card, everything was fine, and then asked why I was making such an issue out of things.

I asked the teller with the goatee if he intended on apologizing for what he'd done. He said there was nothing he should apologize about; after all, he was only trying to help me.

I then turned, and looked the manager directly in the eye, and asked if he was going to reprimand him in any manner. I also pointed out that because he appeared unable to appreciate the harm he'd done, there must have been numerous occasions in the past when he'd done the same thing.

The manager reassured me that wasn't the case. I then remarked that this might be "personal"; I looked the man with the goatee straight in the eye, and asked, "Are you Muslim?"

"No", he answered abruptly, at the same time I was wondering if either the manager, or the teller, knew the meaning of the word, "negligence", and if they did why there was a need for a manager to be in the store.

The manager told the teller to return to his post. The teller had a smile on his face as he turned, and headed back to the check out counter.

"You don't believe I should be compensated for the inconvenience I've endured, and the expense of now having to travel to a Marks & Spencer in London to acquire groceries?"

"I'm sorry about that, but that's up to you." The man appeared totally disconnected from the hardship I faced due to his employee not having a clue how to conduct himself. I told him about the expense I now faced; the train fares from Crawley to Brighton; Brighton to London, and London to Crawley; also the time it would take me to make this roundabout journey, but he expressed not a smidgen of remorse, and did not show the slightest interest in offering compensation for my loss of time, and money.

I told him to take my email address; review the video; then get back to me. I made the assumption at that point that the problem was that he was just slow, or I should say, I wanted to believe the problem was that he was slow, (which is a nicer to think, rather than believing a person is callously indifferent), and, at some later point in time, he would recognize he was wrong.

I adopted a softer tone of voice, and remarked that it was a long trip to London, and due to my now not being able to buy groceries; could he offer something to eat. He collected a sandwich; a bag of chips, and chocolate milk, which he noted was, "Just a few quid".

It was nice of him to point out that I wasn't inconveniencing him in the slightest. He was implying that I shouldn't feel guilty about not being able to pay for the food, because they didn't add up to being all that expensive.

People fail to learn from experiences, because they project the problems they're caused upon others, additionally, people like to remind themselves, and others, that no one's perfect; the suggestion behind this remark is that "mishaps" are unavoidable, which is meant to discourage people from making "an issue" out of incidents that have happened. The impression people want to others to believe, and also believe themselves, is that their actions are as natural as, for example, if a person is tired; he'll fall asleep; in other words, nature is just following its natural course.

I left the store; bought a ticket to London; then sat down on one of the station benches to eat the sandwich I was given. Not long after I thought to myself that I should return to the M & S store, and tell the security guard, who did his part in insuring the wrong that had been done to me was not put right, why I'd done, and said, what I had in the store.

"Would you like to know why I'm here, and what this is all about?" He appeared curious, and eagerly replied, "Certainly!"

"Did you read in The Times newspaper today that the Prime Minister now wants married Muslim women to pass a literacy test?"

"No; I haven't had the time to read the newspaper, yet."

"Well, it's a brilliant way to get rid of all Muslims from Britain. They'll never learn English; I can guarantee that; anymore than that teller I was dealing with can change his behaviour."

"I see what you mean", he said, as a smile crossed his face. He didn't say anything else; which didn't surprise me; being politically correct, which means making sure one doesn't say, or do, something that offends others, is a full time occupation for practically everybody in Britain today; just the thought that one might be tempted at some point in time to be politically incorrect could have a devastating impact on a person's mental equilibrium; Catholics typically go to confession to relieve their mental burden.

I waited a couple of days, but I didn't hear back from the manager, who told me his name is, Marcello Martiradonna; I, therefore, decided to return to the M & S store in Brighton's train station, and gather more information about the people employed there. I wanted, in particular, the names of the people who had gone out of their way to deny me food, although, according to their perception of the matter, they were actually helping me. In order to do this I told them my bank had contacted me, and informed me I'd been a victim of fraud. I thought this would shake them up a bit; they all seemed so complacent, and apathetic, when I was in the store, I found it shocking.

As soon as I entered the store, I noticed the big, burly looking, security guard on duty, walk over to the cashiers, and overheard him utter, in a soft voice, I'm sure he thought I wouldn't hear, "Nigel Shindler's here". Well, I thought to myself, it's nice to know I've left an impression on people. I remember my father often saying to me that one of his favourite lines was;

"You can love me, or hate me, but you won't ignore me".

I approached the security guard, and then asked if I could speak with the store manager.

"I don't know the manager; I'm just security here." This was said in far from a friendly tone of voice.

"Would you happen to know where he is? Surely, you'd have to know the manager."

"As I said, that's not my job."

I was just about to search the store myself, when I saw a thin, young man, walking toward me, who appeared to have an employee card attached to a necklace around his neck.

"Excuse me; could you tell me where the manager is please?"

"That would be me; how can I help you?"

"I was here a few days ago, I left my email address with the manager on duty, and expected to hear back. The thing is, I've since heard from my bank; they're claiming fraud; one of the employees swiped my card without my authorization, and I need the names of the people I was dealing with that day."

"That's not a problem at all."

I couldn't believe my ears; did he really say that? I asked myself.

"My name is Nigel Shindler."

"Oh, yes, I'm aware of what happened. The thing is he hasn't finished his investigation; that's why he hasn't been in touch with you."

"I wasn't aware there was anything further to investigate; the employee confessed. What I couldn't believe was that he, and the manager, claimed he was helping me; how is that possible? The teller was negligent, and I've been more than a little inconvenienced as a result." I told him about the price of the train tickers I'd bought.

He then told me his name was, Kevin Gibbs, and he was the supervisor that day, but Marcello Martiradonna was the store manger, and he was the man of duty the day the incident happened, so I'd already been in contact with the right person to resolve the matter.

Mr Gibbs asked me what time the incident occurred, I told him; he then said he'd review the video tape; make copies of the receipts associated with my card, and also collect the names of the people involved; this would take him probably an hour, he stated.

I thanked him for all his help; we shook hands, and I said I'd return to the store in about an hour. Before doing so, I told him I thought I'd buy a single item, just to make sure my card was O.K. for this store. I found him overly solicitous in agreeing with me; he said I shouldn't have to wait in line; so when I was ready, he'd ring it through himself. Frankly, I thought that if he wanted to make my life easier, he could buy the item himself. I didn't mention this, though; I didn't want to raise any hairs at this point; I still hadn't acquired the additional information I was after.

An hour later I returned to the store, and greeted Mr Kevin Gibbs again. He told me that he hadn't reviewed the video; he gave no explanation as to why he didn't, but he'd made copies of the transactions associated with my card that day. He failed to mention, however, the names of the two tellers, and I pointed that out. He said he wasn't sure about their names, but the manager would provide them once he'd completed his investigation.

Another case of fraud had taken place; he didn't do what he'd said he'd do; was this a contagious disease? I thought to myself. He didn't appear to be ashamed, or embarrassed, by the fact he hadn't kept his word. What I'd noticed with every employee I'd dealt with so far was their inappropriate affect; there was no indication in their tone of voice; facial expressions; general body language, that they were empathizing with my situation.

I described the female teller, and asked whether she was present in store. He then told me she had blond hair; not red hair, as I'd indicated, and she was the "shift manager" on the day the incident in question occurred.

"Really; she behaved like she owned the store; how could you not know her name?" I asked.

"Oh; I just thought of something", Kevin exclaimed, "Her name should be on the board adjacent to the entrance".

We walked over, and looked at the board together.

"No, it's not down here."

What was hard for me to digest was that Kevin Gibbs appeared to be unaware that his behaviour was absolutely ridiculous; the store had to have a list of the employees, and the times they worked, and because he was one of the managers, he should have access to that list.

Kevin reassured me that Marcello would be in contact with me in a short while, and he would have the names I wanted. I decided to be cordial, and I let the matter drop.

I collected some groceries; paid with my credit card, then, who should poke his head from behind the counter just as I'm about to leave? The Arab, who, due to either negligence, or malice, deprived me of food on my previous visit. He didn't have a goatee on this occasion, but he still had the same supercilious smirk. He ever so slightly bowed his head before saying he was sorry about what had happened. I no longer saw him as the primary problem, but instead the people who were paid to supervise, and failed to acknowledge that the people under them were negligent.

I shook his hand, and said; "Forgive, and forget".

I thought he'd construe those words as meaning that his admittance of doing wrong had cleared the matter up; quite the contrary had occurred. What was now quite evident to me was that I'd uncovered a deep rooted psychology problem; the people in the store did not appear to be conscious of the irrationalities, and contradictions, in their behaviour; such being the case, it would be impossible to reason with them.

The only way I thought I might be able to change their behaviour was by placing them in a controlled, enclosed, environment long enough to be de-programmed; typically, this means making it impossible for them to avoid evidence that proves their way of thinking, or what they believe is true, is flawed, or incorrect.

Kevin Gribbs, who'd come across at first as sincere, and eager to help, had not done what I'd asked him to do, and what he claimed he would do, so, I waited a few days to hear from Marcello Martiradonna; with no luck, so I decided to return to Brighton again, and see if I could somehow acquire the names of the people I'd dealt with, and also learn more about what was in their psyches; I was hoping to find something that would explain the irrationalities within their behaviour; I wanted to expose what was going on in their unconscious minds. Their defensive behaviour insures they are not, and cannot, be aware of the implications of their behaviour; thus, negating the possibility that either guilt, or shame, can be inflicted upon them.

A person who is fashioned this way is a narcissist; a person perpetually involved in his own world, which is separate from all others. Their sense of aloneness, and insecurity, because of this would be untenable if they were not capable of deceiving themselves into believing they are the contrary to what they happen to be.

Facebook; twitter, as well as other forms of social media, create the illusion in people's minds that they are getting to know, and are apart of, other people's lives; of course, due to the number of people they consider friends, and the manner in which they communicate, no genuine contact is occurring. The forms of social media I've mentioned are merely ways for people to occupy their lives while insuring no learning is taking place.

Narcissists are the same as altruists in the sense that they wish to derive pleasure from the actions they undertake. Altruists, due to the elevation of their mental development; create, and are, as a consequence, responsible for all the great accomplishments in history. Narcissists cannot follow the same route because their minds still haven't developed, thus, they resort to acts of destruction in their quest to find enjoyment, or excitement; not happiness, in the ordinary sense the word is used.

Happiness is fleeting by nature, and only occurs within altruists when it has been earned; greater the accomplishment; more lasting is the happiness experienced. Narcissists, on the other hand, cannot create something that lasts, so they resort to demolishing things in order to have the sense they have control over their environment; greater the conceived value of the creation they ruin; more elevated is the excitement involved in destroying it. The affect is habituating, so a person needs to destroy more to get the same affect, therefore, their behaviour pattern is the same as an addiction, in fact, it can be conceived as the same as injecting a synthetic product into the body in order to derive a certain affect. Such a person contains no light; the force that creates, and lives in accordance with the Word of God, is absent, the person is, therefore, only one thing, the dark.

Light connects, and bonds, with other forms containing light; the dark has no connection with anything else. One can see evidence of this deep rooted sense of isolation in so many things people do in their daily lives as they go about their affairs.

People kiss, hug, and say, "I love you", to people they hardly know, but it wouldn't matter how long they acquaint themselves with another person, because, due to their level of awareness, they can never form a relationship with another person, which helps to explain why such a large percentage of marriages wind up in divorce.

People who are dark do not have the capacity to compare themselves to anything else, and, of course, because the word, bad, is a relative term, a person who cannot make a comparison between himself and anything else, has no reason to conceive himself as bad, never mind, evil, but what can be taken for granted is that the dark must hide its nature in order to destroy.

When I returned to the M & S in Brighton station my objective was to expose the dark in the employees I'd dealt with; but how, considering the fact that it hides its identity in order to repetitively carry out destructive acts, could I show the true force within the people working in that store? The answer is to convince the dark that it has already been exposed, hence, it has nothing to lose by surfacing, and acting in accordance with its nature; this is why people are capable of doing diabolical things once they've committed a crime, and then discovered they've been caught in the act; picture a man who's just robbed a store, only to discover as he's exiting the store with loot in hand, that a person has been recording his movements with a camera; that person then might drop the loot in order to attack the witness with the camera; what's he got to lose?

When I entered the store on this occasion there was a different manager on duty; he was short, thin, and probably in his late twenties.

"Hello; are you the manager?"

"Yes; how can I help you?" He had a pronounced, distinctive, Italian accent.

"My name's Nigel Shindler."

"Ah, yes; I know you."

"Good, so I don't have to explain why I'm here; unfortunately, I still haven't acquired the names of the employees I was dealing with. My bank is claiming fraud, and I need those names to clear up the matter."

"That's not so easy for me to do?"

"Why's that?"

"My boss was been in touch with other departments; people higher up?"

"Why? What's it got to do with anybody outside this store? Your boss is responsible for the behaviour of the two tellers. How does this concern any other store?"

"Ah, well; you see, this could be damaging to the reputation of Marks & Spencer nation wide?"

"Really; if they employ similar people in each of their stores, I imagine it could be; but that's a separate matter; I need the names I asked for."

"I go get them for you." That was quite a turn around, I thought; one minute names are classified information, the next, he agrees to hand them over.

"Actually, I'll write them down for you now."
He took a pen, and notepad, out of his shirt pocket, and proceeded to write down what I'd asked Kevin Gibbs to provide.

"You know, we are all very well educated here; we have many people here who go to university, and have degrees; all sorts; we employee people from a variety of backgrounds."

"Really; why would anybody with a university degree want to work as a cashier? That makes no sense."

"The problem is their degrees aren't accepted here; they have to go back to school to fulfil requirements."

"I'm not surprised. I know in Canada many people buy their degrees."

"You don't think so highly of immigrants do you?"

"The issue that I have with immigrants is the same issue Angela Merkel has; she's stated that Germany's problems are entirely due to immigrants not learning German, or about German culture; the same problem exists in Britain; if people can't speak English, or learn about British culture; how can they make a contribution?"

"So you are against immigrants; I knew it!!"

"I never said I was against them; I said I have an issue with them. I've been studying this matter for some while. The level of ignorance in this country is appalling; most people can't do the simplest of tasks."

"I think you're a racist."

"Nonsense; there are the rats, and the British."

"Aha, I knew it."

"Know what? "The Rats" is a poem; have you read the poem? I don't think you have, so how could you possibly know what I'm referring to?"

"Because you don't like immigrants, you can't buy food here." He tried taking the basket full of groceries out of my hand.

"I have the right to like, and dislike, whomever I please; I also have the right to offend."

"No you don't!"

"I certainly do."

"You can buy food on this occasion, but I don't want you to come back into my store again"

"I'll come back whenever I please." He called over two men; one, I could tell by the uniform, was the security guard for the store; I couldn't figure out where the other man came from, or what he was doing there, both, however, were trying to look imposing by crossing their arms over their chests, and wearing grim expressions on their faces.

I ignored them, and got in line to buy my groceries. While doing so, the manager, Tiaoo Brandao, went from one person to another spouting, and in a loud voice announced that I'm a "racist"; a "sexist"; and that I hate immigrant. Nobody appeared to care what he was saying, but he continued to represent himself as a spokesman in regard to my beliefs, or rather, my perceptions, to one person after another.

"He thinks immigrants are rats!!" He said to one woman in line.

"I never said that; the rats, is the title of a poem."

I bought the groceries, and just as I was about to leave the store, I said; "I'm a living, walking, cartoon of Mohammed". This last comment appeared to enrage the manager; he followed me outside the store, and said;

"What did you say?"

"Say, about what?"

"You know, if I wasn't on duty know", he showed me the timepiece attached to his necklace; "I'd beat the crap out of you."

"My, aren't you civilized!!"

"Yeh, I'd beat the shit out of you man."

He said as we made our way through the crowded train station until I reached one of the exits.

He didn't appear to have any concern about people hearing what he was saying. I had the impression, from comments he made, that he knew I'm a successful author. I would imagine he was concerned about my broadcasting to the world that he's a buffoon, and he must have thought that since he is one; what's the point in not behaving like one? In his mind, the world already knew, and because he knew he couldn't do anything to stop the inevitable from happening, what's the sense in not acting like one.

Dear Mr Kevin Gibbs

I appreciate the help you've provided thus far in clearing up this matter.

According to the print out you gave me, it covers all receipts for all registers from 4:30 pm to 6:30 pm, on January 21, 2016, yet, it does not show any indication that till 9 was used; the one, to the best of my knowledge, (please check video to verify), Sarah - I still don't have her full, (the Shift Manager), used after the transaction had already been declined three times at another till by; I'm not certain of the name of the person who admitted to Marcello Martiradonna that he took the credit card from my hand, and swiped it twice, although, as he confessed, the transaction was not authorized.

Thank you in anticipation of your response.

Sincerely

Nigel shindler

P.S. - thanks for giving me the good news about Justin Trudeau.

Dear Mr Nigel Shindler,

I am writing to you with regards of the incident on the 18th January 2016.

I have investigated this matter fully with our store internal systems, viewed the CCTV footage and also interviewed the team that was on duty on that evening.

Unfortunately during the transaction for your shopping your card was declined three times and you were no able to purchase the selected items in my store.

I gathered from my team that the team member who served you after your first two failed attempt offered to try swiping the card for you of which you handed to him. Once he swiped the card on to the machine also this transaction was declined. At this point the team member handed the transaction to the team leader on duty and try on a different till and as she was dealing with your transaction you decided not to continue with this.

I have been told by the team leader on duty last night that you visited the store to try to recover the transaction details (attached to this e-mail) and mentioned that your bank believes that there could be a fraudulent activity behind this.

I can assure you that no fraudulent activity will be possible at the till pint without a charge to your card account or a record in my store systems, therefore if you have any charges for the specific transaction on your bank statements I will urge you to let me know so this can be resolved without delays.

As per our conversation on the day I am sincerely sorry for your inconvenience, and so is my team, and I hope that the goodwill gesture I offered to take on the house shows that we have our customer at heart.

Can I please advise you check with your bank if there is any charges from my store for that evening and as mentioned above, and please let me have a copy of the bank statement and I will make sure that this is resolved promptly.

I have notified my line manager of the incident and if there is any follow up to this they will be able to give me and the store support in resolving this matter.

I hope the information given will be sufficient, however If you have any other question please do not hesitate to let me know or alternatively you can contact the M&S Customer Service help desk.

Kind Regards

Marcello Martiradonna

Store Manager
M&S SIMPLY FOOD

Office 01273 748128
Mobile 07763929157

Attachments area
    Preview attachment 22011601 - Nigel Shindler.pdf

22011601 - Nigel Shindler.pdf

Nigel   13:21 (2 minutes ago)
Shindler <songsfromearth@gmail.com>
to Brighton

Dear *Marcello Martiradonna*
    In the email I sent to you I asked for the names of the persons at the tills; whom you refer to as "team member", and "team leader".
       Thank you in anticipation of your cooperation.
          Sincerely
            Dr Nigel Shindler

"The Diagnostic and Statistical Manual of Mental Disorders of the American Psychiatric Association (DSM-III) characterizes narcissism as a personality disorder. Narcissism is usually seen as an infatuation with self so extreme that the interests of others are ignored, others serving merely as mirrors of one's own grandiosity."

"DSM-II characterizes pathological narcissism in terms of an exaggerated concern with power and control, the result of which is interpersonal exploitative. Typical also is an orientation of entitlement, the notion that one is worthy of great admiration, respect, and reward regardless of one's achievements."

"...the pathological narcissist's grandiosity is coupled with great fragility of self-esteem..." – Alfred Adler stated that a Superiority Complex hides, or masks, an Inferiority Complex; on that note, I find it interesting that Muslims want to rule the world, but can't stand the thought of someone drawing a cartoon that isn't to their liking.

Society now consists of more narcissists, percentage wise, that ever before, correspondingly, more rules, regulations, policies, and procedures, are needed to keep all narcissists satisfied; everybody wants their pet-peeves addressed; I've been told the N word, for example, is the most derogatory word in the English language; frankly, I have no interest in bolstering the egos of the narcissists; I say, and do, as I please, and if somebody doesn't like that, they have the option of not listening, or moving away. Offending someone is not a violation of their Rights; treating such beliefs as if they are real is method the Monolith uses to keep the general public focused on trivialities rather than spending their time engaged in activities they can learn from.

Partial quotations from, "Narcissism: Socrates, the Frankfurt School, and Psychoanalytic Theory", by C. Fred Alford, Yale university press, Copyright 1988

| MasterCard | 528061******9756 | Voided Tender | 25.37 |
| MasterCard | 528061******9756 | Voided Tender | 25.37 |

---------- End of Report ----------

Operator: 86790389 Kevin Gibbs

Marcello - Unit Manager
Marchiradonney
Sareh - shift

Brighton.Station @ marks-and-spencer.com

M&S
EST. 1884

Arrivals Lounge

M&S
EST. 1884

Arrivals Lounge

A woman employed at the M &S store in Heathrow kept harassing me to the point I finally told her manager that if she didn't stop, I'd have to call the police. This was actually a set-up; shortly after I encountered a British Transport officer, and a police medic who claimed I was the one creating a scene, and I'd have to leave the Airport immediately, or I'd be arrested.

```
***CARDHOLDER COPY***
Mastercard
TID:  ****0106
CARD: ***********9756 AUTH:  27099
READ: Swiped
Signature Verified

***************************************
        *** Join Sparks ***
***************************************
  You could have earned 27 Sparks today
      Join Sparks in the M&S App
    Or at marksandspencer.com/sparks

    Please retain for your records


        M&S Operated by:
  Select Service Partner UK Limited
          169 Euston Road
          London NW1 2AE

28/12/15  16:30  81268020 191679 106 1268
```

9 99021268106191679 7

```
***CARDHOLDER COPY***
Mastercard
TID:  ****0106
CARD: ***********9756 AUTH:  24376
READ: Swiped
Signature Verified

    Please retain for your records


        M&S Operated by:
  Select Service Partner UK Limited
          169 Euston Road
          London NW1 2AE

26/12/15  11:02  81268020 191192 106 1268
```

9 99021268106191192 1

```
        Marks & Spencer
        Heathrow Term 2

      *** SALE ***
        ***STORE COPY***
Mastercard
TID:  ****0106
CARD: ***********9756 AUTH:  27099
READ: Swiped

PLEASE DEBIT MY ACCOUNT      £2.69

      Please Sign Below
```

```
28/12/15  16:30  81268020 191679 106 1268
```

M&S
EST. 1884

```
        Arrivals Lounge
        Heathrow Airport
           Hounslow
           TW6 1EW
         020 8897 9615
      VAT NO. GB884257978
      www.marksandspencer.com

00747004    MILK DIGESTIVE      £0.90
Items: 1    Balance to Pay      £0.90

Cash Tendered                   £0.90

    Please retain for your records


        M&S Operated by:
  Select Service Partner UK Limited
          169 Euston Road
          London NW1 2AE

27/12/15  16:09  81268020 61304  101 1268
```

The Seven Deadly Sins

What does it mean when it is said that a sin is deadly? What is it that dies?

People commit sins while they are alive, therefore, it can be taken for granted that it isn't the life itself that ceases, but something else. What is that?

The soul; that is what makes Homo-sapiens, human. A person without a soul can function, but he cannot live; which means being conscious, aware, and thoughtful, of the environment that surrounds him.

When the soul is weakened, a person's capacity to recognize the environment is dimmed, and he is less able to detect the manner in which his actions affect any milieu. The result that arises from this taking place too long is that a person may live in a pile of horse manure, but not know how the horse manure got there, and how it affects himself; his health, and overall well-being.

The world is out of control; death, destruction, and despair, exist everywhere; yet, remarkably, very few are aware of how hideous the present condition is, and furthermore, how their actions have contributed to this overall state of affairs. How could such a thing be possible? One might ask; especially given the fact that history has shown that man has the capacity to persevere horrendous ordeals, and overcome tremendous hardships.

The key is to consider the meaning of the word, pride, and how such a thing can be a sin; it doesn't have any relation to the word, proud.

A person may be pleased with his accomplishments, which motivates him to continue doing more, but when pride becomes entangled in any work, at some point someone will pay a price. To provide an example; two artists sit side by side at a table, each has a set of crayons, and pencils, and both decide to create a piece of art. One of the artists makes a picture that attracts a lot of attention, while the other artist is hardly noticed at all. The more popular artist loves the attention, and praise, he's given, and decides one day to arrive earlier than the other artist at the a the studio where they both work, and takes some crayons and pencils that do not belong to him. When the other artist arrives he notices his art materials are gone; the other artist declares he has no idea what happened to the missing art materials a, and certainly had nothing to do with them being taken.

The "art thief" has decided to steal and be deceitful because he believes that due to having more materials he'll produce finer pieces of art, leading to, he hopes, even more attention and praise than he before. He becomes so consumed with wanting attention, as well as praise, that he fails to notice the toll taken by the other artist.

Later when he is reminded of this by something, he rationalizes, and justifies, his actions, by deciding that he is more deserving than the other artist, and, thus, requires more materials.

This pattern continues until one of the artists cannot complete a single piece of art due to the more popular artist taking all the art materials; as a consequence, art lovers are left with only one artist to supply them with pieces of art.

The artist that has becomes a celebrity believes he is actually great at what he does, while failing to acknowledge that he is no longer competing against another artist.

Viewing the evolution of events, it is easy to conclude that to be proud of work you've done is fine, but it is imperative to recognize that every artist has an equal right to create, and shouldn't be hampered from doing so due to being deprived of materials because someone believes him more deserving or entitled.

The worst state of affairs imaginable is when a person loses his life because another has concluded he is deserving of the right to do so. It is most evident at that point that all seven of the deadly sins are deeply ingrained in the person's life, though, efforts to disguise such will also be present.

Try asking your neighbour, an acquaintance, a co-worker, how they acquired various things, accomplished certain tasks; you might get a sketchy, hazy, answer, that doesn't really make a lot of sense. When that happens take it for granted that the capacity to make sense no longer exists, and the soul is gone. All that is left is a person functioning in a manner that serves the purpose of fulfilling a role; absent is a human being who can evolve, and one day realize his Spirit, which is his true nature.

The Greatest trick Satan ever pulled was making Man believe He doesn't Exist

The Preserver part of the Holy Trinity is reported to have said; "But Jesus said, Suffer little children, and forbid them not, to come unto me: for of such is the Kingdom of heaven." Matthew 19:14, (KJV).

One can conclude from this statement that somehow within the scheme of things suffering is good, and keeping yourself like a little child is good also; which most, I'm quite sure, consider to be contrary to how The Lord's approval can be obtained; which some call, grace.

The meaning of the above expression used by Jesus can be understood more fully by examining the encounter that occurs between the Serpent, representing temptation, (the destructive force), anfd Eve. We are actually being informed during this event how the conclusion of the previous Age came about.

Within the Garden of Eden, Adam, and Eve, are given one instruction, some might call it, the solitary Commandment; do not partake of the fruit from The Tree of Knowledge of Good and Evil.

The Serpent is aware that Eve is drawn by touch and sight to the apple, and tells Eve that The Lord is withholding information from Eve; namely, that if she eats the forbidden fruit she will be equal to The Lord in every regard. The temptation must be enormous at this juncture, because if such is true, Eve will not pay a price, bear any consequence, due to the act.

There is a fault in her reasoning, however, that is easy to distinguish; The Lord created her, she is The Lord's creation, therefore, she is from The Lord; if she were equal in stature, there would be two Lords, but she is a part of the one that created her, not separate. When Adam doesn't question where the apple he is presented came from, and decides to eat it, both Adam and Eve are then evicted from the Garden of Eden; this is because they acted against nature, how the universe operates, the Laws it obeys, which is The Lord.

Jesus is reported to have said that a person cannot have two masters, and if one decides to have two, the person will inevitably fall.

Matthew 6:24, (KJV)

"No man can serve two masters: for either he will hate the one, and love the other; or else he will hold to the one, and despise the other. Ye cannot serve

God and mammon."

Outside the Garden of Eden there are trials, troubles, and tribulations, waiting for Adam and Eve; this is not a penalty that has befallen them, a form of punishment, but something Adam and Eve decided to impose upon themselves; overcoming obstacles is the way a person strengths himself.

Surveying the Earth as it is now, it's easy to recognize that a massive portion of the world's population deal with enormous hardship every day; obtaining the mere basic necessities of life is their primary occupation: on a daily basis. Considering that there is far from a shortage of materials in the world to satiate the needs of everyone, the suffering man has placed upon his fellow man is entirely unnecessary, and actually illustrates of the extent to which his activities go against nature.

Eve's weakness was that her cognitive faculty, which resulted in her not seeing through the lies she was being told.

Suffering is something nobody, I'm sure, wants in their life, and pretty much everybody does their utmost to avoid at all times; therefore, if Eve can learn to accurately decipher the weakness within herself, and take the necessary measures to rectify it, the inevitable result is a closer attachment to the Garden of Eden, (symbolizing a state of transcendence).

Children who are brought up well, know right from wrong, and are not inclined to question the obvious, thereby not wasting their time which could otherwise spend on something that can satisfy their eager curious minds. As they explore, mistakes are made, for which they suffer, but then learn, as a consequence, to avoid repeating in the future.

When people who like to question the nature of what surrounds them, encounter those who do not behave the same way, predictably, a conflict arises. The curious child is faced with something that can harm him, which he then takes the required measures to avoid in the future.

The world is full of men and women who do not question what they are told; tragically, many people suffer as a consequence, but no penalty, or punishment, is attached to their behaviour that results in the loss of human life, which, deprives the perpetrator the chance to suffer, and learn from errors in judgment; such a patterns weaken the faculty of free choice/ free will, which could be called, the soul; eventually people become clueless; they haven't the slightest inkling of the damage, destruction, despair, and misery, they cause; not just to others, but also themselves.

The greatest trick Satan ever told was making man believe he doesn't exist; Satan is any Homo-sapien that masquerading as a human, but due to not having a soul, is actually a Savage Primate.

"You are the company you keep", is a common expression A person becomes much like the environment he chooses to inhabit; continually being around liars, makes a person become a liar himself. The monsters people fear are actually themselves.

The United Nations has declared that we all have Human Rights, when any Right is violated by any means, to any degree, one is being attacked; the integrity of one's soul is being jeopardized.

The Lord cannot change what He is, but by viewing the actions of Homo-sapiens, He can more fully conceptualize the nature of Himself, and at the same time not feel as alone; the suffering that results from a person separating himself from nature, serves to remind Him what He is.

After all is said and done, and every Homo-sapien has lost the ability to return to the Garden of Eden; to live in harmony with nature, The Lord then has the opportunity to reflect on His nature, as it is revealed in the book of life, which he created at the beginning of the Age.

Ted Bundy

Many today know the name, Ted Bundy,
And think of this man as a monster.
Unfortunately, what they fail to realize
Is that he is not a man of the past,
Lying dead in a grave;
He's here with us today; everyday.

Mr. Bundy's the iconic killer,
From whom we've learned so much
About how the mind becomes diseased.
Many movies have been made about him;
They fit into the genre people call a Thriller.

Ted Bundy grew up a normal boy,
Or so he tells us in his last interviews,
As his final hours slipped away;
Meters away from where his executioner
Would kill him with a flip of a switch.

He told us his life story;
How he wound up killing so many people.
Previously, it was thought the number was between 30 and 40;
But the actual figure was derived from adding a digit.

The gruesome reality became evermore dark, and insidious;
More than a hundred were slaughtered in a little over four years.
How's it possible people didn't know there was
A mammoth beast of horrific proportions in their midst
Just waiting for the chance to kill?
Could it be they were more concerned with how they'd pay their next bill?

The information he provided about his psyche is incalculable.
We know his drive to murder had nothing to do with acquiring money.
The mixture of sex and violence made him a serial killer;
He warned others about how they could become the same.

Few are aware of this in Britain today,
But they have become hideous creatures just like Ted
The sexual thrill of killing indiscriminately,
Enables them to put butter on their bread,
And sleep on a nice, soft, bed.

The instrument used to orchestrate this diabolical undertaking,
 Is called "brainwashing".
The goal is to have a person behave in accordance
With what others want him to do;
All the while making him believe
He is free to think, and do, whatever he pleases;
That's why people don't recognize what's been done to them.

They often see themselves as the opposite of how they actually are.
They don't see what's wrong with themselves,
Because they avoid the sight of blood on their hands,
And the police look the other way;
So, it's hard for them to see the truth.
They cause untold suffering, misery, and pain,
But all they see is a pay check that the government says they've earned;
Not the bodies decaying in a grave,
Or the ashes of those who've been incinerated.

Ted Bundy drove a Volkswagen Beetle.
Beneath he kept a wrench he used to bash his victims on the head.
He lured them to his vehicle with his charm, and a fake disability.
The weakness of his prey was a willingness to help,
And the sense that he wasn't a threat.
I'm sure most were certain of their perception was correct,
And would have gladly made a bet.

Ted masked the time he spent hunting, killing,
And returning to sights where he desecrated corpses.
By helping prevent suicides on a telephone line,
And joining a university, and taking many courses.

Despite the trail of evidence that time, and time again, led to Ted,
People refused to believe this could be the case;
"There must be some misunderstanding", people concluded.

He fooled just about everybody, including Ann Rule.
She knew so much about crime, and what criminals do,
But Ted remained a stranger to her as they sat side by side
Helping save the lives of those who saw no cause to live.

Unbeknownst to the people in parliament,
And businesses practically everywhere,
The people that murder in much the same manner as Ted
Are the ones that have no problem taking bribes,
And continually tell a whole bunch of lies.

The Killer Beside You

Cujo isn't a dog that barks at the moon;
He's the man around the corner that stomps on your foot,
When you don't give him enough room.

You've invited him into your home,
Despite knowing he patrols the neighbourhood
Like he owns the whole world.
He believes he has the right to dictate how others think, speak, and feel.
He's a laugh to look at; a joke to listen to;
Regardless, Cujo steadfastly believes he's the real deal.

He calls himself a Muslim,
And we're led to believe he's quite holy,
But if you contradict anything he has to say,
You might be stoned to death the same day.

You can never be sure if he's got a gun tucked under his fur coat,
However, if he approaches, don't run; be brave;
Remember, Britain is your home,
And you're not prepared to leave on a boat.

Cujo, considers the Koran to be the word of God;
To a large extent it's utter nonsense;
You'd think it was written by a dog made sick
From the bite of an animal with rabies;
He considers it an honour to kill members of his own family,
Even those with babies.
Cujo is upset, and confused.
He doesn't know why,
But he's sure the man to blame is the one that brought him food;

Not just for him, but his whole brood.
The first chance he gets, he'll make him suffer;
He'll use his sharp fangs to bite deep into his flesh,
And wound the hand the once fed him:
At the same time he'll spread the disease he has no idea he's got.

Ask Cujo why he does these things,
And he'll respond with a growl,
While saliva drips from the corners of his mouth,
Then he'll say that he behaves in accordance with
A book that originated further to the south;
Written, supposedly, by a man named Mohamed,
Who spoke to an angel named Gabriel:
All of this he accepts as true, and not some sort of fable.

The neighbours that know a little about Cujo's nature,
Have no interest in getting too close to him;
He's been told that if he comes near,
They'll hop in their cars, and flee.

Cujo doesn't like the idea of a meal getting away,
Or missing the opportunity to cause a great deal of pain,
So he threatens to blow them all up if they decide to behave this way.

"How can you consider yourself religious, and good,
If you could even consider doing such a thing?"
The British gentleman asks;
All the while conscious that it's impossible to reason

With a dog that's lost its' mind
"If I die as well, my God will judge me most kind."
"Really; how did you come to that conclusion?"
"Because 72 virgins will be waiting for me in heaven."
"How do you know that's true?"

Cujo begins to growl, and get upset.
He doesn't like to think;
It unsettles his stomach,
And makes him quite hungry.
He believes that if he maims, or kills,
Those that give him his daily bread,
He will one day stand beside Mohamed,
Instead of in hell,
Along with those the Messenger chose to behead.

When the simple minded fool is perturbed by things he cannot figure out,
He waits for nightfall,
When sleep can give him peace;
Such is the nature of the beast.

Note for feminists; the word, "he", refers to male, and female.

The evil that exists in the world today is due to the deliberate spread of chaos manufactured by those who decided to make themselves rich by taking control of people's minds. They did this by using information gathered during the Second World War that has, to a large degree, been kept hidden from the general public, and obtaining the wealth that once belonged to Jews slaughtered during the Holocaust; I call them "Hitler's Silent Partners".

Those who control America and Israel, whom I call the "Monolith", repeatedly create situations intended to have a destabilizing affectin certain regions. There is still no peace between Israel and the Arab World, for example, because that would not be of financial beneft to the Monolith.

The hoodlums the Kennedy's tried to expose, and incarcerate, still use violence; threats, and intimidation, to get what they want. They don't just use alcohol now to stupefy the public, but a vast array of drugs, and have people wear uniforms that display the word "police" to give people the impression crime is being fought, while they make sure a portion of the drugs the police seize is put back into circulation, and a percentage of proceeds from crimes they confiscate wind up in bank accounts.

The public is kept miss-informed, and ill-informed, by the media which submits to the dictates of the Monlith, and are, consequantly, unable to process the information dispensed due to the proliferation of the assimilation education system which is designed to make sure that by the time "students" enter university they are bereft of any desire to form attachments tosuch books as "Plato's Republic", but rather an electronic device they carry with them wherever they go to call, tweet, and text, their "friends", and due to their minds being kept in an infantile state, they use immediately whenever the desire to gratify their senses should arise, with no regard concerning the penalty paid by another for having done so.

The members of the Monolith are not capable of originality; the world lost all the Jews during the Holocaust; which explains why there has been no development of technology since the Second World War. All the computer gadgetry we have is due to a miniaturization of technology that was available in the 1950's; wires, diodes, transistors, and resistors, which channel electrical currents one way or the other; signified by the numbers 1, and 0.

Serial killer

A **serial killer** is a person who murders three or more people, usually due to abnormal psychological gratification, with the murders taking place over more than a month and including a significant break (a "cooling off period") between them. Some sources, such as the FBI, disregard the "three or more" criterion and define serial killing as "a series of two or more murders, committed as separate events, usually, but not always, by one offender acting alone".

Although psychological gratification is the usual motive for serial killing, and most serial killings involve sexual contact with the victim, the FBI states that the motives of serial killers can include anger, thrill, financial gain, and attention seeking. The murders may be attempted or completed in a similar fashion, and the victims may have something in common: age group, appearance, gender, or race, for example.

Serial killing is not the same as mass murdering, nor is it spree killing, in which murders are committed in two or more locations in a short time. However, cases of extended bouts of sequential killings over periods of weeks or months with no apparent "cooling off period" or "return to normalcy" have caused some experts to suggest a hybrid category of "spree-serial killer".

Etymology

The English term and concept of "serial killer" are commonly attributed to former FBI Special agent Robert Ressler in 1974. Author Ann Rule postulates in her book *Kiss Me, Kill Me* (2004) that the English-language credit for coining the term *serial killer* goes to LAPD detective Pierce Brooks, who created the ViCAP system in 1985. However, in his book *Serial Killers: The Method and Madness of Monsters* (2004), criminal justice historian Peter Vronsky argues that while Ressler might have coined the term *serial homicide* within law, at Bramshill Police Academy in Britain, the terms *serial murder* and *serial murderer* appear in 1966 in John Brophy's book *The Meaning of Murder*. Moreover, Vronsky reports that the term *serial killer* does not appear in Ann Rule's book on Ted Bundy, *The Stranger Beside Me* (1980), when the term was not yet in popular use. In his more recent study, Vronsky states that the construct "serial killing" first enters into American popular usage with the appearance of the term in the *New York Times* in the spring of 1981, to describe Atlanta serial killer Wayne Williams. Subsequently throughout the 1980s, the term was used in the pages of the *New York Times* on 233 occasions, but by the end of the 1990s, in the publication's second decade, the use of the term escalated to 2,514 times in the nation's "newspaper of record".

The German term and concept were coined by the influential Ernst Gennat, who described Peter Kürten as a *serienmörder* (literally "serial murderer") in his article "Die Düsseldorfer Sexualverbrechen" in 1930.

Characteristics

General

Some commonly found characteristics of serial killers include the following:

- They may exhibit varying degrees of mental illness and/or psychopathy, which may contribute to their homicidal behavior.
  - For example, someone who is mentally ill may have psychotic breaks that cause them to believe they are another person or are compelled to murder by other entities.
  - Psychopathic behavior that is consistent with traits common to some serial killers include sensation seeking, a lack of remorse or guilt, impulsivity, the need for control, and predatory behavior. Unlike people with major mental disorders such as schizophrenia, psychopaths can seem normal and often quite charming, a state of adaptation that psychiatrist Hervey Cleckley called the "mask of sanity".
- They were often abused—emotionally, physically and/or sexually— by a family member.
- Serial killers may be more likely to engage in fetishism, partialism or necrophilia, which are paraphilias that involve a strong tendency to experience the object of erotic interest almost as if it were a physical representation of the symbolized body. Individuals engage in paraphilias which are organized along a continuum; participating in varying levels of fantasy perhaps by focusing on body parts (partialism), symbolic objects which serve as physical extensions of the body (fetishism), or the anatomical physicality of the human body; specifically regarding its inner parts and sexual organs (one example being necrophilia).
- A disproportionate number exhibit one, two, or all three of the Macdonald triad of predictors of future violent behavior:
  - Many are fascinated with fire setting.
  - They are involved in sadistic activity; especially in children who have not reached sexual maturity, this activity may take the form of torturing animals.

- More than 60 percent, or simply a large proportion, wet their beds beyond the age of 12. However, recent authorities (see citations in the Enuresissection of the Macdonald triad article) question or deny the statistical significance of this figure; subsequent research suggests that bed-wetting may not be relevant.

- They were frequently bullied or socially isolated as children or adolescents. For example, Henry Lee Lucas was ridiculed as a child and later cited the mass rejection by his peers as a cause for his hatred of everyone. Kenneth Bianchi was teased as a child because he urinated in his pants, suffered twitching, and as a teenager was ignored by his peers.

- Some were involved in petty crimes, such as dishonesty, fraud, theft, vandalism, or similar offenses.

- Often, they have trouble staying employed and tend to work in menial jobs. The FBI, however, states, "Serial murderers often seem normal; have families and/or a steady job." Other sources state they often come from unstable families.

- Studies have suggested that serial killers generally have an average or low-average IQ, although they are often described, and perceived, as possessing IQs in the above-average range. A sample of 174 IQs of serial killers had a median IQ of 93; only serial killers who used bombs had an average IQ above the population mean.

There are exceptions to these criteria, however. For example, Harold Shipman was a successful professional (a General Practitioner working for the NHS). He was considered a pillar of the local community; he even won a professional award for a children's asthma clinic and was interviewed by Granada Television's *World in Action*. Dennis Nilsen was an ex-soldier turned civil servant and trade unionist who had no previous criminal record when arrested. Neither was known to have exhibited many of the tell-tale signs. Vlado Taneski, a crime reporter, was a career journalist who was caught after a series of articles he wrote gave clues that he had murdered people. Russell Williams was a successful and respected career Royal Canadian Air Force Officer who was convicted of the murder of two women, along with fetish burglaries and rapes

German serial killer Fritz Haarmannwith police detectives, November 1924

Many serial killers have faced similar problems in their childhood development. Hickey's Trauma Control Model explains how early childhood trauma can set the child up for deviant behavior in adulthood; the child's environment (either their parents or society) is the dominant factor determining whether or not the child's behavior escalates into homicidal activity.

Family or lack thereof, is the most prominent part of a child's development because it is what the child can identify with on a regular basis. "The serial killer is no different from any other individual who is instigated to seek approval from parents, sexual partners, or others." This need for approval is what influences children to attempt to develop social relationships with their family and peers, but if they are rejected or neglected, they are unable to do so. This results in the lowering of their self-esteem and helps develop their fantasy world, in which they are in control. According to the Hickey's Trauma Control Model the development of a serial killer is based on an early trauma followed by facilitators (e.g., alcohol, drugs, pornography, or other factors that constitute a facilitator, depending on individual circumstances) and disposition (an inability to attach being one common factor).

Family interaction also plays an important role in a child's growth and development. "The quality of their attachments to parents and other members of the family is critical to how these children relate to and value other members of society." Wilson and Seaman (1990) conducted a study on incarcerated serial killers, and what they felt was the most influential factor that contributed to their homicidal activity. Almost all of the serial killers in the study had experienced some sort of environmental problems during their childhood, such as a broken home caused by divorce, or a lack of discipline in the home. It was common for the serial killers to come from a family that had experienced divorce, separation, or the lack of a parent. Furthermore, nearly half of the serial killers had experienced some type of physical or sexual abuse, and even more had experienced emotional neglect. When a parent has a drug or alcohol problem, the attention in the household is on the parents rather than the child. This neglect of the child leads to the lowering of their self-esteem and helps develop a fantasy world in which they are in control. Hickey's Trauma Control Model supports how the neglect from parents can facilitate deviant behavior, especially if the child sees substance abuse in action. This then leads to disposition (the inability to attach), which can further lead to homicidal behavior, unless the child finds a way to develop substantial relationships and fight the label they receive. If a child receives no support from those around him or her, then he or she is unlikely to recover from the traumatic event in a positive way. As stated by E. E. Maccoby, "the family has continued to be seen as a major—perhaps *the* major—arena for socialization".

Chromosomal make up

There have been recent studies looking into the possibility that an abnormality with one's chromosomes could be the trigger for serial killers. Two serial killers, Bobby Joe Long and Richard Speck, came to attention for reported chromosomal abnormalities. Long had an extra X chromosome. Speck was erroneously reported to have an extra Y chromosome; in fact, his karyotype was performed twice and was normal each time. Hellen Morrison, an American forensic psychiatrist, said in an interview that while researchers do not have an exact gene identity, the fact that the majority of serial killers are men leads researchers to believe there is "a change associated with the male chromosome make up."

Fantasy

Children who do not have the power to control the mistreatment they suffer sometimes create a new reality to which they can escape. This new reality becomes their fantasy that they have total control of and becomes part of their daily existence. In this fantasy world, their emotional development is guided and maintained. According to Garrison (1996), "the child becomes sociopathic because the normal development of the concepts of right and wrong and empathy towards others is retarded because the child's emotional and social development occurs within his self-centered fantasies. A person can do no wrong in his own world and the pain of others is of no consequence when the purpose of the fantasy world is to satisfy the needs of one person" (Garrison, 1996). Boundaries between fantasy and reality are lost and fantasies turn to dominance, control, sexual conquest, and violence, eventually leading to murder. Fantasy can lead to the first step in the process of a dissociative state, which, in the words of Stephen Giannangelo, "allows the serial killer to leave the stream of consciousness for what is, to him, a better place".

Criminologist Jose Sanchez reports, "the young criminal you see today is more detached from his victim, more ready to hurt or kill ... The lack of empathy for their victims among young criminals is just one symptom of a problem that afflicts the whole society." Lorenzo Carcaterra, author of *Gangster* (2001), explains how potential criminals are labeled by society, which can then lead to their offspring also developing in the same way through the cycle of violence. The ability for serial killers to appreciate the mental life of others is severely compromised, presumably leading to their dehumanization of others. This process may be considered an expression of the intersubjectivity associated with a cognitive deficit regarding the capability to make sharp distinctions between other people and inanimate objects. For these individuals, objects can appear to possess animistic or humanistic power while people are perceived as objects. Before he was executed, serial killer Ted Bundy stated media violence and pornography had stimulated and increased his need to commit homicide, although this statement was made during last-ditch efforts to appeal his death sentence. However, correlation is not causation (a disturbed physiological disposition, psychosis, lack of socialization, or aggressiveness may contribute to both fantasy creation and serial killing without fantasy creation generally contributing to serial killing for instance). There are exceptions to the typical fantasy patterns of serial killers, as in the case of Dennis Rader, who was a loving family man and the leader of his church.

Organized, Disorganized, and Mixed

The FBI's *Crime Classification Manual* places serial killers into three categories: *organized*, *disorganized*, and *mixed* (i.e., offenders who exhibit organized and disorganized characteristics). Some killers descend from being organized into disorganized as their killings continue, as in the case of psychological decompensation, or vice versa, as when a previously disorganized killer identifies one or more specific aspects of the act of killing as his/her source of gratification and develops a *modus operandi* structured around that aspect / those aspects.

Organized serial killers often plan their crimes methodically, usually abducting victims, killing them in one place and disposing of them in another. They often lure the victims with ploys appealing to their sense of sympathy. Others specifically target prostitutes, who are likely to go voluntarily with a stranger. These killers maintain a high degree of control over the crime scene and usually have a solid knowledge of forensic science that enables them to cover their tracks, such as burying the body or weighing it down and sinking it in a river. They follow their crimes in the news media carefully and often take pride in their actions, as if it were all a grand project. Often, organized killers have social and other interpersonal skills sufficient to enable them to develop both personal and romantic relationships, friends and lovers and sometimes even attract and maintain a spouse and sustain a family including children. Among serial killers, those of this type are in the event of their capture most likely to be described by acquaintances as kind and unlikely to hurt anyone. Bundy and John Wayne Gacy are examples of organized serial killers. In general, the IQs of organized serial killers tend to be near normal range, with a mean of 94.7. Organized nonsocial offenders tend to be on the higher end of the average, with a mean IQ of 99.2.

Disorganized serial killers are usually far more impulsive, often committing their murders with a random weapon available at the time, and usually do not attempt to hide the body. They are likely to be unemployed, a loner, or both, with very few friends. They often turn out to have a history of mental illness, and their modus operandi (M.O.) or lack thereof is often marked by excessive violence and sometimes necrophilia and/or sexual violence. Disorganized serial killers have been found to have a slightly lower mean IQ than organized serial killers, at 92.8.

Medical professionals

Some people with a pathological interest in the power of life and death tend to be attracted to medical professions or acquiring such a job. These kinds of killers are sometimes referred to as "angels of death" or angels of mercy. Medical professionals will kill their patients for money, for a sense of sadistic pleasure, for a belief that they are "easing" the patient's pain, or simply "because they can". One such killer was nurse Jane Toppan, who admitted during her murder trial that she was sexually aroused by death. She would administer a drug mixture to patients she chose as her victims, lie in bed with them and hold them close to her body as they died.

Another medical profession serial killer is Genene Jones. It is believed she killed 11 to 46 infants and children while working at Bexar County Medical Center Hospital in San Antonio, Texas. She is currently serving a 99 year sentence for the murder of Chelsea McClellan and the attempted murder of Rolando Santos, and is eligible for parole in 2017 due to a law in Texas at the time of her sentencing to reduce prison overcrowding.

Female Serial Killers

Female serial killers are rare compared to their male counterparts. Sources suggest that female serial killers represented less than one in every six known serial murderers in the U.S. between 1800 and 2004 (64 females from a total of 416 known offenders), or that around 15% of U.S. serial killers have been women, with a collective number of victims between 427 and 612. *Lethal Ladies'* authors, Amanda L. Farrell, Robert D. Keppel, and Victoria B. Titterington state that "the Justice Department indicated 36 female serial killers have been active over the course of the last century."

According to *The Journal of Forensic Psychiatry & Psychology,* there is evidence that 16 % of all serial killers are women.

Kelleher and Kelleher (1998) created several categories to describe female serial killers. They used the classifications of widow, angel *of death, sexual predator, revenge, profit or crime, team killer, question of sanity, unexplained,* and *unsolved.* In using these categories, they observed that most women fell into the categories of black widow or team killer. Although motivations for female serial killers can include attention seeking, addiction, or the result of psychopathological behavioral factors, female serial killers are commonly categorized as murdering men for material gain, usually being emotionally close to their victims, and generally needing to have a relationship with the victim, hence the traditional cultural image of the "black widow." In describing murderer Stacey Castor, forensic psychiatrist James Knoll offered a psychological perspective on what defines a "black widow" type. In simple terms, he described it as a woman who kills two or more husbands or lovers for material gain. Though Castor was not officially defined as a serial killer, it is likely that she would have killed again.

One "analysis of 86 female serial killers from the U.S. found that the victims tended to be spouses, children or the elderly". Other studies indicate that since 1975, increasingly strangers are marginally the most preferred victim of female serial killers, or that only 26% of female serial killers kill for material gain only. Sources state that "[e]ach killer will have her own proclivities, needs and triggers." A review of the published literature on female serial murder stated that "sexual or sadistic motives are believed to be extremely rare in female serial murderers and psychopathic traits and histories of childhood abuse have been consistently reported in these women." A study by Eric W. Hickey (2010) of 64 female serial killers in the U.S. indicated that sexual activity was one of several motives in 10% of the cases, enjoyment in 11% and control in 14%, and that 51% of all U.S. female serial killers murdered at least one woman and 31% murdered at least one child. In other cases, women have been involved as an accomplice with a male serial killer as a part of a serial killing team. A 2015 study published in *The Journal of Forensic Psychiatry & Psychology* found that the most common motive for female serial killers was for financial gain and almost 40 percent of them had experienced some sort of mental illness.

Peter Vronsky in *Female Serial Killers* (2007) maintains that female serial killers today often kill for the same reason males do: as a means of expressing rage and control. He suggests that sometimes the theft of the victims' property by the female "black widow" type serial killer appears to be for material gain, but really is akin to a male serial killer's collecting of totems (souvenirs) from the victim as a way of exerting continued control over the victim and reliving it. By contrast, Hickey states that although popular perception sees "black widow" female serial killers as something of the Victorian past, in his statistical study of female serial killer cases reported in the U.S. since 1826, approximately 75% occurred since 1950.

The methods that female serial killers use for murder are frequently covert or low-profile, such as murder by poison (the preferred choice for killing). Other methods used by female serial killers include shootings (used by 20%), suffocation (16%), stabbing (11%), and drowning (5%). They commit killings in specific places, such as their home or a health-care facility, or at different locations within the same city or state. A notable exception to the typical characteristics of female serial killers is Aileen Wuornos, who killed outdoors instead of at home, used a gun instead of poison, killed strangers instead of friends or family, and killed for personal gratification. The most prolific female serial killer in all of history is allegedly Elizabeth Báthory. Countess Elizabeth Báthory de Ecsed (Báthory Erzsébet in Hungarian, August 17, 1560 – August 21, 1614) was a countess from the renowned Báthory family. After her husband's death, she and four collaborators were accused of torturing and killing hundreds of girls and young women, with one witness attributing to them over 600 victims, though the number for which they were convicted was 80. Elizabeth herself was neither tried nor convicted. In 1610, however, she was imprisoned in the Csejte Castle, where she remained bricked in a set of rooms until her death four years later.

A 2010 article by Perri and Lichtenwald addressed some of the misperceptions concerning female criminality. In the article, Perri and Lichtenwald analyze the current research regarding female psychopathy, including case studies of female psychopathic killers featuring Münchausen syndrome by proxy, cesarean section homicide, fraud detection homicide, female kill teams, and a female serial killer.

Ethnicity and Serial Killer Demographics in the U.S.

The racial demographics regarding serial killers are often subject to debate. In the United States, the majority of reported and investigated serial killers are white males, from a lower-to-middle-class background, usually in their late twenties to early thirties. However, there are African American, Asian, and Hispanic (of any race) serial killers as well, and, according to the FBI, based on percentages of the U.S. population, whites are not more likely than other races to be serial killers. Criminal profiler Pat Brown says serial killers are usually reported as white because the media typically focuses on "All-American" white and pretty female victims who were the targets of white male offenders, that crimes among minority offenders in urban communities, where crime rates are higher, are under-investigated, and that minority serial killers likely exist at the same ratios as white serial killers for the population. She believes that the myth that serial killers are always white might have become "truth" in some research fields due to the over-reporting of white serial killers in the media.

According to some sources, the percentage of serial killers who are African American is estimated to be between 13 and 22 percent. Another study has shown that 16 percent of serial killers are African American, what author Maurice Godwin describes as a "sizeable portion". A 2014 Radford/FGCU Serial Killer Database annual statistics report showed that for the decades 1900–2010, the percentage of White serial killers was 52.1% while the percentage of African American serial killers was 40.3%. Popular racial stereotypes about the lower intelligence of African-Americans, and the stereotype that serial killers are white males, may explain the media focus on serial killers that are white and the failure to adequately report on those that are black. Similarly, in a 2005 article Anthony Walsh, professor of criminal justice at Boise State University, argued a review of post-WWII serial killings in America finds that the prevalence of African-American serial killers has typically been drastically underestimated in both professional research literature and the mass media. As a paradigmatic case of this media double-standard, Walsh cites news reporting on white killer Gary Heidnik and African-American killer Harrison Graham. Both men were residents of Philadelphia, Pennsylvania; both imprisoned, tortured and killed several women; and both were arrested only months apart in 1987. "Heidnik received widespread national attention, became the subject of books and television shows, and served as a model for the fictitious Buffalo Bill in *Silence of the Lambs*", writes Walsh, while "Graham received virtually no media attention outside of Philadelphia, despite having been convicted of four more murders than Heidnik".

Motives

The motives of serial killers are generally placed into four categories: *visionary*, *mission-oriented*, *hedonistic* and *power or control*; however, the motives of any given killer may display considerable overlap among these categories.

Visionary

Visionary serial killers suffer from psychotic breaks with reality, sometimes believing they are another person or are compelled to murder by entities such as the Devil or God. The two most common subgroups are "demon mandated" and "God mandated".

Herbert Mullin believed the American casualties in the Vietnam War were preventing California from experiencing the Big One. As the war wound down, Mullin claimed his father instructed him via telepathy to raise the number of "human sacrifices to nature" in order to delay a catastrophic earthquake that would plunge California into the ocean. David Berkowitz ("Son of Sam") is also an example of a visionary killer. He claimed a demon transmitted orders through his neighbor's dog and instructed him to commit murder.

Mission-oriented

Mission-oriented killers typically justify their acts as "ridding the world" of a certain type of person perceived as undesirable, such as homosexuals, prostitutes, or people of different ethnicity or religion; however, they are generally not psychotic. For example, the Zebra killers in the San Francisco Bay Area specifically targeted Caucasians. Some see themselves as attempting to change society, often to cure a societal ill.

Hedonistic

This type of serial killer seeks thrills and derives pleasure from killing, seeing people as expendable means to this goal. Forensic psychologists have identified three subtypes of the hedonistic killer: "lust", "thrill" and "comfort".

Lust

Sex is the primary motive of lust killers, whether or not the victims are dead, and fantasy plays a large role in their killings. Their sexual gratification depends on the amount of torture and mutilation they perform on their victims. The sexual serial murderer has a psychological need to have absolute control, dominance, and power over their victims, and the infliction of torture, pain, and ultimately death is used in an attempt to fulfill their need. They usually use weapons that require close contact with the victims, such as knives or hands. As lust killers continue with their murders, the time between killings decreases or the required level of stimulation increases, sometimes both.

Kenneth Bianchi, one of the "Hillside Stranglers", murdered women and girls of different ages, races and appearance because his sexual urges required different types of stimulation and increasing intensity. Jeffrey Dahmer, who was repeatedly diagnosed with borderline, searched for his perfect fantasy lover—beautiful, submissive and eternal. As his desire increased, he experimented with drugs, alcohol, and exotic sex. His increasing need for stimulation was demonstrated by the dismemberment of victims, whose heads and genitals he preserved, and by his attempts to create a "living zombie" under his control (by pouring acid into a hole drilled into the victim's skull). Dahmer once said, "Lust played a big part of it. Control and lust. Once it happened the first time, it just seemed like it had control of my life from there on in. The killing was just a means to an end. That was the least satisfactory part. I didn't enjoy doing that. That's why I tried to create living zombies with … acid and the drill." He further elaborated on this, also saying, "I wanted to see if it was possible to make—again, it sounds really gross—uh, zombies, people that would not have a will of their own, but would follow my instructions without resistance. So after that, I started using the drilling technique." He experimented with cannibalism to "ensure his victims would always be a part of him".

Thrill

The primary motive of a thrill killer is to induce pain or terror in their victims, which provides stimulation and excitement for the killer. They seek the adrenaline rush provided by hunting and killing victims. Thrill killers murder only for the kill; usually the attack is not prolonged, and there is no sexual aspect. Usually the victims are strangers, although the killer may have followed them for a period of time. Thrill killers can abstain from killing for long periods of time and become more successful at killing as they refine their murder methods. Many attempt to commit the perfect crime and believe they will not be caught. Robert Hansen took his victims to a secluded area, where he would let them loose and then hunt and kill them. In one of his letters to San Francisco Bay Area newspapers, the Zodiac Killer wrote "[killing] gives me the most thrilling experience it is even better than getting your rocks off with a girl". Coral Watts was described by a surviving victim as "excited and hyper and clappin' and just making noises like he was excited, that this was gonna be fun" during the 1982 attack. Slashing, stabbing, hanging, drowning, asphyxiating, and strangling were among the ways Watts killed.

Comfort (profit)

Material gain and a comfortable lifestyle are the primary motives of comfort killers. Usually, the victims are family members and close acquaintances. After a murder, a comfort killer will usually wait for a period of time before killing again to allow any suspicions by family or authorities to subside. They often use poison, most notably arsenic, to kill their victims. Female serial killers are often comfort killers, although not all comfort killers are female. Dorothea Puente killed her tenants for their Social Security checks and buried them in the backyard of her home. H. H. Holmes killed for insurance and business profits. Professional killers ("hitmen") may also be considered comfort serial killers. Richard Kuklinski charged tens of thousands of dollars for a "hit", earning enough money to support his family in a middle-class lifestyle (Bruno, 1993).

Some, like Puente and Holmes, may be involved in and/or have previous convictions for theft, fraud, non payment of debts, embezzlement and other crimes of a similar nature. Dorothea Puente was finally arrested on a parole violation, having been on parole for a previous fraud conviction.

Power/control

The main objective for this type of serial killer is to gain and exert power over their victim. Such killers are sometimes abused as children, leaving them with feelings of powerlessness and inadequacy as adults. Many power- or control-motivated killers sexually abuse their victims, but they differ from hedonistic killers in that rape is not motivated by lust (as it would be with a lust murder) but as simply another form of dominating the victim. Ted Bundy is an example of a power/control-oriented serial killer. He traveled around the United States seeking women to control.

Media

Many serial killers claim that a violent culture influenced them to commit murders. During his final interview, Bundy stated that hardcore pornography was responsible for his actions. Others idolize figures for their deeds or perceived vigilante justice, such as Peter Kürten, who idolized Jack the Ripper, or John Wayne Gacy and Ed Kemper, who both idolized the actor John Wayne. Many movies, books, and documentaries have been written about serial killers, detailing the lives and crimes that have been committed. The movie Bundy, which was released in 2002, focuses on serial killer Ted Bundy's personal life in college, leading up to his execution. Another movie, Dahmer, was released in the same year, and tells the story of Jeffrey Dahmer. Serial killers are also portrayed in fictional media, often times as having substantial intelligence and looking for difficult targets, despite the contradiction with the psychological profile of serial killers. Killers who have a strong desire for fame or to be renowned for their actions desire media attention as a way of validating and spreading their crimes; fear is also a component here, as some serial killers enjoy causing fear. An example is the BTK Killer, who sought attention from the press during his murder spree.

Theories

Biological and sociological

Theories for why certain people commit serial murder have been advanced. Some theorists believe the reasons are biological, suggesting serial killers are born, not made, and that their violent behavior is a result of abnormal brain activity. Holmes and Holmes believe that "until a reliable sample can be obtained and tested, there is no scientific statement that can be made concerning the exact role of biology as a determining factor of a serial killer personality." The "Fractured Identity Syndrome" (FIS) is a merging of Charles Cooley's "looking glass self" and Erving Goffman's "virtual" and "actual social identity" theories. The FIS suggests a social event, or series of events, during one's childhood or adolescence results in a fracturing of the personality of the serial killer. The term "fracture" is defined as a small breakage of the personality which is often not visible to the outside world and is only felt by the killer.

"Social Process Theory" has also been suggested as an explanation for serial murder. Social process theory states that offenders may turn to crime due to peer pressure, family, and friends. Criminal behavior is a process of interaction with social institutions, in which everyone has the potential for criminal behavior. A lack of family structure and identity could also be a cause leading to serial murder traits. A child used as a scapegoat will be deprived of their capacity to feel guilt. Displaced anger could result in animal torture, as identified in the Macdonald triad, and a further lack of basic identity.

Military

A dishonorably discharged Marine, Charles participated in the kidnapping, sadistic torture, rape and murder of numerous victims

The "military theory" has been proposed as an explanation for why serial murderers kill, as some serial murderers have served in the military or related fields. According to Castle and Hensley, 7% of the serial killers studied had military experience. This figure may be a proportional under-representation when compared to the number of military veterans in a nation's total population. For example, according to the United States census for the year 2000, military veterans comprised 12.7% of the U.S. population; in England, it was estimated in 2007 that military veterans comprised 9.1% of the population. Though by contrast, about 2.5% of the population of Canada in 2006 consisted of military veterans.

There are two theories that can be used to study the correlation between serial killing and military training: *Applied learning theory* states that serial killing can be learned. The military is training for higher kill rates from servicemen while training the soldiers to be desensitized to taking a human life. *Social learning theory* can be used when soldiers get praised and accommodated for killing. They learn, or believe that they learn, that it is acceptable to kill because they were praised for it in the military. Serial killers want accreditation for the work that they have done.

In both military and serial killing, the offender or the soldier may become desensitized to killing as well as compartmentalized; the soldiers do not see enemy personnel as "human" and neither do serial killers see their victims as humans. The theories do not imply that military institutions make a deliberate effort to produce serial killers; to the contrary, all military personnel are trained to recognize when, where, and against whom it is appropriate to use deadly force, which starts with the basic *Law of Land Warfare*, taught during the initial training phase, and may include more stringent policies for military personnel in law enforcement or security. They are also taught ethics in basic training. **(Wikipedia)**

Any Woman can get Bacterial Vaginosis

Having bacterial vaginosis can increase your chance of getting an STD.

What is bacterial vaginosis?

Bacterial vaginosis (BV) is an infection caused when too much of certain bacteria change the normal balance of bacteria in the vagina.

How common is bacterial vaginosis?

Bacterial vaginosis is the most common vaginal infection in women ages 15-44.

How is bacterial vaginosis spread?

We do not know about the cause of BV or how some women get it. BV is linked to an imbalance of "good" and "harmful" bacteria that are normally found in a woman's vagina.

We do know that having a new sex partner or multiple sex partners and douching can upset the balance of bacteria in the vagina and put women at increased risk for getting BV.

However, we do not know how sex contributes to BV. BV is not considered an STD, but having BV can increase your chances of getting an STD. BV rarely affects women who have never had sex.

You cannot get BV from toilet seats, bedding, or swimming pools.

How can I avoid getting bacterial vaginosis?

Doctors and scientists do not completely understand how BV is spread, and there are no known best ways to prevent it.

The following basic prevention steps may help lower your risk of developing BV:

I'm pregnant. How does bacterial vaginosis affect my baby?

Pregnant women can get BV. Pregnant women with BV are more likely to have babies who are born premature (early) or with low birth weight than women who do not have BV while pregnant. Low birth weight means having a baby that weighs less than 5.5 pounds at birth.

Treatment is especially important for pregnant women.

How do I know if I have bacterial vaginosis?

Many women with BV do not have symptoms. If you do have symptoms, you may notice a thin white or grey vaginal discharge, odour, pain, itching, or burning in the vagina. Some women have a strong fish-like odour, especially after sex. You may also have burning when urinating; itching around the outside of the vagina, or both.

How will my doctor know if I have bacterial vaginosis?

A health care provider will look at your vagina for signs of BV and perform laboratory tests on a sample of vaginal fluid to determine if BV is present.

Can bacterial vaginosis be cured?

BV will sometimes go away without treatment. But if you have symptoms of BV you should be checked and treated. It is important that you take all of the medicine prescribed to you, even if your symptoms go away. A health care provider can treat BV with antibiotics, but BV can recur even after treatment. Treatment may also reduce the risk for STDs.

Male sex partners of women diagnosed with BV generally do not need to be treated. However, BV may be transferred between female sex partners.

What happens if I don't get treated?

BV can cause some serious health risks, including

- Increasing your chance of getting HIV if you have sex with someone who is infected with HIV;
- If you are HIV positive, increasing your chance of passing HIV to your sex partner;
- Making it more likely that you will deliver your baby too early if

you have BV while pregnant;

- Increasing your chance of getting other STDs, such as chlamydia and gonorrhoea. These bacteria can sometimes cause pelvic inflammatory disease (PID), which can make it difficult or impossible for you to have children.

What is PID?

Pelvic inflammatory disease is an infection of a woman's reproductive organs. It is a complication often caused by some STDs, like chlamydia and gonorrhea. Other infections that are not sexually transmitted can also cause PID.

How do I get PID?

You are more likely to get PID if you

- Have an STD and do not get treated;
- Have more than one sex partner;
- Have a sex partner who has sex partners other than you;
- Have had PID before;
- Are sexually active and are age 25 or younger;
- Douche;
- Use an intrauterine device (IUD) for birth control.

How can I reduce my risk of getting PID?

The only way to avoid STDs is to not have vaginal, anal, or oral sex.

If you are sexually active, you can do the following things to lower your chances of getting PID:

- Being in a long-term mutually monogamous relationship with a partner who has been tested and has negative STD test results;
- Using latex condoms the right way every time you have sex.

How do I know if I have PID?

There are no tests for PID. A diagnosis is usually based on a combination of your medical history, physical exam, and other test results. You may not realize you have PID because your symptoms may be mild, or you may not experience any symptoms. However, if you do have symptoms, you may notice

- Pain in your lower abdomen;

- Fever;
- An unusual discharge with a bad odour from your vagina;
- Pain and/or bleeding when you have sex;
- Burning sensation when you urinate; or
- Bleeding between periods.

You should

- Be examined by your doctor if you notice any of these symptoms;
- Promptly see a doctor if you think you or your sex partner(s) have or were exposed to an STD;
- Promptly see a doctor if you have any genital symptoms such as an unusual sore, a smelly discharge, burning when peeing, or bleeding between periods;
- Get a test for chlamydia every year if you are sexually active and younger than 25 years of age.

    Have an honest and open talk with your health care provider if you are sexually active and ask whether you should be tested for other STDs.

Breast Feeding

- If breastfeeding rates were to go up from 35 per cent to 75 per cent in our neonatal units, it would save the NHS an estimated £6 million, largely by reducing the incidence of a particular kind of bowel disease. Pushing feeding rates up to 45 per cent among mothers in general would save £11 million in treating childhood diseases, and around £26 million in breast-cancer treatments for the mothers.
- And we haven't even got on to the other pluses: breast milk is free (although, admittedly, a lactating mothers' equine appetite doesn't come cheap). The milk changes in composition according to your baby's needs. It's convenient – no bottles or mixing or sterilising equipment needed. It's at the perfect temperature. And it provides the opportunity for a bonding cuddle.
- Not that I want to induce guilt in women who really can't breastfeed. It is not an easy thing to do, especially at first, and I often bristle at the "Breastapo" for saying that mothers who don't breastfeed are simply not trying hard enough. For many, breastfeeding can hurt horribly – cabbage leaves were my morphine – and when things aren't going smoothly, there is the added anguish and guilt at not being able to satisfy your precious baby's basic need

to be fed.

- In the age of austerity, many mums also feel the pressure to be back at work quickly. There are few jobs which can easily absorb the constant interruptions of a breastfeeding schedule, and even fewer workwear options that can easily absorb a regurgitated feed. It is no coincidence that the countries at the top of the lactation league are those with very long, generously paid, usually state-sponsored maternity leave.

- Lest I sound too negative, there is heartening evidence that British mothers do start off with the best intentions: 81 per cent begin by feeding their children from the breast. But that just makes it even more shocking that only a fifth last the recommended six months. So there is a real case for paying for greater practical and social support, via the NHS and midwives, for mothers who find breastfeeding a tough, lonely and painful road to travel.

- But there are also mothers who never even start – and this is where our efforts need to be particularly sharpened up. In some areas of the country, such as the poorer parts of Derbyshire and Yorkshire, breastfeeding rates can be as low as 20 per cent. In the most deprived communities, only one in eight mums is still going 12 weeks after birth; in some communities, it is perceived as a disgusting, unsexy, unmodern way to feed a baby.

- So, last year, academics at Sheffield University started a voucher scheme, in which breastfeeding mothers in deprived areas would qualify for £200 of shopping vouchers, redeemable in chains such as Poundstretcher. At the time, I called it "nudge theory for nipples" – and, despite some controversy over the "bribing" of mothers, the scheme does appear to have worked. Half of the women eligible to join the Nosh (Nourishing Start for Health) project signed up, and the majority stuck it out for eight weeks (data on whether they lasted the full six months is still being collated).

- Overall, breastfeeding rates in the areas where vouchers were offered were around 34 per cent. That alone, says the lead researcher, Dr Clare Relton, makes the scheme a judicious investment for the state. And interestingly, the financial inducement did not – as widely feared – affect the relationship between mother and midwife.

- The final results will be published shortly in the Lancet, but the scheme is already being extended to around 4,000 new mums, with researchers from the Universities of Dundee and Brunel joining up (though I doubt vouchers for Claridge's will be among those on offer).

- In some ways, breastfeeding is an easier target for advocates of "nudge theory" than obesity: mothers need to be motivated for only six months, while the obese have to stay motivated over a lifetime in order to maintain a healthy weight. Reports suggest that the vouchers did indeed serve as that well-timed little push to steer mothers away from the formula carton. The vouchers, incidentally, went on nappies, baby clothes, toys and occasionally groceries.

- Some will still call them a bribe – but does that actually matter? Public and private institutions have always used financial inducements, and penalties, to change behaviour: tax cuts; tax relief; incentives to lose weight; duties on alcohol, cigarettes and petrol; traffic fines; bonuses; early bird discounts. If we penalise people for doing things that are bad for them, is it really immoral to reward them for doing things that are good for them? The benefits of breastfeeding, after all, go not just to the mother and child, but to the public purse.

- Boosting breastfeeding rates is a pragmatic national calculation – unlike the one made by Mr Farage when he suggested that mothers form a lactating huddle in the corner. After all, living in a free, tolerant society means that our attention is occasionally drawn to things we'd rather not see: I'm not particularly taken with builders' bottoms, tattoos or extreme facial piercings – or blokeish, beer-swilling politicians, for that matter. But I am perfectly free to avert my gaze.

- It was telling that not a single person taking afternoon tea in Claridge's complained about the mother who was asked to cover up. I have rarely seen a woman ostentatiously whipping her bosom out in order to assuage a hungry infant – we're normally too exhausted to set about such flamboyant, feminist gestures.

- But even if mothers up and down the land do occasionally flash a nipple, can't we just respect their right to breastfeed without embarrassment?

- Given its undoubted benefits, it's those who are purse-lipped about public breastfeeding – or who frown on the state for encouraging it – who should be asked to stand in the corner.

- **Sponsored By Standard Life Investments.**

Breastfeeding: good for you, your baby and the NHS.

"Increase in breastfeeding could save NHS £40m a year," The Independent reports after a recent economic modelling study projected a reduction in childhood diseases and breast cancer rates would lead to considerable savings for the health service.

The roven key benefits, and potential savings, associated with breastfeeding a baby include;

a reduced risk of bowel infection(gastroenteritis), lower respiratory tract infection (bronchiolitis), middle ear infection (otitis media) and an uncommon, but serious, condition called necrotising enterocolitis (bowel tissue death).

Breastfeeding also brings benefits to the mother, such as a reduced risk of breast cancer.

Increasing breastfeeding rates in neonatal units from 35% to 75% could save £6 million per year by reducing the incidence of necrotising enterocolitis, according to the study.

In the general population, if the percentage of women who breastfed for at least four months increased from 7% to 45%, the NHS would save £11 million per year from a reduction in the types of common infant conditions described above.

Similarly, the same increase could result in NHS savings of around £21 million related to breast cancer alone over the course of a first-time mothers' lifetime.

While the figures presented in the study are only estimates, it would certainly seem breastfeeding is not only good for mother and baby: it is also good for the NHS.

How many infants born in the United States are breastfed?

The CDC National Immunization Survey is a nationally representative sample of the U.S. population that provides the percent of U.S. children who are breastfed by birth year. Rates are provided on breastfeeding initiation, duration, and exclusivity.

For additional breastfeeding statistics, see Data and Statistics.

Are growth charts available to assess growth in breastfed infants?

In the United States, the 2006 WHO growth standard charts are recommended for use with both breastfed and formula fed infants and children, from birth to aged 2 years to monitor growth. The WHO growth charts reflect growth patterns among children who were predominantly breastfed for at least 4 months and still breastfeeding at 12 months. These charts describe the growth of healthy children living in well-supported environments in six countries throughout the world including the United States. The WHO growth charts show how infants and children should grow rather than simply how they do grow in a certain time and place and are therefore recommended for all infants. The WHO growth charts are relevant to the U.S. infant and young child population as U.S. children were included in the WHO study sample and their growth tracks along the median of the pooled international sample.

The WHO growth charts establish the growth of the breastfed infant as the norm for growth. Healthy breastfed infants typically put on weight more slowly than formula fed infants in the first year of life. Formula fed infants gain weight more rapidly after about 3 months of age. Differences in weight patterns continue even after complementary foods are introduced. The Centers for Disease Control and Prevention (CDC) and the American Academy of Pediatrics recommend that health care providers in primary care settings use the 2000 CDC growth reference charts for children and teens aged 2 to 20 years to monitor growth in the United States.[1] The 2000 CDC growth reference charts include the weight-for-age, stature-for-age, and BMI (Body Mass Index)-for-age charts for boys and girls aged 2 to 20 years.

When should a baby start eating solid foods such as cereals, vegetables, and fruits?

Breast milk alone is sufficient to support optimal growth and development for approximately the first 6 months after birth. For these very young infants, the American Academy of Pediatrics (AAP) states that water, juice, and other foods are generally unnecessary. Even when babies enjoy discovering new tastes and textures, solid foods should not replace breastfeeding, but merely complement breast milk as the infant's main source of nutrients throughout the first year. Beyond one year, as the variety and volume of solid foods gradually increase, breast milk remains an ideal addition to the child's diet.

How long should a mother breastfeed?

The American Academy of Pediatrics (AAP) recommends that breastfeeding should continue for at least 12 months, and thereafter for as long as mother and baby desire.

The World Health Organization recommends continued breastfeeding up to 2 years of age or beyond.

What can happen if someone else's breast milk is given to another child?

HIV and other serious infectious diseases can be transmitted through breast milk. However, the risk of infection from a single bottle of breast milk, even if the mother is HIV positive, is extremely small. For women who do not have HIV or other serious infectious diseases, there is little risk to the child who receives her breast milk. See Diseases and Conditions for more information. **(Wikipedia)**

Gerald Durrell

Gerald "Gerry" Malcolm Durrell, OBE (7 January 1925 – 30 January 1995) was an English naturalist, zookeeper, conservationist, author, and television presenter. He founded what is now called the Durrell Wildlife Conservation Trust and the Jersey Zoo (now Durrell Wildlife Park) on the Channel Island of Jersey in 1958, but is perhaps best remembered for writing a number of books based on his life as an animal collector and enthusiast. He was the youngest brother of novelist Lawrence Durrell

Books

Durrell's books, both fiction and non-fiction, have a wry, loose style that poked fun at himself as well as those around him. Perhaps his best-known work is *My Family and Other Animals* (1956), which tells of his idyllic, if oddball, childhood on Corfu. Later made into a TV series, it is delightfully deprecating about the whole family, especially elder brother Lawrence, who became a famous novelist. Despite Durrell's jokes at the expense of "brother Larry", the two were close friends all their lives.

Gerald Durrell always insisted that he wrote for royalties to help the cause of environmental stewardship, not out of an inherent love for writing.

Gerald Durrell describes himself as a writer in comparison to his brother Lawrence: The subtle difference between us is that he loves writing and I don't. To me it's simply a way to make money which enables me to do my animal work, nothing more. **(Wikipedia)**

Learning to care for the world, means taking an interest,
And to do that you need to
Develop your mind.

Diagnosis - Hyper- democracacy + Assimilation education system = narcissistic personality disorder, ("God Complex"); mass psychogenic illness

Prognosis - There is no cure. This illness reaches the deepest, most primitive, regions of the brain; the "Life Instinct", "Death Instinct"; which explains why people have no insight into their behaviour; the act of introspection would lead to the mind's infrastructure collapsing. Peole can never reveal what they are; what they do, because that would threaten their existence; others would know of their violent, aggressive nature, and then be aware that they are a threat to their survival; consequently, everybody hides their nature,
In 40 years Britain will be a garden fit for Jews to live in.
The two ingredients that form the disease must first be eradicated.
Only Jews will reside in Bitian 40 years from now; everybody else will be dead.

Without a "Moral Guide" man throughout history has displayed a tendency to give into his most basic instincts. It is quite extraordinary that in the world today those who are enormously rich steadfastly remain on the lowest level of Abraham Maslow's "Hierarchy of Needs".

Inertia is man's finest companion; for most, only once given a swift kick in the ass, will they shift their weight even an inch. The "assimilation education system", does not expect pupils to get "passing grades" to graduate from one grade level to another. Universities insist courses be paid for, and "students" be present for a percentage of classes, in order to acquire a piece of paper that entitles them to obtain enormous salaries for doing a horrendously meaningless task.

Men and women abuse themselves by wasting their God given abilities, and, consequently, destroy others at the same time. They make of themselves sadomasochistic monsters. Man becomes a masochist due to not being able to manage the liberty of individuality. He feels overwhelmed, alone, and afraid, and, thus, he decides consciously, or unconsciously, to get rid of the burden of freedom by submitting to a person, or power, he believes is overwhelmingly strong, (the "primary power" is substituted for a "secondary power"); tis is why "cults" exist, and fascist, and totalitarian, dictatorships, and why people submit to the dictates of a "system", regardless of the immortality involved.

People persistently endeavour to waste resources, in the environment, and also themselves; consequently, in order to live the opulent lifestyles they do, they, quite literally, devour others, by deriving a sector of their own population the basic resources needed to survive.

God Complex

Description

A person with a god complex may refuse to admit the possibility of their error or failure, even in the face of irrefutable evidence, intractable problems or difficult or impossible tasks. A person with a god complex is also highly dogmatic in their views, meaning the person speaks of their personal opinions as though they are unquestionably correct.

Someone with a god complex may exhibit no regard for the conventions and demands of society, and may request special consideration or privileges.

God complex is not a clinical term or diagnosable disorder and does not appear in the *Diagnostic and Statistical Manual of Mental Disorders (DSM)*.

The first person to use the term god-complex was Ernest Jones (1913-51). His description, at least in the contents page of *Essays in Applied Psycho-Analysis*, describes the god complex as belief that one is a god.

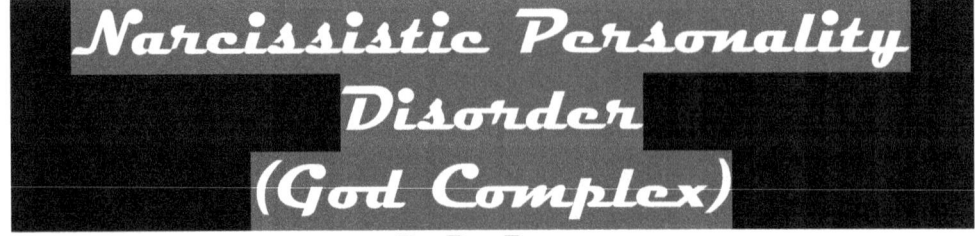

Narcissistic Personality Disorder (God Complex)

By Dr. Bayer

Disorder (NPD) is characterized by the Diagnostic and Statistical Manual of Mental Disorders as "a pervasive pattern of grandiosity, need for admiration, and a lack of empathy."

Think of this combination of traits: "a pervasive pattern of grandiosity, need for admiration, and a lack of empathy."

In my practice, I personally witness various levels of narcissistic behaviour, where the adulation is generated internally, resulting in an obsessive preoccupation with issues of status, power, and potency. As the individual excels, the narcissism is accentuated, sometimes driven to epic proportions—hence the omnipotence of the "God Complex."

I had one particularly troubled patient, who earned an astounding $300 million annually leading a major Wall Street hedge fund. He was 43 years old, hailed from an old line Massachusetts Mayflower family and grew up on an elaborate estate in Connecticut, while his family also had a large Carnegie Hill co-op in Manhattan.

His astounding success on Wall Street had enabled him to withdraw even further from reality, to the point where he was practically incapable of normal human discourse.

My patient was ensconced in a nine-floor brownstone on the Upper East Side of Manhattan and obsessed with home interior design and multi-thousand dollar doorknobs.

He retained a co-ed naked Brazilian housecleaning service, where he was entitled to sexual relations with the cleaning staff (male and female) at his whim. The sex of the partner was irrelevant, as long as he had a convenient conduit for orgasm, which he used to fend off debilitating depression. Such lack of empathy and dehumanization are also manifestations of narcissism.

Though this behaviour—and his irreconcilable lack of empathy—are certainly repulsive, they pale in comparison to his frank disdain and contempt for all other humans who were not wealthy. He truly considered them "peasants and pond-scum," which were among his favourite characterizations of all non-peers.

He was not able to embrace the *Gyroscope* methodology, nor was he in touch with enough of his inner pain. He is exceptionally well defended, and convinced of his perceptions. I was frankly confused with regard to why he consulted me. He terminated treatment willingly, and entered a sexual addiction rehab in Mississippi as per my strong suggestion. **(Wikipedia)**

Father
Dear Father

Time churns; sometimes it swallows you whole;
Without, it seems, offering one the knowledge of where to go.

Could it be, the wise man asks,
That the answer lies at the top of a grassy knoll?

Men, who are wise, tread in the footsteps of others when it is necessary;
But they also have the power to veer away from the trodden path
Whenever they choose.
It is during those times that he truly experiences what it means to be a man.

Any man that can do this
Knows this act is due to a power instilled in his heart,
Created by the "man" that allowed that heart to beat.
I could say this understanding is neat,
But so many do the same,

And this is what causes great men to experience pain.
Is that a shame?

How can we experience joy without experiencing things we despise?
Disdain, cause us grief, and has us wandering in depths
Where only darkness can lurk?

Long suffering is what eventually enlightens. Is this true?
We'll all find out eventually I'm sure!

Oh look yonder! The knoll spoken of before is in sight.
Now we can see a chance to make a change,
That could really bring a worthy gain.

Let's travel there!
Then maybe from the perch that lies on top
We can view the past, where we have been,
Then, I'm sure; we'll be in awe of all we have seen.

What delights the eye fills the soul,
Lifts us up to heights no man thought he could possibly go.

Does man evolve?
Or does he become increasing more aware that as he continues to change,
He comes closer to realizing we are all the same?

May we be still.
When life is over, we will find a stall to rest,
And recover from the wearying journey we call life.
The greatest reward is to know our work is done.
Oh, this is what it feels like to be home!

Let no one deter you from following your will.
You have the power within you to be still.
When that time arrives you know you have not a care in the world.
Everything has been looked after; not by you, but another.

Your purpose has been served; your role you've played to the max.
Sit back, relax! Now you can enjoy the Show.
You've earned the price of the ticket.
Not with gold; its worth is nothing but a grand illusion!

Knowledge brings happiness.
One day I'm sure all of us will realize this is all that matters.
Time, after all, is nothing but immaterial.
Always has been, always will be,
Forever more.
Ah-men. Hallelujah!

The message is in the script.
The script is the message.

John Edward Douglas is a former special agent and unit chief with the U.S. Federal Bureau of Investigation (FBI), one of the first criminal profilers, and criminal psychology author.

Profiling

Douglas examined crime scenes and created profiles of the perpetrators, describing their habits and attempting to predict their next moves. In cases where his work helped to capture the criminals, he built strategies for interrogating and prosecuting them as well. At the time of criminal profiling's conception, Douglas claimed to have been doubted and criticized by his own colleagues until both police and the FBI realized that he had developed an extremely useful tool for the capture of criminals.

Since his retirement from the FBI in 1995, Douglas has gained international fame as the author of a series of books detailing his life tracking serial killers, and has appeared numerous times on television. Douglas has also written textbooks for criminal profiling classes. He is the author, along with Mark Olshaker, of several books. His books are considered to be some of the most insightful works written on the minds, motives, and operation of serial killers, and the methods and lives of those who track them.

The End Is Near,
Have No Fear

Suffering, definitely leads to a heightened awareness.
Gathered are the mightiest hands that direct this course;
Unknown to practically all, is this tremendous force.

It corrects wrongs, it justifies righteousness.
For the good of all it ensures paths are inevitably followed.
Elaborate, complex, intricate, but adherence is done with care;
To the point of being fastidious.

Why should all this be brought about?
The shared opinion of those who are scholarly,
Is that the destiny was decided at the beginning.

Only a few among many would greet this place.
God's goal, probably, was merely to view Himself;
His strength is indomitable.
Those who have persevered, suffered almost intolerable cruelty,
Are the ones deserving of greeting Him; He is the greatest force.

People believe He has disguised Himself.
   Truth be told, He is everywhere, He is within each of us.
   How silly it is to frown, after all, He is here;
   Do look around.

The Rats (1974) is a horror novel written by British writer James Herbert. This was Herbert's first novel and included graphic depictions of death and mutilation. A film adaptation was made in 1982, called *Deadly Eyes*. A 1985 adventure game for the Commodore 64 and ZX Spectrum based on the book was published by Hodder & Stoughton Ltd and produced by Five Ways Software Ltd.

The Rats was followed by two sequels, *Lair* and *Domain*.

Herbert became inspired to write *The Rats* in early 1972, whilst watching Tod Browning's *Dracula*; specifically, after seeing the scene where Renfield describes a nightmare he had involving hordes of rats. Linking the film to childhood memories he had of rats in the London suburbs, Herbert stated in later interviews that he wrote the book primarily as a pastime; "It seemed like a good idea at the time, I was as naive as that." The manuscript was typed by Herbert's wife Eileen, who sent it off after nine months to nine different publishers.

The novel opens introducing the reader to an alcoholic vagrant, resting in an abandoned and forgotten lock-keeper's house by a canal. As he is ruminating over the injustices inflicted upon him in his life, he is suddenly set upon by a pack of dog-sized rats and is devoured alive.

Harris, a young, east London art teacher notices that one of his students has a bloodied bandage around his hand. When he enquires as to what caused the damage, the student answers that he was attacked by a rat. Meanwhile, a baby girl and her dog are killed by the giant rats, now aided by packs of smaller black rats. The girl's mother rescues her daughter's mutilated body, but not before sustaining bites as well. Harris takes the student to the hospital and sees the grieving mother with her dead child. According to the doctor, the number of seemingly unprovoked rat attacks has strangely increased.

The next rat attack occurs at the remains of a bombsite, where a group of squabbling vagrants are slaughtered. Harris is visited at work by the Minister of Health, Mr. Foskins, who reveals that the bitten student, and all the other surviving victims of rat attacks, died of a mysterious disease 24 hours after being bitten. Foskins asks Harris to keep the existence of the disease a secret and lead an exterminator named Ferris to the area where the student had been bitten. Accompanied by Ferris, Harris goes to the canal described by the student and sights a group of giant rats. Harris attempts to contact the police while Ferris follows the rats who then attack and kill him.

The rat attacks become increasingly more daring, as more and more public places are attacked. A tube station is assaulted, leaving few survivors. Next, Harris' own school is attacked, resulting in the death of the headmaster. With the existence of the rats' disease now becoming public knowledge, a meeting is held, in which a young researcher by the name of Stephen Howard comes up with the idea of using a virus to infect the rats. The virus is injected into several puppies which are left in areas of the attacks. This results in the deaths of thousands of rats, which crawl to the surface to die. A few weeks later however, the rats adapt to the virus, yet at the same time, losing the toxicity of their bites. The rats brutally attack a cinema and overrun the London Zoo. Based on the fact that rats communicate with each other using ultrasound, a plan is formulated to use ultrasonic machines to lure the rats into poison gas chambers.

Foskins is dismissed as Health Minister and reveals to Harris that he's been investigating possible clues as to the rats origins and comes to the conclusion that they were smuggled from the tropics by a Zoologist living near a canal. Pursuing the disgraced health minister past waves of entranced rats; Harris finds the abandoned house and enters it. He goes into the cellar and finds Foskins' corpse being devoured by rats of unusually great size. He kills them after a bloody battle and discovers the rats' leader hidden in the shadows; a white, hairless and obese rat with two heads. Harris kills the creature and leaves.

The epilogue indicates that one rat survived the purge by being trapped in the basement of a grocery shop. There, it gives birth to a new litter, including a new white rat. **(Wikipedia)**

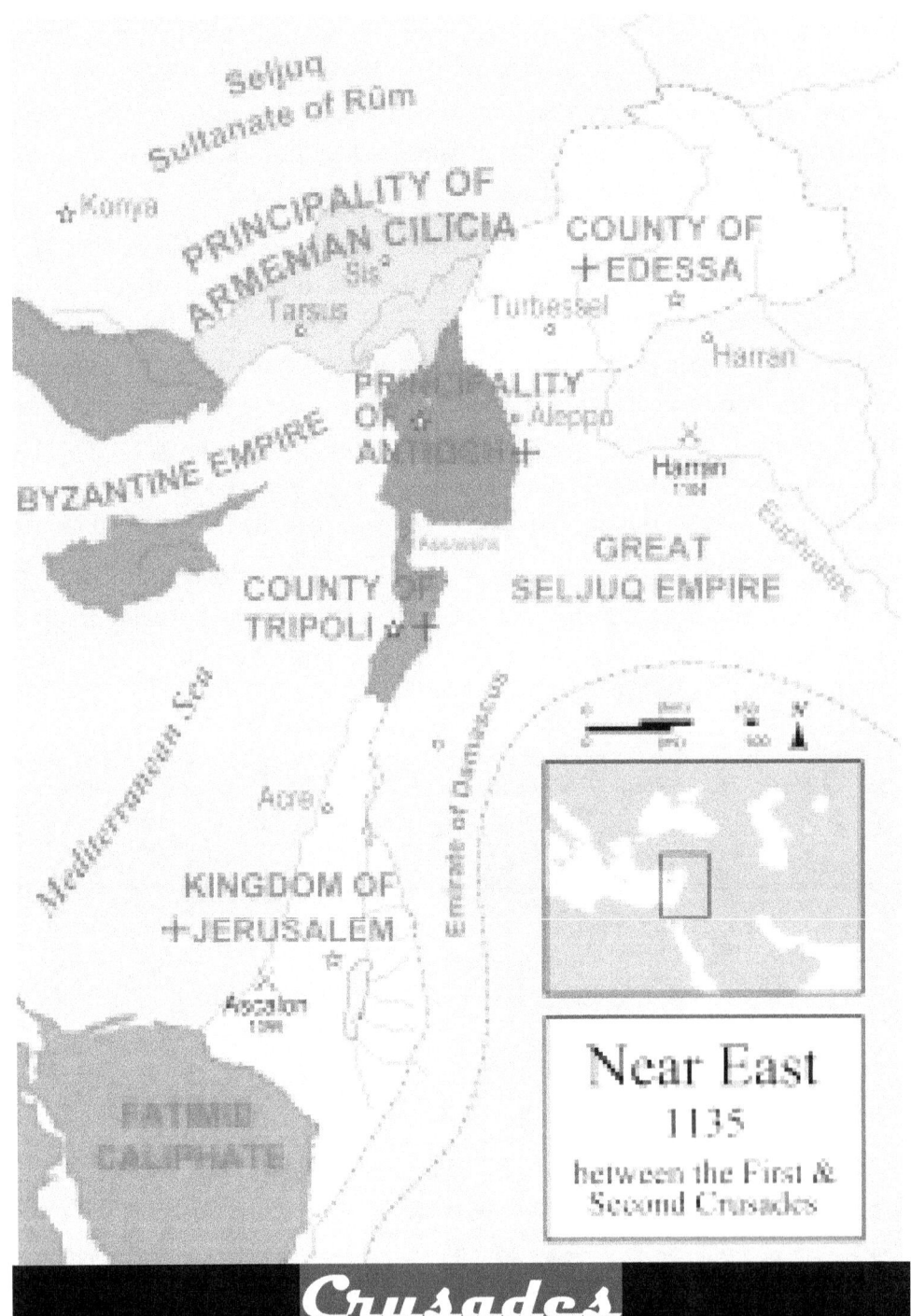

Near East
1135
between the First &
Second Crusades

Crusades

The Crusades were a series of intermittent military campaigns sanctioned by various Popes in the Middle Ages, from 1096 to 1487. In 1095, Byzantine Emperor Alexios I sent an ambassador from Constantinople to Pope Urban II in Italy requesting military support in the conflict with the eastward invading Turks.

The Pope responded promptly by calling Catholic soldiers to join the First Crusade. The immediate goal was to guarantee pilgrims access to the holy sites in the Land that were under Muslim control. His long-range goal was to reunite the Eastern and Western branches of

Christendom after their split in 1054, with the pope as head of the united Church. A complex 200-year struggle ensued.

Hundreds of thousands of people from many different classes and nations of Western Europe became crusaders by taking a public vow and receiving plenary indulgences from the church. Some crusaders were peasants hoping for Apotheosis at Jerusalem. Pope Urban II claimed that anyone who participated was forgiven of their sins. In addition to demonstrating devotion to God, as stated by him, participation satisfied feudal obligations and provided opportunities for economic and political gain. Crusaders often pillaged the countries through which they traveled, and contrary to their promises the leaders retained much of this territory rather than returning it to the Byzantines.

The People's Crusade prompted the murder of thousands of Jews, known as the Rhineland massacres. Constantinople was sacked during the Fourth Crusade, rendering the reunification of Christendom impossible for that time. Due to the weakening that resulted from the siege, the remnants of the Byzantine Empire finally fell to the Ottomans in 1453. Western European potentates mounted no coherent response when the last Catholic stronghold in the region, Acre, fell in 1291.

Opinions concerning the conduct of crusaders have varied from laudatory to highly critical. The impact of the crusades was profound; they reopened the Mediterranean to commerce and travel, enabling Genoa and Venice to flourish. Crusader armies would trade with the local populations while travelling, and Orthodox Byzantine emperors often organized markets for crusaders moving through their territory. The Crusades consolidated the collective identity of the Latin Church under papal leadership, and were a source of heroism, chivalry, and piety. This consequently spawned medieval romance, philosophy, and literature. However, the crusades reinforced the connection between Western Catholicism, feudalism, and militarism, which was counter to the Peace and Truce of God that Urban had promoted.

Terminology

Madrid Skylitzes illuminated manuscript depicting Byzantine Greeks punishing ninth-century Cretan Saracens

Crusade is a modern term derived from the French *croisade* and Spanish *cruzada*; by 1750, forms of the word "crusade" had established themselves in English, French, and German. The *Oxford English Dictionary* records its first use in English in 1757 by William

Shenstone. When a crusader swore a vow (*votus*) to reach Jerusalem, they received a cloth cross (*crux*) to be sewn on their clothing. This "taking of the cross" became associated with the entire journey, and crusaders saw themselves as undertaking an *iter* (journey) or *peregrinatio* (armed pilgrimage). The inspiration for this "messianism of the poor" was an expected mass apotheosis at Jerusalem.

The *numbering of the Crusades* is debated, with some historians counting seven major Crusades and a number of minor ones from 1096 to 1291. Others consider the Fifth Crusade of Frederick II as two crusades, making the crusade launched by Louis IX in 1270 the Eighth. Sometimes the Eighth Crusade is considered two, the second of which is the Ninth Crusade.

In the pluralistic view of the Crusades developed during the 20th century, "Crusade" encompasses all papal-sanctioned military campaigns in Southwestern Asia or in Europe. A key distinction between the Crusades and other holy wars was that the authorization for the Crusades came directly from the pope, who claimed to be working on behalf of Christ. This takes into account the view of the Roman Catholic Church and medieval contemporaries, such as Saint Bernard of Clairvaux, which gives equal precedence to military campaigns undertaken for political reasons and to combat paganism and heresy. This broad definition includes the persecution of heretics in Southern France, the political conflict between Christians in Sicily, the Christian re-conquest of Iberia, Hussite Wars, and the conquest of pagans in the Baltic. A narrower view is that the Crusades were a defensive war in the Levant against Muslims to free the Holy Land from Muslim rule.

Popes periodically declared political crusades as a means of conflict resolution amongst Roman Catholics; the first of these was declared by Pope Innocent III against Markward of Anweiler in 1202. Others include a crusade against the Stedingers, several (declared by a number of popes) against Emperor Frederick II and his sons, and two crusades against opponents of King Henry III of England who received the same privileges as participants in the Fifth Crusade.

A common term for Muslim was *Saracen*; before the 16th century, the words "Muslim" and "Islam" were rarely used by Europeans. In Greek and Latin, "Saracen" originated in the early first millennium to refer to non-Arab peoples inhabiting the desert areas around the Roman province of Arabia. The term evolved to include Arab tribes, and by the 12th century it was an ethnic and religious marker synonymous with "Muslim" in Medieval Latin literature. *Frank* and *Latin* were used during the Crusades for Western Europeans, distinguishing them from *Greeks*.

During the 16th-century Reformation and Counter-Reformation, historians saw the Crusades through the lens of their own religious beliefs. Protestants saw them as a manifestation of the evils of the papacy, and Catholics viewed them as forces for good. Enlightenment historians tended to view the Middle Ages in general, and the Crusades in particular, as the efforts of barbarian cultures driven by fanaticism. By the early Romantic period in the 19th century, that harsh view of the Crusades and their era had softened; scholarship later in the century emphasized specialization and detail.

Eighteenth-century Enlightenment scholars and modern Western historians have expressed moral outrage at the conduct of the crusaders. Steven Runciman wrote during the 1950s, "High ideals were besmirched by cruelty and greed ... the Holy War was nothing more than a long act of intolerance in the name of God". The 20th century produced three important histories of the Crusades: by Runciman, Rene Grousset and a multi-author work edited by K. M. Stetton. During that century, two definitions of the Crusades developed; one includes all papal-led efforts in Western Asia and Europe, but historian Thomas Madden wrote: "The crusade, first and foremost, was a war against Muslims for the defense of the Christian faith .... They began as a result of a Muslim conquest of Christian territories." Madden wrote that the goal of Pope Urban was that "[t]he Christians of the East must be free from the brutal and humiliating conditions of Muslim rule."

After the 1291 fall of Acre, European support for the Crusades continued despite criticism by contemporaries (such as Roger Bacon, who believed them ineffective: "Those who survive, together with their children, are more and more embittered against the Christian faith").According to historian Norman Davies, the Crusades contradicted the Peace and Truce of God supported by Urban and reinforced the connection between Western Christendom, feudalism, and militarism. The formation of military religious orders scandalized the Orthodox Byzantines, and crusaders pillaged countries they crossed on their journey east. Violating their oath to restore land to the Byzantines, they often kept the land for themselves. The early People's Crusade instigated a pogrom in the Rhineland and the massacre of thousands of Jews in Central Europe; during the late 19th century, this crusade was used by Jewish historians to support Zionism. The Fourth Crusade resulted in the sacking of Constantinople, effectively ending any chance of reconciling the East–West Schism and leading to the fall of the Byzantine Empire to the Ottomans. Enlightenment historians criticized the Crusades' misdirection—that of the Fourth in particular, which attacked a Christian power (the Byzantine Empire) instead of Islam. David Nicolle called the Fourth Crusade controversial in its "betrayal" of

Byzantium, and in *The History of the Decline and Fall of the Roman Empire* Edward Gibbon wrote that the crusaders' efforts would have been more effective improving their own countries.

The Great Seljuk Empire at its greatest extent (1092)

After Muslim forces defeated the Byzantines at the Battle of Yarmouk in 636, control of Palestine passed through the Umayyad and Abbasid Dynasties and the Fatimids. Tolerance, trade, and political relationships between the Arabs and the Christian states of Europe ebbed and flowed until 1072, when the Fatimids lost control of Palestine to the rapidly-expanding Great Seljuk Empire. Although the Fatimid caliph al-Hakim bi-Amr Allah ordered the destruction of the Church of the Holy Sepulchre, his successor allowed the Byzantine Empire to rebuild it. The Muslim rulers allowed pilgrimages by Catholics to sacred sites. Resident Christians were considered Dhimmi and intermarriage was not uncommon. Cultures and creeds coexisted and competed, but the frontier conditions were inhospitable to Catholic pilgrims and merchants. The disruption of pilgrimages by the conquering Seljuk Turks prompted support for the Crusades in Western Europe.

Fifteenth-century illustrated French translation of Boccaccio's *De Casibus Virorum Illustrium*, showing Seljuk emperor Alp Arslan ritually humiliating Romanos IV in 1071 after Manzikert; Alp Arslan allowed Romanos to return to Constantinople, where he was killed by the Byzantines.

The Byzantine Empire expanded at the end of the 10th century, with Basil II spending most of his half-century reign in conquest. Although he left a growing treasury, he neglected domestic affairs and ignored the cost of incorporating his conquests into the Byzantine *ecumene*. None of Basil's successors were militarily or politically talented, and the task of governing the Empire increasingly devolved to the civil service. Their efforts to spend the Byzantine economy back into prosperity triggered inflation. To balance an increasingly unstable budget, Basil's standing army was dismissed and his *thematic* troops replaced by *tagmata*. After the 1071 defeat of the Byzantine army at the Battle of Manzikert, the Seljuk Turks held nearly all of Anatolia and the empire descended into frequent civil wars.

The reconquest of the Iberian Peninsula from the Muslims began during the 8th century, reaching its turning point with the 1085 recapture of Toledo. Although at the 1095 Council of Clermont Urban II compared the Iberian wars to his First Crusade, it was not until Pope Callixtus II's 1123 encyclical that they attained crusade status. After the encyclical, the papacy declared Iberian crusades in 1147, 1193, 1197, 1210, 1212, 1221, and 1229. Crusader privileges were also given to those aiding the major military orders (the Templars and Hospitallers) and the Iberian orders which eventually merged with the two main orders: the Order of Calatrava and the Order of Santiago. From 1212 to 1265, the Iberian Christian kingdoms drove the Muslims to the Emirate of Granada in the far south of the peninsula. In 1492 the emirate was conquered, and Muslims and Jews were expelled from the peninsula.

An aggressive, reformist papacy clashed with the Eastern Empire and Western secular monarchs, leading to the 1054 East–West Schism and the Investiture Controversy (which began around 1075 and continued during the First Crusade). The papacy began to assert its independence from secular rulers, marshaling arguments for the proper use of armed force by Catholics. The result was intense piety, an interest in religious affairs, and religious propaganda advocating a just war to reclaim Palestine from the Muslims. The majority view was that non-Christians could not be forced to accept Christian baptism or be physically assaulted for having a different faith, although a minority believed that vengeance and forcible conversion were justified for the denial of Christian faith and government. Participation in such a war was seen as a form of penance which could counterbalance sin. In Europe, the Germans were expanding at the expense of the Slavs and Sicily was conquered by Norman adventurer Robert Guiscard in 1072.

Illumination from the *Livre des Passages d'Outre-mer* (c. 1490) of Urban II at the Council of Clermont (from the Bibliothèque Nationale)

Emperor Alexios I Komnenos requested military aid (probably mercenaries to reinforce his *tagmata*) from Pope Urban II at the 1095 Council of Piacenza to fight the Seljuks, exaggerating the danger facing the Eastern Empire to secure his required troops. On 27 November 1095 at the Council of Clermont, attended by nearly 300 French clerics, Urban raised the issues of the problems in the East and the struggle of the Eastern Roman Empire against the Muslims. Five major sources of information exist on the council: the anonymous *Gesta Francorum* (*The Deeds of the Franks*, dated about 1100–1101; Fulcher of Chartres, who attended the council; Robert, who may have been present, and Baldric, archbishop of Dol and Guibert de Nogent (who were not). The accounts, written retrospectively, differ greatly. In his 1106–7 *Historia Iherosolimitana*, Robert the Monk wrote that Urban asked western Christians to aid the Byzantine Empire because "*Deus vult*" ("God wills it") and promised absolution to participants; according to other sources, the pope promised an indulgence. In the accounts, Urban emphasizes reconquering the Holy Land more than aiding the emperor and lists gruesome offences allegedly committed by Muslims. The crusade was preached across France; Urban wrote to those "waiting in Flanders" that the Turks, in addition to ravaging the "churches of God in the eastern regions", seized "the Holy City of Christ, embellished by his passion and resurrection—and blasphemy to say it—have sold her and her churches into abominable slavery". Although the pope did not explicitly call for the reconquest of Jerusalem, he called for military "liberation" of the Eastern Churches and appointed Adhemar of Le Puy to lead the crusade (which began on 15 August, commemorating the Assumption of Mary).

History

First Crusade (1096–1099) and immediate aftermath

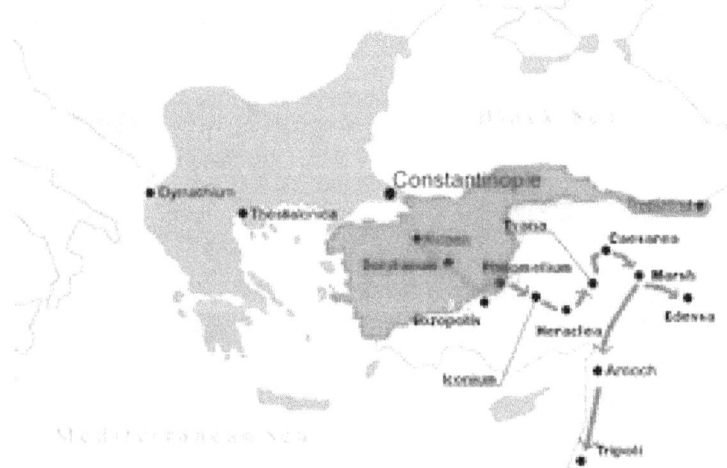

Route of the First Crusade through Asia

Pope Urban II in Rome in 1095 received an ambassador from Byzantine Emperor Alexius I in Constantinople seeking urgent help against the Turkish threat. The pope acted promptly and called a crusade with the goals of securing access to the Holy Sites. Historian Paul Everett Pierson says he also "hoped that if the crusaders aided the Eastern Church by defeating the Turks, the Church would be reunited under his leadership. Inspired by Pope Urban II's preaching, Peter the Hermit led as many as 20,000 people, mostly peasants, to the Holy Land shortly after Easter 1096. When they arrived in Germany in spring 1096, units of crusaders commenced the Rhineland massacres in the cities of Speyer, Worms, Mainz and Cologne, despite the efforts by Catholic bishops to protect the Jews. Major leaders included Emichoand Peter the Hermit. The range of anti-Jewish activity was broad, extending from limited, spontaneous violence to full-scale military attacks on the Jewish communities of Mainz and Cologne. This was the first major outbreak of anti-Jewish violence in Europe and was cited by 19th-century Zionists as showing the need for a Jewish state. When the group finally reached the Byzantine Empire, Emperor Alexios urged them to wait for the western nobles, but they insisted upon proceeding and fell to a Turkish ambush outside Nicaea, from which only about 3,000 people escaped.

The official crusader armies departed from France and Italy in August and September 1096. The bulk of the army divided into four parts, which travelled separately to Constantinople. With non-combatants included, the western forces may have contained as many as 100,000 people. The armies journeyed eastward by land toward Constantinople, where they received a wary welcome from the Byzantine Emperor. The main army, mostly comprising French and Norman knights under baronial leadership, pledged to restore lost territories to the empire and marched south through Anatolia. The leaders of the First Crusade included Godfrey of Bouillon, Robert Curthose, Hugh of Vermandois, Baldwin of Bouillon,Tancred de Hauteville, Raymond of Toulouse, Bohemond of Taranto, Robert II, Count of Flanders, and Stephen, Count of Blois. The king of France and Henry IV, Holy Roman Emperor, were in conflict with the Pope and did not participate.

The crusader armies initially fought the Turks at the lengthy Siege of Antioch, which began in October 1097 and lasted until June 1098. When they entered Antioch, the crusaders massacred the Muslim inhabitants and pillaged the city. However, a large Muslim army led by Kerbogha immediately besieged the victorious crusaders, who were now inside Antioch. Bohemond of Taranto successfully rallied the crusader army and defeated Kerbogha on 28 June. Bohemond and his men retained control of the city, despite his pledge to Alexios. Most of the remaining crusader army marched south, moving from town to town along the coast, finally reaching Jerusalem on 7 June 1099 with only a fraction of their original forces.

Jews and Muslims fought together to defend Jerusalem against the invading Franks, but the crusaders entered the city on 15 July 1099. They proceeded to massacre the remaining Jewish and Muslim civilians and also pillaged or destroyed mosques or the city itself. In his *Historia Francorum qui ceperunt Iherusalem*, Raymond D'Aguilers exalted actions which would be considered atrocities from a modern viewpoint. As a result of the First Crusade, four primary crusader states were created: Edessa, Antioch, Tripoli, and Jerusalem. On a popular level, the First Crusade unleashed a wave of impassioned, pious Catholic fury which was expressed in the massacres of Jews that accompanied the crusades and the violent treatment of the "schismatic" Orthodox Christians of the east.

Following the First Crusade was a second, less successful crusade known as the Crusade of 1101, in which Turks led by Kilij Arslan defeated the crusaders in three separate battles.

12th Century

In the early 12th-century, smaller scale crusading continued. Pope Calixtus II promoted the Venetian Crusade of 1122–1124; Count Fulk V of Anjou visited in 1120 and 1129 and Conrad III of Germany in 1124, leading to recognition of the Knights by Pope Honorius II. In 1135 Pope Innocent II's grant of crusading indulgences to those who opposed papal enemies is seen by some historians as the beginning of politically motivated crusades. The crusader states were initially secure, but Imad ad-Din Zengi, who was appointed governor of Mosul in 1127, captured Aleppo in 1128 and Edessa (Urfa) in 1144. These defeats led Pope Eugenius III to call for another crusade on 1 March 1145. The new crusade was supported by various preachers, most notably by Bernard of Clairvaux. Armies from France and Germany, under King and Conrad III, respectively, marched to Jerusalem in 1147 and also besieged Damascus, but failed to win any major victories. Meanwhile, a group of crusaders from northern Europe stopped in Portugal and allied with the king of Portugal, Afonso I, retaking Lisbon from the Muslims in 1147. A detachment from this group of crusaders helped Count Raymond Berenguer IV of Barcelona conquer the city of Tortosa the following year.

In the Holy Land, both the kings of France and Germany had returned to their countries by 1150 without any changes. Bernard of Clairvaux, who had encouraged the Second Crusade in his preachings, was upset with the violence and slaughter directed toward the Jewish population of the Rhineland. In 1172, Henry the Lion, Duke of Saxony, made a pilgrimage that is sometimes considered a crusade. At the same time, Saxons and Danes fought against Wends in the Wendish Crusade.

The Wends defeated the Danes; the Saxons did not make any considerable contributions to the crusade. The crusades continued, although no official papal bulls were issued authorizing new crusades. Henry restarted efforts to conquer the Wends in 1160, and they were defeated in 1162.

Detail of a miniature of King Philip II of France arriving in the Holy Land

Saladin created a united opposition force and presented a new threat to the Latin states. Following his victory at the Battle of Hattin, he easily overwhelmed the disunited crusaders in 1187 and retook Jerusalem on 29 September of that year. Terms were arranged and the city surrendered; Saladin entered the city on 2 October. According to Benedict of Peterborough, Pope Urban III died of deep sadness on 19 October 1187 upon hearing news of the defeat. On 29 October Pope Gregory VIII issued a papal bull, *Audita tremendi*, proposing the Third Crusade. Planning to recapture Jerusalem, Frederick I, Holy Roman Emperor, Philip II of France, and Richard I of England organized their forces. Frederick died en route to Jerusalem; few of his men reached the Holy Land. The other two armies arrived successfully but were beset by political quarrels. Philip returned to France, leaving most of his forces behind. Richard conquered the island of Cyprus from the Byzantines in 1191 because the shipwreck survivors including his sister were taken prisoner by the island's ruler, Isaac Komnenos. He then recaptured the city of Acre after a long siege. The crusader army travelled south along the Mediterranean coast, defeated the Muslims near Arsuf, and recaptured the port city of Jaffa. They were near Jerusalem, but supply shortages forced them to end the crusade without taking Jerusalem. Richard left the following year after negotiating a treaty with Saladin. The terms allowed unarmed Catholics to make pilgrimages to Jerusalem and permitted merchants to trade. Henry VI, Holy Roman Emperor, initiated the German Crusade in 1197 to fulfill the promises made by his father, Frederick. Led by Conrad of Wittelsbach, Archbishop of

Mainz, the army landed at Acre and captured the cities of Sidon and Beirut. However, most of the crusaders returned to Germany after Henry died.

Nineteenth-century depiction of two Livonian Knights

When Pope Celestine III called for a crusade against Northern European pagans in 1193, Bishop Berthold of Hanover led a large army to defeat and his death in 1198. In response to the defeat, Pope Innocent III issued a papal bull declaring a crusade against the mostly-pagan Livonians. Albrecht von Buxthoeven, who was consecrated as bishop in 1199, arrived the following year with a large force and established Riga as the seat of his bishopric in 1201. In 1202 he formed the Livonian Knights to help convert the pagans to Catholicism and, more importantly, to protect German commerce. The Livonians were conquered and converted between 1202 and 1209. In 1217 Pope Honorius III declared a crusade against the Prussians, and Konrad of Masovia gave Chelmno to the Teutonic Knights in 1226 as a base for the crusade. In 1236 the Livonian Knights were defeated by the Lithuanians at Saule, and in 1237 Pope Gregory IX merged the remainder of the military order into the Teutonic Knights as the Livonian Order.

By 1249 the Teutonic Knights completed their conquest of the Old Prussians, whom they ruled as lords of the German emperor. They then conquered and converted the Lithuanians, a process which lasted into the 1380s. The order tried unsuccessfully to conquer Orthodox Russia, particularly the Republics of Pskov and Novgorod (with the support of Pope Gregory IX), as part of the Northern Crusades. In 1240 the Novgorod army defeated the Swedes in the Battle of the Neva, and two years later they defeated the Livonian Order in the Battle on the Ice.

Innocent III began preaching what became the Fourth Crusade in 1200 in France, England, and Germany, primarily in France. It was a vehicle for the political ambitions of Doge Enrico Dandolo of Venice (a vassal state of Byzantium at the time) and German King Philip of Swabia, who was married to Irene of Byzantium. Dandolo saw an opportunity to expand Venice's possessions in the Near East and break loose from Byzantine

Pope Innocent III excommunicating the Albigensians (left), and an Albigensian massacre by crusaders

vassalage; Philip saw the crusade as a chance to restore his exiled nephew, Alexios IV Angelos, to the throne of Byzantium. Although the crusaders contracted with the Venetians for a fleet and provisions to transport them to the Holy Land, they were unable to pay when too few knights arrived in Venice. They agreed, therefore, to divert the crusade to Constantinople and share what could be looted as payment. As collateral, the crusaders seized the Christian city of Zara on 24 November 1202 and were excommunicated by the appalled Innocent. They met limited resistance in their initial siege of Constantinople, sailing down the Dardanelles and breaching the sea walls. Alexios IV Angelos was strangled after a palace coup, robbing them of success, and they repeated the siege in April 1204. This time the city was sacked, churches pillaged, and many citizens killed; the crusaders divided the empire into Latin fiefs and Venetian colonies. In the latter, the defence of La Cava and Nicosia was emphasized. In April 1205 the crusaders were defeated by the Bulgars and remaining Greeks at Adrianople, where Kaloyan of Bulgaria captured and imprisoned new Latin emperor Baldwin of Flanders. While deploring its methods, the papacy initially supported the apparent forced reunion of the Eastern and Western churches. The Fourth Crusade effectively left two Roman Empires in the East: a Latin "Empire of the Straits" which existed until 1261 and a Byzantine enclave ruled from Nicaea, which regained control in the absence of the Venetian fleet. Venice was the sole beneficiary in the long run.

Although the Albigensian Crusade was launched in 1208 to eliminate the Cathars of Occitania (present-day southern France), the decades-long struggle had as much to do with the desire of northern France to extend its control southwards as it did with battling heresy. The Cathars were ultimately driven underground, and southern France lost its independence. In 1221 Pope Honorius III called on King Andrew II to subjugate the heretics in Bosnia, and Hungarian forces responded to additional papal calls in 1234 and 1241; the latter campaign ended with the Mongol invasion of Hungary in 1241. The Bosnian church was theologically Catholic, but its schism with the Roman Catholic Church extended well past the end of the middle Ages. Innocent III declared that a new crusade would begin in 1217, and summoned the Fourth Council of the Lateran in 1215. Most of the crusaders came from Germany, Flanders, and Frisian, with a large army from Hungary led by Andrew II and additional forces led by Duke Leopold VI. Andrew and Leopold arrived in Acre in October 1217, but little was accomplished and Andrew returned to Hungary in January 1218. After the arrival of more crusaders, Leopold and king of Jerusalem John of Brienne laid siege to Damietta in Egypt; they captured it in November 1219. Further efforts by the papal legate, Pelagius, to move further into Egypt were fruitless. Blocked by Ayyubid Sultan Al-Kamil's forces, the crusaders were forced to surrender. Al-Kamil forced the

return of Damietta, agreed to an eight-year truce, and the crusaders left Egypt.

Frederick II (left) meets al-Kamil (right) in a manuscript illumination from Giovanni Villani's *Nuova Cronica*

After repeatedly breaking his vow to crusade, Emperor Frederick II was excommunicated. He finally sailed from Brindisi, landing at Acre in September 1228 following a stop in Cyprus. Frederick agreed to a peace treaty with Al-Kamil which allowed Latin Christians to rule most of Jerusalem and a strip of territory from Acre to Jerusalem, with the Muslims controlling their sacred areas in Jerusalem. In return Frederick pledged to protect Al-Kamil against all enemies, even if they were Christian. Following this crusade was an effort by King Theobald I of Navarre in 1239 and 1240, originally summoned in 1234 by Gregory IX to assemble in July 1239 at the end of a truce. In addition to Theobald, Peter of Dreux, Hugh, Duke of Burgundy and other French nobles participated. They arrived in Acre in September 1239; after a November defeat, Theobald arranged a treaty with the Muslims which returned territory to the crusading states but caused disaffection among the crusaders. Theobald returned to Europe in September 1240; Richard of Cornwall, younger brother of King Henry III of England, took the cross and arrived in Acre the following month. After enforcing Theobald's treaty, Richard left the Holy Land for Europe in May 1241.

During the summer of 1244 a Khwarezmian force summoned by al-Kamil's son, al-Salih Ayyub, stormed and took Jerusalem. The Franks allied with Ayyub's uncle Ismail and the emir of Homs, and their combined forces went into battle at La Forbie in Gaza. The crusader army and its allies were defeated within forty-eight hours by the Khwarezmian army. King Louis IX of France organized a crusade after taking the cross in December 1244, preaching and recruiting from 1245 to 1248. Louis' forces set sail from France in May 1249, landing in Egypt near Damietta on 5 June 1249. After the Nile floodwaters receded, the army marched into the interior in November, and by February was near Mansura. They were defeated, and Louis was captured as he retreated towards Damietta. He was ransomed for 800,000 bezants, and a ten-year truce was agreed. Louis went to Syria, remaining there until 1254 to solidify and fortify the kingdom of Jerusalem.

In 1256 the Venetians were evicted from Tyre, prompting the War of Saint Sabas over territory in Acre claimed by Genoa and Venice; although the Venetians conquered the disputed territory (destroying Saint Sabas' fortifications), they could not expel the Genoese. During a 14-month blockade, Genoa allied with Philip of Monfort, John of Arsuf, and the Knights Hospitaller and Venice was supported by the Count of Jaffa and the Knights Templar. By 1261 the Genoese were expelled but Pope Urban IV, concerned about the impact of the war on defence against the Mongols, organized a peace council. The conflict resumed in 1264 when the Genoese received aid from Michael VIII Palaiologos, Emperor of Nicaea, and Venice unsuccessfully tried to conquer Tyre. Both sides used Muslim soldiers (primarily Turcopoles) against their Christian foes, and the Genoese forged an alliance with the Egyptian sultan Baibars. The war significantly impaired the kingdom's ability to withstand external threats. Except for religious buildings, most fortified buildings in Acre were destroyed; at one point, the city looked as if it had been ravaged by a Muslim army.

According to Rothelin, continuator of William *History*, 20,000 men died in the conflict (when the crusader states were chronically short of soldiers). The war ended in 1270, and in 1288 Genoa regained its quarter in Acre.

In 1266 Louis IX' brother Charles seized Sicily, previously-controlled parts of the eastern Adriatic, Corfu, Butrinto, Avlona, and Suboto. The Treaty of Viterbo was agreed with the exiled Baldwin II of Constantinople and William II Villehardouin; the heirs of both Latin princes would marry Charles' children, and if there were no heirs Charles would receive the empire and principality. Charles turned his brother's crusade to his own advantage, persuading Louis to direct the Eighth Crusade against Charles' rebel vassals in Tunis. Louis' death, illness among the crusaders and a fleet-devastating storm forced Charles to postpone his designs on Constantinople. Michael VIII Palailogos was alarmed by Charles' planned crusade to restore the Latin Empire, which had fallen in 1261, and Charles' expansion in the Mediterranean. Michael delayed Charles by beginning negotiations with Pope Gregory X for union of the Greek and the Latin churches. At the Second Council of Lyon a union of the churches was declared, with Charles and Philip of Courtenay compelled to form a truce with Byzantium. This union would later prove unacceptable to the Greeks. Michael also provided Genoa with funds to encourage revolt in Charles' northern Italian territories. In 1268 Charles executed Conradin, great-grandson of Isabella I of Jerusalem and principal pretender to the throne of Jerusalem, when he seized Sicily from the Holy Roman Empire. Charles purchased the rights to Jerusalem from Maria of Antioch, the only surviving grandchild of Queen Isabella, creating a claim rivaling that of Hugh III of Cyprus (Isabella's great-grandson).

Charles spent his life trying to amass a Mediterranean empire, and he and Louis saw themselves as God's instruments to uphold the papacy. Ignoring his advisers, in 1270 Louis IX again attacked the Arabs in Tunis. The weather was hot, and his army was devastated by disease. Louis died, ending the last major attempt to take the Holy Land. From 1265 to 1271, mamluks led by Baibars drove the Franks to a few small coastal outposts. The future Edward I of England vowed to crusade with Louis IX, but he was delayed and did not arrive in North Africa until November 1270. After Louis' death, Edward went to Sicily and then to Acre in May 1271. His forces were small, however, and he was displeased with the truce between Baibars and King Hugh of Jerusalem. Edward learned of his father's death and his succession to the throne in December 1272, but he did not return to England until 1274 (although he accomplished little in the Holy Land). The 1281 election of a French pope, Martin IV, brought the full power of the papacy into line behind Charles. He campaigned unsuccessfully in Albania and Achaea before preparing to launch his crusade (with 400 ships, carrying 27,000 mounted knights) against Constantinople. Michael VIII Palailogos allied with Peter III of Aragon to foment an uprising, the Sicilian Vespers, during which the crusader fleet was abandoned and burnt. The Sicilians appealed to Peter, who was proclaimed king, and the Capetian House of Anjou was exiled from Sicily. Martin excommunicated Peter and called for a crusade against Aragon before Charles died in 1285, allowing Henry II of Cyprus to reclaim Jerusalem. One factor in the crusaders' decline was the disunity and conflict among Latin Christian interests in the eastern Mediterranean. Martin compromised the papacy by supporting Charles of Anjou, and botched secular "crusades" against Sicily and Aragon tarnished its spiritual lustre. The collapse of the papacy's moral authority and the rise of nationalism rang the death knell for crusading, ultimately leading to the Avignon Papacy and the Western Schism. The Crusade of Aragón was declared by Martin against Peter III in 1284 and 1285, with Peter supporting anti-Angevin forces in Sicily after the Sicilian Vespers and Martin supporting Charles of Anjou. Pope Boniface VIII proclaimed a crusade against Frederick III of Sicily (Peter's youngest son) in 1298, but was unable to prevent Frederick's coronation and recognition as king of Sicily.

The mainland Crusader states of the *outremer* were extinguished with the fall of Tripoli in 1289 and Acre in 1291. Most remaining Latin Christians left for destinations in the *Frankokratia* or were killed or enslaved. Minor crusading efforts lingered into the 14th century; Peter I of Cyprus captured and sacked Alexandria in 1365 in what became known as the Alexandrian Crusade, although his motivation was as much commercial as religious. Louis II led the 1390 Barbary Crusade against Muslim pirates in North Africa; after a ten-week siege, the crusaders signed a ten-year truce.

14th and 15th Centuries

A number of crusades were launched during the 14th and 15th centuries to counter the expansion of the Ottoman Empire; the first (in 1396) was led by Sigismund of Luxemburg, king of Hungary. Many French nobles joined Sigismund's forces, including the crusade's military leader John the Fearless (son of the Duke of Burgundy). Although Sigismund advised the crusaders to focus on defence when they reached the Danube, they besieged the city of Nicopolis. The Ottomans defeated them in the Battle of Nicopolis on 25 September, capturing 3,000 prisoners. The Hussite Crusades, also known as the Hussite Wars and the Bohemian Wars, involved military action against the followers of Jan Hus in Bohemia from 1420 to about 1431. Crusades were declared five times during that period: in 1420, 1421, 1422, 1427, and 1431. These expeditions forced the Hussite forces, who disagreed on many doctrinal points, to unite to drive out the invaders. The wars ended in 1436 with the ratification of the Compactata of Iglau by the Church.

Polish-Hungarian King Władysław Warneńczyk invaded the recently conquered Ottoman territory, reaching Belgrade in January 1444; a negotiated truce was repudiated by Sultan Murad II within days of its ratification. Further efforts by the crusaders ended in the Battle of Varna on 10 November, a decisive Ottoman victory which led to the withdrawal of the crusaders. This withdrawal, following the last Western attempt to aid the Byzantine Empire, led to the 1453 fall of Constantinople. John Hunyadi and Giovanni da Capistrano organized a 1456 crusade to lift the Ottomon siege of Belgrade. In April 1487, Pope Innocent VIII called for a crusade against the Waldensians of Savoy, the Piedmont, and the Dauphiné in southern France and northern Italy. The only efforts actually undertaken, resulting in little change, were in the Dauphiné.

Crusader States

The First Crusade established the first four crusader states in the Eastern Mediterranean: the County of Edessa (1098–1149), the Principality (1098–1268), the Kingdom of Jerusalem (1099–1291), and the County of Tripoli (1104—although Tripoli was not conquered until 1109—to 1289). The Armenian Kingdom of Cilicia originated before the Crusades, but it received kingdom status from Pope Innocent III and later became fully westernized by the House of Lusignan. According to historian Jonathan Riley-Smith, these states were the first examples of "Europe overseas". They are generally known as *outremer*, from the French *outre-mer* ("overseas" in English).

The Fourth Crusade established a Latin Empire in the east and allowed the partition of Byzantine territory by its participants. The Latin emperor controlled one-fourth of the Byzantine territory, Venice three-eighths (including three-eighths of the city of Constantinople), and the remainder was divided among the other crusade leaders. This began the period of Greek history known as *Frankokratia* or *Latinokratia* ("Frankish [or Latin] rule"), when Catholic Western European nobles—primarily from France and Italy—established states on former Byzantine territory and ruled over the Orthodox Byzantine. The *Partitio terrarum imperii Romaniae* is a valuable record of early-13th-century Byzantine administrative divisions (*episkepsis*) and family estates.

Finance

Crusades were expensive; as the number of wars increased, their costs escalated. Pope Urban II called upon the rich to help First Crusade lords such as Duke Robert of Normandy and Count Raymond of St. Gilles, who subsidized knights in their armies. The total cost to King Louis IX of France of the 1284–1285 crusades was estimated at 1,537,570 *livres*, six times the king's annual income. This may be conservative, since records indicate that Louis spent 1,000,000 *livres* in Palestine after his Egyptian campaign. Rulers demanded subsidies from their subjects, and alms and bequests prompted by the conquest of Palestine were additional sources of income. The popes ordered that collection boxes be placed in churches and, beginning in the mid-twelfth century, granted indulgences in exchange for donations and bequests.

Military Orders

The military orders, especially the Templars and the Hospitallers, played a major role in providing support for the Crusader States, for they provided decisive forces of highly trained and motivated soldiers at critical moments. The Hospitallers and the Templars became international organizations, with depots across Western Europe and the East. The Teutonic Knights focused on the Baltic, and the Spanish military orders of Santiago, Calatrava, Alcantara, and Montesa concentrated on the Iberian Peninsula. The Hospitallers (Knights of the Order of the Hospital of St John of Jerusalem) had been founded in Jerusalem before the First Crusade but greatly enlarged its mission once the Crusades began. After the fall of Acre they relocated to Cyprus, conquering and ruling Rhodes (1309–1522) and Malta (1530–1801). The Poor Knights of Christ and its Temple of Solomon were founded in 1118 to protect pilgrims en route to Jerusalem. They became wealthy and powerful through banking and real estate. In 1322 the King of France suppressed the order, ostensibly for sodomy, magic and heresy but probably for financial and political reasons.

Roles of Women, Children, and Class

Women were intimately connected to the Crusades; they aided in recruitment, took over the crusaders' responsibilities in their absence, and provided financial and moral support. Historians contend that the most significant role played by women in the West was in maintaining the *status quo*. Landholders left for the Holy Land, leaving control of their estates to regents who were often wives or mothers. Since the Church recognized that risk to families and estates might discourage crusaders, special papal protection was a crusading privilege. A number of aristocratic women participated in crusades, such as Eleanor of Aquitaine (who joined her husband, Louis VII). Non-aristocratic women also served in positions such as washerwomen. More controversial was women taking an active role (counter to their femininity); accounts of fighting women were primarily by

Muslim historians, who portrayed Christian women who killed as barbarous and ungodly.

The Children's Crusade was said to have been a Catholic movement in France and Germany in 1212 who tried to reach the Holy Land. The traditional narrative is probably conflated from some factual and mythical notions of the period including visions by a French or German boy, an intention to peacefully convert Muslims in the Holy Land to Christianity, a band of several thousand youths set out for Italy, and children being sold into slavery. A study published in 1977 casts doubt on the existence of these events, and many historians came to believe that they were not (or not primarily) children but multiple bands of "wandering poor" in Germany and France, some of whom tried to reach the Holy Land and others who never intended to do so.

Three crusading efforts were made by peasants during the mid-1250s and the early 14th century. The first, the Shepherds' Crusade of 1251, was preached in northern France. After a meeting with Blanche of Castile, it became disorganized and was disbanded by the government. The second, in 1309, occurred in England, northeastern France, and Germany; as many as 30,000 peasants arrived at Avignon before it was disbanded. The third, in 1320, became a series of attacks on clergy and Jews and was forcibly suppressed. However, this "crusade" is primarily seen as a revolt against the French monarchy. The Jews had been allowed to return to France, after being expelled in 1306; any debts owed to the Jews before their expulsion were collected by the monarchy, drove the *Pastoureaux* (by which this movement is called).

Legacy

Western Europeans in the East adopted native customs, saw themselves as citizens of their new home and intermarried. This led to a people and culture descended from the remaining European inhabitants of the crusader states, particularly French Levantines in Lebanon, Palestine, and Turkey.

Traders from the maritime republics of the Mediterranean, continued to live in Constantinople, Smyrna, and other parts of Anatolia, and the eastern Mediterranean coast during the middle Byzantine and Ottoman eras. These people, known as Levantines or Franco-Levantines, are Roman Catholic.

They are now concentrated in the Istanbul districts of Galata, Beyoğlu and Nişantaşı; the İzmir districts of Karşıyaka, Bornova and Buca, and in Mersin (where they were influential in creating and reviving an operatic tradition. After the British occupied parts of Ottoman Syria in the aftermath of World War I, the term "Levantine" was used pejoratively for inhabitants of mixed Arab and European descent and for Europeans (usually French, Italian or Greek) who adopted local dress and customs.

The Crusades influenced the attitude of the Western Church towards warfare, with the frequent calling of crusades habituating the clergy to violence. They also sparked a debate about the legitimacy of seizing land and possessions from pagans on purely religious grounds which would resurface during the Age of Discovery in the 15th and 16th centuries. The needs of crusading stimulated secular governmental developments, not all of which were positive; resources used in crusading could have been used by developing states for local and regional needs.

Its power and prestige rose by the Crusades, the papal curia had greater control of the western Church and extended the system of papal taxation through the ecclesiastical structure of the West. The system of indulgences grew significantly in late medieval Europe, sparking the Protestant Reformation in the early 16th century. Although the Albigensian Crusade intended to eliminate Catharism in Languedoc, France acquired lands with closer cultural and linguistic ties to Catalonia. The crusade also had a role in the creation and institutionalization of the Dominican Orderand the Medieval Inquisition. The persecution of Jews in the First Crusade is part of the long history of anti-Semitism in Europe. The need to raise, transport and supply large armies led to flourishing trade between Europe and the *outremer*. Genoa and Venice flourished, with profitable trading colonies in crusader states in the Holy Land and (later) in captured Byzantine territory. **(Wikipedia)**

The Israeli Relinquishment of the Temple Mount

The most relevant factual basis for disproving the "Al-Aksa is in danger" libel is, as noted, the de facto Israeli relinquishment of the Temple Mount, for which I could find no precedent in any other country or religion. The birthfather of this relinquishment, which for years has been called "the status quo on the Temple Mount," was Moshe Dayan, who served as Israeli defense minister during the Six-Day War. The thrilling liberation of the Western Wall and the Temple Mount was documented in detail in dozens of publications that appeared after the war. Even the cry of paratroop commander Mordechai Gur into his field radio – "The Temple Mount is in our hands!" – entered the pantheon of national symbols of the State of Israel. And yet, the reality that Israel devised on the Temple Mount, and the heavy limitations it imposed on itself there, were very far from the euphoria of the liberation itself and the overwhelming encounter with the place where the two Temples of the Jewish people had stood in the past, long the focal point of its spiritual life.

After the Six-Day War, the reality that Israel devised on the Temple Mount, and the heavy limitations it imposed on itself there, contravened in many ways everything that believing Jews pray for every day.

The reality that the State of Israel created at the site indeed contravened in many ways everything that believing Jews, keepers of the Torah and the commandments, pray for and mention in their prayers every day: "that the Temple be rebuilt speedily in our days....And there we will serve You in reverence, as in the days of old and as in former years."

Dayan's first act on the Temple Mount, only a few hours after IDF Chief Rabbi Shlomo Goren blew the shofar and gave the Shehecheyanu blessing beside the Western Wall, was to immediately remove the Israeli flag that the paratroopers had raised on the mount.

Dayan's second act was to clear out the paratroop company that was supposed to remain permanently stationed in the northern part of the mount. Dayan rejected the insistent pleas of the head of Central Command, Uzi Narkiss, who tried to prevent him from taking this measure. Narkiss reminded Dayan that Jordan, too, had stationed a military contingent on the mount to maintain order, and that long ago the Romans had done the same, deploying a garrison force in the Antonia Fortress that Herod had built near the mount. But Dayan was not persuaded. He told Narkiss that it seemed to him the place would have to be left in the hands of the Muslim guards. (3)

Despite harsh criticism from religious and nationalist circles, (4) Dayan, just a few hours after his first public announcement to the Israeli people about the holy places and particularly the Temple Mount, succinctly stated: "We have returned to the holiest of our places, never to be parted from them again....We did not come to conquer the sacred sites of others or to restrict their religious rights, but rather to ensure the integrity of the city and to live in it with others in fraternity."

Here Dayan behaved as the successor of David Ben-Gurion, who already during the War of Independence in July 1948, when it appeared that the Jewish forces were about to conquer the Old City, ordered David Shaltiel, Haganah commander in Jerusalem, to "prepare a special force, loyal and disciplined...that will use without mercy a machinegun against any Jew who tries to rob or desecrate a holy place, Christian or Muslim." Ben-Gurion also recommended that Shaltiel mine the entrances to the holy places so as to prevent harm to them.

Nineteen years later, a few hours after Dayan's decision, he summoned Prime Minister Levi Eshkol and the heads of the religious communities, and promised them that the places that were holy to them would not be harmed. Eshkol, for his part, announced to the chief rabbis of Israel that they would be responsible for arrangements in the vicinity of the Western

Wall, and promised the religious leaders of the Christian and Muslim communities that they would continue to determine the arrangements at the places holy to them: the Church of the Holy Sepulchre and the Temple Mount.

Moshe Dayan's most significant act on the Temple Mount, which was widely criticized, was to forbid Jewish prayer there, unlike the arrangements at the Machpelah Cave in Hebron where there is also a functioning mosque.

Dayan's most significant act on the Temple Mount, which sparked controversy over the years and was widely criticized, was to forbid Jewish prayer and worship there, unlike the arrangements that emerged at the Machpelah Cave in Hebron where there is also a functioning mosque. Dayan decided to leave the mount and its management in the hands of the Muslim Wakf, while at the same time insisting that Jews would be able to visit it (but not pray at it!) without restriction. Dayan thought, and years later even committed the thought to writing, that since for Muslims the mount is a "Muslim prayer mosque" while for Jews it is no more than "a historical site of commemoration of the past...one should not hinder the Arabs from behaving there as they now do." The Israeli defense minister believed that Islam must be allowed to express its religious sovereignty – as opposed to national sovereignty – over the mount; that the Arab-Israeli conflict must be kept on the territorial-national level; and that the potential for a conflict between the Jewish religion and the Muslim religion must be removed. In granting Jews the right to visit the mount, Dayan sought to placate the Jewish demands for worship and sovereignty there. In giving religious sovereignty over the mount to the Muslims, he believed he was defusing the site as a center of Palestinian nationalism.

The basic elements of the status quo that Dayan designed on the Temple Mount have remained the same up to the present. Despite countless attempts by Jews to pray on the mount, the state has upheld the prohibition on Jewish prayer there. According to the Protection of Holy Places Law (1967), the religious affairs minister is indeed authorized to exercise his power and lay down regulations for Jewish and Muslim prayer on the mount; but those who have held this post have avoided doing so, conforming with the governmental decree. The Supreme Court as well, to which Jews have appealed numerous times to change this policy and allow Jews to pray at their holiest of places, has backed the government's policy for considerations of "maintaining order and public security." The court has determined that the right to pray is not enforceable without regulations, and that implementing the right without such regulations would pose a grave danger to public peace. In its ruling in the case of *The Temple Mount Faithful v. Tzahi Hanegbi* (the internal security minister at the time), the court clarified that every Jew has the right to ascend the Temple Mount, to pray on it, and to commune with his Creator. That is part of the freedom of religious worship; that is part of the freedom of expression. At the same time, this right, like other basic rights, is not an absolute right, and in a place at which the likelihood of damage to the public peace and even to human life is almost certain – this can justify limiting the freedom of religious worship and also limiting the freedom of expression.

Even the rabbinical establishment has long assented to this policy de facto for its own reasons, which are rooted in Halakhah (Jewish religious law). The prohibition on Jews entering the Temple Mount is anchored in the Halakhic status of Jews in our times, who are regarded as "defiled by contact with the dead." At present, unlike in ancient times, there is no possibility of being purified from this defilement. Not all the rabbis have agreed with this prohibition, and recent years have seen a great increase in the number of rabbis who have changed their stance and permitted Jews to enter the mount. At the same time, the Israeli Chief Rabbinate, which is the decisive institutional actor when it comes to Halakhah, has so far stuck to its position that Jews may not enter the mount. Almost all the adjudicators in the haredi (ultra-Orthodox) world think the same, and so do many of the leading religious-Zionist adjudicators.

An even wider consensus is embodied in the almost comprehensive Halakhic ruling that it is forbidden at present to build the Third Temple, for which Jews yearn in their prayers. This opinion is common to both rabbis who now permit entry to the Temple Mount and those who prohibit it. The rabbis categorically forbid building the Temple, whether the proposal entails building it in place of the mosques or within the mount compound but without harming them. The possibility of building the Temple is negated for several reasons; the main ones are:

1. The view that building the Temple will be allowed only with the coming of the Messiah.

2. Many believe that the Third Temple will not be built in human times but will descend, complete, from the heavens.

3. A good many more view the contemporary generation as lacking a sufficient level of spirituality, purity, and maturity to be worthy of the Temple.

4. The Halakhic obstacle to the entry of Jews to the Mount, and the absence of the "red heifer," whose ash, according to Jewish sources, served in ancient times to purify Jews defiled by death.

5. The fear of an interreligious clash between Islam and Judaism involving harm to Jews and Jewish religious targets all over the world.

Seemingly, the logic of many Halakhic adjudicators concerning the Temple Mount over the years was summed up by the former deputy president of the Israel Supreme Court, Menachem Elon, in his ruling on *The Temple Mount Faithful v. the Attorney-General*. Elon explained that "this special attitude in the world of Judaism, that the more sacred the place or

the issue is, there is a special duty not to draw near to it or enter it, does not entail distancing or avoidance but, rather, nearness and veneration." He also quoted statements in this spirit by Rabbi Avraham Yitzchak Hacohen Kook, who discussed the issue of the Temple and the Temple Mount at length. Similar words were written more recently by Rabbi Shlomo Aviner, head of the Ateret Cohanim yeshiva, in his book Shalhevetya: "Our ownership and our belonging are revealed in the fact that we do not approach this place, and our national genius is evident in the fact that we show the whole world that: there is a place that we do not enter...
the distance does not separate; on the contrary;
it connects."

For his part, Rabbi Yuval Sherlo, head of the *hesder* (combining military service and religious study) yeshiva in Petah Tikva and one of the leading rabbis of religious Zionism, takes a more complex position. Like hundreds of other Zionist rabbis, Sherlo advocates Jewish prayer on the Temple Mount but does not countenance harming the mosques. He recognizes the value of the mount and is not prepared to relinquish the Jewish connection to it, while also appreciating the obstacles to fully realizing that connection. Sherlo is in favor of studying the issues of holiness and the Temple, and of "internalizing the constant feeling that something is lacking for us," but he also emphasizes that "the building of the Temple begins at a different place" – from the standpoint of "I will build a Temple in my heart, a place for doing justice, charity, morality, and law between a man and his fellow and amending the world and society."

Over the years the State of Israel has adhered—mainly with the help of its security mechanisms, the Shin Bet (Israel Security Agency) and the police—to Dayan's status quo. Furthermore, Israel initiated or accepted two major changes on the Temple Mount, to the benefit of the Muslim side.

First, notwithstanding Dayan's original decision, for many years the police have not allowed free entry by Jews to the Temple Mount, even for mere visits. The police restrict the number of Jews, particularly religious Jews, who can enter. Only a few dozen religious Jews are allowed to be there at once, and they are shadowed by Wakf guards and policemen who keep an eye on them, check their belongings to make sure they have not "smuggled" onto the mount a *tallit* (prayer shawl), *tefillin* (phylacteries), or prayer book, while warily ascertaining that their lips are not moving in prayer. Only after such a contingent of religious Jews has left is another group of a few dozen allowed to enter. The hours of entry for Jews to the mount are also restricted and meager, and in times of riots and tensions the site is closed to them altogether.

Second, in the mid-1990s two large underground recesses on the Temple Mount were modified, greatly expanding the area available for Muslim prayer: the underground recess at the south-eastern corner of the mount, which is called Solomon's Stables (for the Muslims, the Al-Marwani Mosque), and the recess under the Al-Aksa Mosque, which is called Ancient Al-Aksa.

The archaeological management of the Temple Mount is also carried out under difficult limitations. These stem from the Wakf's position that it is the sovereign, the ruler, and the decision-maker for the site. The State of Israel shows deference toward this position even though officially it does not agree with it. For example, there were years in which archaeological management was not permitted at all. Quite often, rehabilitation, renovation, and building work on the mount is performed (in coordination with the Israeli government) by foreign governments and bodies, such as Jordan or Egypt, while the State of Israel "tiptoes" around these projects. In the 1994 peace treaty with Jordan, the Israeli government also recognized Jordan's future senior status regarding the Muslim holy places in Jerusalem including the mount (Al-Haram al-Sharif); at such time as peace treaties and final status agreements will be signed with the rest of the Arab world.

The fact that the official and actual policy of the State of Israel leaves the management of the Temple Mount in the hands of the Muslim Wakf is not recognized in the Muslim world today.

The fact that the official and actual policy of the State of Israel, as embodied in decisions of the Chief Rabbinical Council, the government, and the Supreme Court, leaves the management of the Temple Mount in the hands of the Muslim Wakf is not recognized in the Muslim world today. On the contrary, Palestinian and Muslim elements portray the activities of nongovernmental and nonmainstream Jewish elements, some of them extreme and marginal, who seek an immediate renewal of Temple worship and even the destruction of the mosques, as reflecting the official and actual position of the State of Israel.

The reality, of course, is different. The State of Israel acts to foil such plans. Over the years extreme, nonofficial Jewish elements have tried to damage the Temple Mount and its mosques, and Muslim extremists have made use of it for purposes of terror and incitement. These attempts have been thwarted by the iron hand of the Israeli security authorities: the police, the Shin Bet, and the Mossad. It is only this overall security responsibility for the site that the State of Israel has maintained exclusively. Nevertheless, all keys to the gates of the Temple Mount compound, except for the Mughrabi Gate (on its western side), are exclusively in the hands of the Wakf. Only the keys to the Mughrabi Gate are held jointly by the Wakf and Israel, with each side having a copy.

Israel has also passed the Protection of Holy Places Law. It stipulates that these places will be protected against desecration and any other harm, and against anything that could detract from the different religions' freedom of access to these places, or injure feelings connected to them. The punishments for transgressors are severe: seven years in prison for whoever desecrates a holy place, five years in prison for "whosoever does anything likely to violate the freedom of access of the members of the different religions to the places sacred to them or their feelings with regard to those places."

Taking all this into account, the claim that the state and its institutions have formulated a plot to destroy the Temple Mount mosques, and establish the Third Temple in their stead, is absurd and invalid. The State of Israel has indeed adhered to the Jewish heritage, honours Jewish history, and sees itself as committed to its ancient roots, a context in which the Temple Mount and the Temple are central. Regarding the mount, however, this involves an ideological and spiritual heritage, not a practical one; a profound bond and commitment, but only on the level of consciousness. At the same time, the State of Israel does just about everything, in both its statements and its actions, to make clear that it has no intentions of building the Third Temple or destroying the Temple Mount mosques. All this has in no way prevented the many-faceted "Al-Aksa is in danger" libel from developing and taking hold of the imaginations and hearts of tens of millions of Muslims.

Israel is the finest example I can think of right now of a country in need of unity; how this can be brought about is quite simple; the reason why this hasn't happened since the founding of the State in 1948 is due to the absence of a unifying; the presence of the entity I signify as the "Jew".

Israel is not a "Jewish" state, nor is it a "Muslim" state; both terms are patently ridiculous due to the perception that a patch of land being holy is an example of "idolatry"; Moses had a similar problem with his "followers" while the "lost generation" travelled from the nation they'd been enslaved, Egypt, to Canaan.

The West Bank, and the Gaza Strip, should be annexed, and become a part of Israel, for the same reason the Golan Heights was annexed, shortly after the land of Israel was originally created. The Jews were given international recognition as a separate people once the horrors of the Holocaust were exposed, and as such were deserving of safe place to reside.

The claim made by leaders of the Arab World that the continent of Europe transplanted the problem they created on their soil, is absurd, and merits the label of being preposterous. They have shown no greater a capacity for compassion, and empathy, as those who participated in the methodical eradication of Jewish people during the reign of the Nazi Party.

Brainwashing (verb)

1. pressurize (someone) into adopting radically different beliefs by using systematic and often forcible means.

"people are **brainwashed into** believing family life is the best"
synonyms: indoctrinate, condition, re-
educate, persuade, propagandize, influence,inculcate, drill;
pressurize
"women of the nineties have been brainwashed into thinking they should go back to work"

Mind Control

Brainwashing (verb)

2. pressurize (someone) into adopting radically different beliefs by using systematic and often forcible means.

"people are **brainwashed into** believing family life is the best"
synonyms: indoctrinate, condition, re-
educate, persuade, propagandize, influence,inculcate, drill;
pressurize
"women of the nineties have been brainwashed into thinking they should go back to work"

Mind control

(also known as **brainwashing, reeducation, brainsweeping, coercive persuasion, thought control,** or **thought reform**) is a theory that human subjects can be indoctrinated in a way that causes "an impairment of autonomy, an inability to think independently, and a disruption of beliefs and affiliations. In this context, brainwashing refers to the involuntary reeducation of basic beliefs and values".

Theories of brainwashing and of mind control were originally developed to explain how totalitarian regimes systematically indoctrinate prisoners of war through propaganda and torture techniques. These theories were later expanded and modified by psychologists including Margaret Singer and Philip Zimbardo to explain a wider range of phenomena, especially conversions to some new religious movements (NRMs). The suggestion that NRMs use mind control techniques has resulted in scientific and legal debate; with Eileen Barker, James Richardson, and other scholars, as well as legal experts, rejecting at least the popular understanding of the concept.

Newer theories have been proposed by scholars including: Robert Cialdini, Robert Jay Lifton, Daniel Romanovsky, Kathleen Taylor, and Benjamin Zablocki. The concept of mind control is sometimes involved in legal cases, especially regarding child custody; and is also a major theme in both science fiction and in criticism of modern corporate culture.

The Korean War and Brainwashing

Origin of the concept

The *Oxford English Dictionary* records the earliest known English-language usage of *brainwashing* in an article by newspaperman Edward Hunter, in *Miami News*, published on 7 October 1950. Hunter, an outspoken anticommunist and said to be a CIA agent working undercover as a journalist, wrote a series of books and articles on the theme of Chinese brainwashing, and the word brainwashing quickly became a stock phrase in Cold War headlines.

The Chinese term 洗腦 (*xǐ nǎo*, literally "wash brain") was originally used to describe methodologies of coercive persuasion used under the Maoist government in China, which aimed to transform individuals with a reactionary imperialist mindset into "right-thinking" members of the new Chinese social system. The term punned on the Taoist custom of "cleansing/washing the heart/mind" (洗心, *xǐ xīn*) before conducting certain ceremonies or entering certain holy places.

Hunter and those who picked up the Chinese term used it to explain why, during the Korean War (1950-1953); some American prisoners of war cooperated with their Chinese captors, even in a few cases defecting to the enemy side. British radio operator Robert W. Ford and British army Colonel James Carne also claimed that the Chinese subjected them to brainwashing techniques during their war-era imprisonment.

The U.S. military and government laid charges of "brainwashing" in an effort to undermine detailed confessions made by military personnel to war crimes, including biological warfare. After Chinese radio broadcasts claimed to quote Frank Schwable, Chief of Staff of the First Marine Air Wing admitting to participating in germ warfare, United Nations commander Gen. Mark W. Clark asserted: "Whether these statements ever passed the lips of these unfortunate men is doubtful. If they did, however, too familiar are the mind-annihilating methods of these Communists in extorting whatever words they want.... The men themselves are not to blame, and they have my deepest sympathy for having been used in this abominable way."

In the 1950s many American movies were filmed that featured brainwashing of POWs, including *The Rack*, *The Bamboo Prison*, *Toward the Unknown*, and *The Fearmakers*. Fraser A. Sherman comments: "The possibility that advanced psychological techniques could reprogram people's minds became a permanent part of pop culture." *Forbidden Area* told the story of Soviet secret agents who had been brainwashed (through classical conditioning) by their own government so they wouldn't reveal their true identities. In 1962 *The Manchurian Candidate* "put brainwashing front and center" and featured a plot by the Soviet government to take over the United States by use of a brainwashed presidential candidate.

The concept of brainwashing became associated with the research of Russian psychologist Ivan Pavlov; which mostly involved dogs, not humans, as subjects. In *The Manchurian Candidate* the head brainwasher is Dr. Yen Lo, of the Pavlov Institute.

Korean War Brainwashing Debunked

In 1956, after reexamining the concept of brainwashing following the Korean War, the U.S. Army published a report entitled *Communist Interrogation, Indoctrination, and Exploitation of Prisoners of War* which called brainwashing a "popular misconception." The report states "exhaustive research of several government agencies failed to reveal even one conclusively documented case of 'brainwashing' of an American prisoner of war in Korea."

US POWs captured by North Korea were brutalized with starvation, beatings, forced death marches, exposure to extremes of temperature, binding in stress positions, and withholding of medical care, but the abuse had no relation to indoctrination "in which [North Korea was] not particularly interested." In contrast American POWs in the custody of North Korea's Chinese Communist allies did face a concerted interrogation and indoctrination program. However, "systematic, physical torture was not employed in connection with interrogation or indoctrination," the report states.

The "most insidious" and effective Chinese technique according to the US Army Report was a convivial display of false friendship, which persuaded some GIs to make anti-American statements, and in a few isolated cases, refuse repatriation and remain in China:

"[w]hen an American soldier was captured by the Chinese, he was given a vigorous handshake and a pat on the back. The enemy 'introduced' himself as a friend of the 'workers' of America ... in many instances the Chinese did not search the American captives, but frequently offered them American cigarettes. This display of friendship caught most Americans totally off-guard and they never recovered from the initial impression made by the Chinese. ... [A]fter the initial contact with the enemy, some Americans seemed to believe that the enemy was sincere and harmless. They relaxed and permitted themselves to be lulled into a well-disguised trap [of cooperating with] the cunning enemy."

Two academic studies of the repatriation of American prisoners of war by Robert Jay Lifton and by Edgar Schein concluded that brainwashing (called "thought reform" by Lifton and "coercive persuasion" by Schein), if it occurred, had at best a transient effect. In 1961, they both published books expanding on these findings. Schein published *Coercive Persuasion* and Lifton published *Thought Reform and the Psychology of Totalism*.

CIA Mind Control Program

In 1999, forensic psychologist Dick Anthony concluded that the CIA had invented the concept of "brainwashing" as a propaganda strategy to undercut communist claims that American POWs in Korean communist camps had voluntarily expressed sympathy for communism. He argued that the books of Edward Hunter (whom he identified as a secret CIA "psychological warfare specialist" passing as a journalist) pushed the CIA brainwashing theory onto the general public. Succumbing to their own propaganda for twenty years, starting in the early 1950s, the CIA and the Defense Department conducted secret research (notably including Project MKULTRA) in an attempt to develop practical brainwashing techniques; the results are unknown. (See also Sidney Gottlieb.)

American Psychological Association rejection of brainwashing theory

Margaret Singer, who also spent time studying the political brainwashing of Korean prisoners of war, in her book *Cults in Our Midst*, describes six conditions which would create an atmosphere in which thought reform is possible. In 1983, the American Psychological Association (APA) asked Singer to chair a taskforce called the APA Task Force on Deceptive and Indirect Techniques of Persuasion and Control (DIMPAC) to investigate whether brainwashing or "coercive persuasion" did indeed play a role in recruitment by such movements.

Before the taskforce had submitted its final report, the APA submitted on 10 February 1987 an *amicus curiæ* brief in an ongoing court case related to brainwashing. Although the amicus curiæ brief written by the APA denies the credibility of the brainwashing theory, the APA submitted the brief under "intense pressure by a consortium of pro-religion scholars (a.k.a. NRM scholars)". The brief repudiated Singer's theories on "coercive persuasion" and suggested that brainwashing theories were without empirical proof. Afterward the APA filed a motion to withdraw its signature from the brief, since Singer's final report had not been completed.

On 11 May 1987, the APA's Board of Social and Ethical Responsibility for Psychology (BSERP) rejected the DIMPAC report because the report "lacks the scientific rigor and evenhanded critical approach necessary for APA imprimatur", and concluded that "after much consideration, BSERP does not believe that we have sufficient information available to guide us in taking a position on this issue." Benjamin Zablocki and Alberto Amitrani interpreted the APA's response as meaning that there was no unanimous decision on the issue either way, suggesting also that Singer retained the respect of the psychological community after the incident.

Two critical letters from external reviewers Benjamin Beit-Hallahmi and Jeffery D. Fisher accompanied the rejection memo. The letters criticized "brainwashing" as an unrecognized theoretical concept and Singer's reasoning as so flawed that it was "almost ridiculous." After her findings were rejected, Singer sued the APA in 1992 for "defamation, frauds, aiding and abetting and conspiracy" and lost. After that time U.S. courts consistently rejected testimonies about mind control and manipulation, stating that such theories were not part of accepted mainline science according to the Frye Standard of 1923.

New Religious Movements

In the 1970s, the anti-cult movement applied mind control theories to explain seemingly sudden and dramatic religious conversions to various new religious movements (NRMs). The media was quick to follow suit, and social scientists sympathetic to the anti-cult movement, who were usually psychologists, developed more sophisticated models of brainwashing. While some psychologists were receptive to these theories, sociologists were for the most part skeptical of their ability to explain conversion to NRMs.

Theories and religious conversion

Over the years various theories of conversion and member retention have been proposed that link mind control to some new religious movements (NRMs), particularly those religious movements referred to as "cults" by their critics. Philip Zimbardo discusses mind control as "the process by which individual or collective freedom of choice and action is compromised by agents or agencies that modify or distort perception, motivation, affect, cognition and/or behavioral outcomes", and he suggests that any human being is susceptible to such manipulation.

Debate over theories as applied to NRMs

James Richardson observes that if the new religious movements (NRMs) had access to powerful brainwashing techniques, one would expect that NRMs would have high growth rates, yet in fact most have not had notable success in recruitment. Most adherents participate for only a short time, and the success in retaining members is limited. For this and other reasons, sociologists of religion including David Bromley and Anson Shupe consider the idea that "cults" are brainwashing American youth to be "implausible." In addition, Thomas Robbins, Massimo Introvigne, Lorne Dawson, Melton, Marc, and Saul Levine, amongst other scholars researching NRMs, have argued and established to the satisfaction of courts, relevant professional associations and scientific communities that there exists no generally accepted scientific theory, based upon methodologically sound research, that supports the brainwashing theories as advanced by the anti-cult movement.

Benjamin Zablocki responds that it is obvious that brainwashing occurs, at least to any objective observer; but that it isn't "a process that is directly observable." The "real sociological issue", he states, is whether "brainwashing occurs frequently enough to be considered an important social problem". Zablocki disagrees with scholars like Richardson, stating that Richardson's observation is flawed. According to Zablocki, Richardson misunderstands brainwashing, conceiving of it as a recruiting process, instead of a retaining process. Zablocki adds that the sheer number of former cult leaders and members who attest to brainwashing in interviews (performed in accordance with guidelines of the National Institute of Mental Health and National Science Foundation) is too large to be a result of anything other than a genuine phenomenon.

Zablocki also points out that in the two most prestigious journals dedicated to the sociology of religion, the number of articles "supporting the brainwashing perspective" have been zero, while over one hundred such articles have been published in other journals "marginal to the field". From this fact, Zablocki concludes that the concept brainwashing has been blacklisted unfairly from the field of sociology of religion.

Eileen Barker criticizes mind control theories because they function to justify costly interventions such as deprogramming or exit counseling. She has also criticized some mental health professionals, including Singer, for accepting expert witness jobs in court cases involving NRMs. Her 1984 book, *The Making of a Moonie: Choice or Brainwashing?* describes the religious conversion process to the Unification Church (whose members are sometimes informally referred to as "Moonies") which had been one of the best known groups said to practice brainwashing. Barker spent close to seven years studying Unification Church members. She interviewed in depth and/or gave probing questionnaires to church members, ex-members, "non-joiners," and control groups of uninvolved people from similar backgrounds, as well as parents, spouses, and friends of members. She also attended numerous Unification Church workshops and communal facilities. Barker writes that she rejects the "brainwashing" theory as an explanation for conversion to the Unification Church, because, as she wrote, it explains neither the many people who attended a recruitment meeting and did not become members, nor the voluntary disaffiliation of members.

Joost Meerloo, a Dutch psychiatrist, was an early leading proponent of the concept of brainwashing. His view was influenced by his experiences during the German occupation of his country in the Second World War and his work with the Dutch government and the American military in the interrogation of accused Nazi war criminals. He later immigrated to the United States and taught at Columbia University. His best-selling 1956 book, *The Rape of the Mind*, concludes by saying: "The modern techniques of brainwashing and menticide-those perversions of psychology-can bring almost any man into submission and surrender. Many of the victims of thought control, brainwashing, and menticide that we have talked about were strong men whose minds and wills were broken and degraded. But although the totalitarians use their knowledge of the mind for vicious and unscrupulous purposes, our democratic society can and must use its knowledge to help man to grow, to guard his freedom, and to understand himself." ("Menticide" is a neologism coined by Meerloo meaning: "Killing of the mind.")

In Italy there has been controversy over the concept of *plagio*, a crime consisting in an absolute psychological—and eventually physical—domination of a person. The effect of such domination is the annihilation of the subject's freedom and self-determination and the consequent negation of his or her personality. The crime of plagio has rarely been prosecuted in Italy, and only one person was ever convicted. In 1981, Italy the Court found the concept to be imprecise, lacking coherence, and liable to arbitrary application.

By the twenty-first century, the concept of brainwashing had spread to other fields and was being applied "with some success" in criminal defense, child custody, and child sexual abuse cases. In some cases "one parent is accused of brainwashing the child to reject the other parent, and in child sex abuse cases where one parent is accused of brainwashing the child to make sex abuse accusations against the other parent" (possibly resulting in or causing parental alienation).

In his 2000 book, *Destroying the World to Save It: Aum Shinrikyo, Apocalyptic Violence, and the New Global Terrorism*, Robert Lifton applied his original ideas about thought reform to Aum Shinrikyo and the War on Terrorism, concluding that in this context thought reform was possible without violence or physical coercion. He also pointed out that in their efforts against terrorism Western governments were also using some mind control techniques, including thought-terminating clichés.

In 2003 Dick Anthony[who?] asserted in the *Washington Post* that "no reasonable person would question that there are situations where people can be influenced against their best interests, but those arguments are evaluated on the basis of fact, not bogus expert testimony." Dismissing the idea of mind control, he has defended NRMs, and argued that involvement in such movements may often have beneficial, rather than harmful effects: "There's a large research literature published in mainstream journals on the mental health effects of new religions. For the most part the effects seem to be positive in any way that's measurable."

In her 2004 book, *Brainwashing: The Science of Thought Control*, neuroscientist and physiologist Kathleen Taylor put forth the theory that the neurological basis for reasoning and cognition in the brain and the self itself are changeable. She describes the physiology behind neurological pathways which include webs of neurons containing dendrites, axons, and synapses; and explains that certain brains with more rigid pathways will be less susceptible to new information or creative stimuli. She uses neurological science to demonstrate that brainwashed individuals have more rigid pathways, and that that rigidity can make it unlikely that the individual will rethink situations or be able to later reorganize these pathways. She explains that repetition is an integral part of brainwashing techniques because connections between neurons become stronger when exposed to incoming signals of frequency and intensity. She argues that people in their teenage years and early twenties are more susceptible to persuasion. Taylor explains that brain activity in the temporal lobe, the region responsible for artistic creativity, also causes spiritual experiences in a process known as lability.

In his 2007 book, *Influence: The Psychology of Persuasion*, social psychologist Robert Cialdini argues that mind control is possible through the covert exploitation of the unconscious rules that underlie and facilitate healthy human social interactions. He states that common social rules can be used to prey upon the unwary. Using categories, he offers specific examples of both mild and extreme mind control—both one on one and in groups—notes the conditions under which each social rule is most easily exploited for false ends, and offers suggestions on how to resist such methods.

In 2009 historian Daniel Romanovsky wrote about what he called "Nazi brainwashing" of the people of Belarus by the occupying Germans during the Second World War, which took place through both mass propaganda and intense re-education, especially in schools. He notes that very soon most people had adopted the Nazi view of the Jews, that they were an inferior race and were closely tied to the Soviet government, views that had not been at all common before the occupation.

Mind control has often been an important theme in science fiction and fantasy stories. Terry O'Brien comments: "Mind control is such a powerful image that if hypnotism did not exist, then something similar would have to have been invented: the plot device is too useful for any writer to ignore. The fear of mind control is equally as powerful an image." A subgenre is "corporate mind control", in which a future society is run by one or more business corporations which dominate society using advertising and mass media to control the population's thoughts and feelings.

Modern corporations are said to practice mind control to create a work force which shares the same common values and culture. Critics have linked "corporate brainwashing" with globalization, saying that corporations are attempting to create a world-wide monocultural network of producers, consumers, and managers. In his 1992 book, *Democracy in an Age of Corporate Colonization*, Stanley A. Deetz says that modern "self awareness" and "self improvement" programs provide corporations with even more effective tools to control the minds of employees than traditional brainwashing. Modern educational systems have also been criticized, by both the left and the right, for contributing to corporate brainwashing.

(Wikipeda)

The Stepford Wives

The Stepford Wives is a 1972 satirical thriller novel by Ira Levin. The story concerns Joanna Eberhart, a photographer and young mother who begins to suspect that the frighteningly submissive housewives in her new idyllic Connecticut neighborhood may be robots created by their husbands.

The book has had two feature film adaptations, both using the same title as the novel: the critically acclaimed 1975 version, and the critically panned 2004 remake. Edgar J. Scherick produced the 1975 version, as well as all three of the made-for-television sequels. Scherick was posthumously credited as producer on the 2004 remake.

The term "Stepford wife," which is often used in popular culture, stems from the novel and is usually a reference to a submissive and docile wife. In the late 20th and early 21st centuries, it was sometimes used in reference to any woman, even an accomplished professional woman, who had subordinated her life and/or career to her husband's interests and who affected submission and devotion to him even in the face of the husband's public problems and disgrace.

In a March 27, 2007 letter to *The New York Times*, Levin said that he based the town of Stepford on Wilton, Connecticut, where he lived in the 1960s. Wilton is a "step" from Stamford, a major city lying 15 miles away.

The premise involves the married men of the fictional town of Stepford, Connecticut and their fawning, submissive, impossibly beautiful wives. The protagonist is Joanna Eberhart, a talented photographer newly arrived from New York City with her husband and children, eager to start a new life. As time goes on, she becomes increasingly disturbed by the zombie-like, submissive wives of Stepford, especially when she sees her once independent-minded friends—fellow new arrivals to Stepford—turn into mindless, docile housewives overnight. Her husband, who seems to be spending more and more time at meetings of the local men's association, mocks her fears.

As the story progresses, Joanna becomes convinced that the wives of Stepford are being poisoned or brainwashed into submission by the men's club. She visits the library and reads up on the pasts of Stepford's wives, finding out that some of the women were once feminist activists and very successful professionals, while the leader of the men's club is a former Disney engineer and others are artists and scientists, capable of creating lifelike robots. Her friend Bobbie helps her investigate, going so far as to write to the EPA to inquire about possible environmental toxins in Stepford. However, eventually, Bobbie is also transformed into a docile housewife and has no interest in her previous activities.

At the end of the novel, Joanna decides to flee Stepford but when she gets home she finds that her children have been taken. She asks her husband to let her leave but he takes her car keys. She manages to escape from the house on foot and several of the men's club members track her down. They corner her in the woods and she accuses them of creating robots out of the town's women. The men deny the accusation and ask Joanna if she would believe them if she saw one of the other women bleed. Joanna agrees to this and they take her to Bobbie's house. Bobbie's husband and son are upstairs, with loud rock music playing—as if to cover screams. The scene ends as Bobbie brandishes a knife at her former friend.

In the story's epilogue, Joanna has become another Stepford wife gliding through the local supermarket and has given up her career as a photographer, while Ruthanne (a new resident in Stepford) appears poised to become the conspiracy's next victim. (Wikipedia)

Individuation

The principle of **individuation** or *principium individuation* describes the manner in which a thing is identified as distinguished from other things.

The concept appears in numerous fields and is encountered in works of Carl Jung, Gilbert Simondon, Bernard Stiegler, Friedrich Nietzsche, Arthur Schopenhauer, David Bohm, Henri Bergson, Gilles Deleuze, and Manuel De Landa.

Usage

The word *individuation* occurs with different meanings and connotations in different fields.

In philosophy

Philosophically, "individuation" expresses the general idea of how a thing is identified as an individual thing that "is not something else". This includes how an individual person is held to be distinct from other elements in the world and how a person is distinct from other persons.

In Jungian psychology

In Jungian psychology, also called analytical psychology, individuation is the process in which the individual self develops out of an undifferentiated unconscious – seen as a developmental psychic process during which innate elements of personality, the components of the immature psyche, and the experiences of the person's life become integrated over time into a well-functioning whole.

In the media industry

The media industry has begun using the term *individuation* to denote new printing and online technologies that permit mass customization of the contents of a newspaper, a magazine, a broadcast program, or a website so that its contents match each individual user's unique interests. This differs from the traditional mass-media practice of producing the same contents for all readers, viewers, listeners, or online users.

Communications theorist Marshall McLuhan alluded to this trend when discussing the future of printed books in an electronically interconnected world.

Carl Jung

According to Jungian psychology, individuation is a process of psychological integration. "In general, it is the process by which individual beings are formed and differentiated [from other human beings]; in particular, it is the development of the psychological individual as a being distinct from the general, collective psychology."

Individuation is a process of transformation whereby the personal and collective unconscious are brought into consciousness (e.g., by means of dreams, active imagination, or free association) to be assimilated into the whole personality. It is a completely natural process necessary for the integration of the psyche. Individuation has a holistic healing effect on the person, both mentally and physically.

In addition to Jung's theory of complexes, his theory of the individuation process forms conceptions of a phylogenetically acquired unconscious filled with mythic images, a non-sexual libido, the general types of extraversion and introversion, the compensatory and prospective functions of dreams, and the synthetic and constructive approaches to fantasy formation and utilization.

"The symbols of the individuation process . . . mark its stages like milestones, prominent among them for Jungians being the shadow, the wise old man . . . and lastly the anima in man and the animus in woman." Thus, "There is often a movement from dealing with the persona at the start . . . to the ego at the second stage, to the shadow as the third stage, to the anima or animus, to the Self as the final stage. Some would interpose the Wise Old Man and the Wise Old Woman as spiritual archetypes coming before the final step of the Self."

Gilbert Simondon

In *L'individuation psychique et collective*, Gilbert Simondon developed a theory of individual and collective individuation in which the individual subject is considered as an effect of individuation rather than a cause. Thus, the individual atom is replaced by a never-ending ontological process of individuation.

Simondon also conceived of "pre-individual fields" which make individuation possible. Individuation is an ever-incomplete process, always leaving a "pre-individual" left over, which makes possible future individuations. Furthermore, individuation always creates both an individual subject and a collective subject, which individuate themselves concurrently.

Bernard Stiegler

The philosophy of Bernard Stiegler draws upon and modifies the work of Gilbert Simondon on individuation and also upon similar ideas in Friedrich Nietzsche and Sigmund Freud. During a talk given at the Tate Modern art gallery in 2004, Stiegler summarized his understanding of individuation. The essential points are the following:

- The *I*, as a psychic individual, can only be thought in relationship to *we*, which is a collective individual. The *I* is constituted in adopting a collective tradition, which it inherits and in which a plurality of *I*'s acknowledge each other's existence.

- This inheritance is an adoption, in that I can very well, as the French grandson of a German immigrant, recognize myself in a past which was not the past of my ancestors but which I can make my own. This process of adoption is thus structurally factual.

- The *I* is essentially a process, not a state, and this process is an individuation — it is a process of psychic individuation. It is the tendency to *become one*, that is, to become indivisible.

- This tendency never accomplishes itself because it runs into a counter-tendency with which it forms a metastable equilibrium. (It must be pointed out how closely this conception of the dynamic of individuation is to the Freudian theory of drives and to the thinking of Nietzsche and Empedocles.)

- The *we* is also such a process (the process of collective individuation). The individuation of the *I* is always inscribed in that of the *we*, whereas the individuation of the *we* takes place only through the individuations, polemical in nature, of the *I*'s which constitute it.

- That which links the individuations of the *I* and the *we* is a pre-individual system possessing positive conditions of effectiveness that belong to what Stiegler calls retentional apparatuses. These retentional apparatuses arise from a technical system which is the condition of the encounter of the *I* and the *we* — the individuation of the *I* and the *we* is in this respect also the individuation of the technical system.

- The technical system is an apparatus which has a specific role wherein all objects are inserted — a technical object exists only insofar as it is disposed within such an apparatus with other technical objects (this is what Gilbert Simondon calls the technical group).

- The technical system is also that which founds the possibility of the constitution of retentional apparatuses, springing from the processes of grammatization growing out of the process of individuation of the technical system. And these retentional apparatuses are the basis for the dispositions between the individuation of the *I* and the individuation of the *we* in a single process of psychic, collective, and technical individuation composed of three branches, each branching out into process groups.

- This process of triple individuation is itself inscribed within a vital individuation which must be apprehended as
  - the vital individuation of natural organs,
  - the technological individuation of artificial organs,
  - and the psycho-social individuation of organizations linking them together.

- In the process of individuation, wherein knowledge as such emerges, there are individuations of mnemo-technological subsystems which over determine, *qua* specific organizations of what Stiegler calls tertiary retentions, the organization, transmission, and elaboration of knowledge stemming from the experience of the sensible. **(Wikipedia)**

C.G. Jung; Quotes

"The meeting of two personalities is like the contact of two chemical substances: if there is any reaction, both are transformed."

"Everything that irritates us about others can lead us to an understanding of ourselves."

"Your visions will become clear only when you can look into your own heart. Who looks outside, dreams; who looks inside, awakes."

"I am not what happened to me, I am what I choose to become."

"You are what you do, not what you say you'll do."

"Knowing your own darkness is the best method for dealing with the darkness's of other people."

"Loneliness does not come from having no people about one, but from being unable to communicate the things that seem important to oneself, or from holding certain views which others find inadmissible."

"The pendulum of the mind oscillates between sense and nonsense, not between right and wrong."

"Until you make the unconscious conscious, it will direct your life and you will call it fate."

"As a child I felt myself to be alone, and I am still, because I know things and must hint at things which others apparently know nothing of, and for the most part do not want to know."

Spears

Isaiah 2:4

And He will judge between the nations, And will render decisions for many peoples; And they will hammer their swords into plowshares and their spears into pruning hooks Nation will not lift up sword against nation, And never again will they learn war.

2 Samuel 2:23

However, he refused to turn aside; therefore Abner struck him in the belly with the butt end of the spear, so that the spear came out at his back. And he fell there and died on the spot. And it came about that all who came to the place where Asahel had fallen and died, stood still.

2 Samuel 18:14

Then Joab said, "I will not waste time here with you." So he took three spears in his hand and thrust them through the heart of Absalom while he was yet alive in the midst of the oak.

2 Samuel 23:21

He killed an Egyptian and impressive man. Now the Egyptian had a spear in his hand, but he went down to him with a club and snatched the spear from the Egyptian's hand and killed him with his own spear.

What Should A Jew Do?

The Jew boy walks home, from the Yeshiva, which is his school.
The path is made of cobbled stone.
Many a rock from here he has thrown.

Fields and meadows lay each side.
The panorama is wide open.
Few places exist that can serve as a secure place to hide.

Nearby, is the village where he resides;
A collection of shacks, huts, bungalows, roads, and dirt paths,
That serves as carriage routes.
Here men, usually, don't bother to wear suits.
They work as laborers.
Many must carry on their bent backs heavy, burdensome, canvas sacks.

The women feed chickens, as well as goats,
And share tales, stories, and Yiddish jokes.
Their hands are often covered with grime, and smell of filthy, dirty, socks,
But when relatives and friends arrive to share toasts,
They are the ones that serve as their loving hosts.

Each home is bordered with an old rickety wooden fence.
In many a yard there is a special area,
Where there is placed a comfortable, beautiful, bench.

The men gather here to chat and talk.
Women and young girls serve traditional dishes.
Each one is given his own special fork.

Don't base a judgment merely on how a place looks;
Much might appear to be in a state of squalor.
So many here shout and holler;
But here is where many devote much time to reading holy books.

The state one must live in,
Says little about the depth of the knowledge a person has within.
The willingness to learn and study hard,
Can lead to a better understanding of the time when all did begin

The boy walks tall, with his chin held high.
For all his life he has decided never to tell a lie.
Truth is golden, and should be valued above all else.
This is how we should measure the value of ones' self.

On his death bed, where he must muster his last sigh,
Remembering he has always striven to be a good Jew,
With an open heart that reflects comfort and grace,
He will then express his love and devotion toward others;
Then wish his dear ones a sincere goodbye.

The Torah and Talmud are held tightly in one hand.
The other sways with the greatest of ease.
He imagines a gentle caressing breeze, as well as palm trees and soft sand.
One day he hopes he will be able to return to his Homeland.

Once he reaches the villages' central square,
He notices the stores are vacant, and empty;
The market stalls have been left bare.
All appears still, but nearby awaits a treacherous lair.
Several youngsters are waiting for him there, hiding behind a rusty iron gate.
Their intent is to wait till he passes,
Then whips and chains will be used to give lashes.

Whistling a tune a Rabbi once taught him,
He neglects to survey what is around
His mind is concerned with the coming Sabbath,
When he will wear a shoal, the sacred gown.

Deprived of even a moment to prepare,
The troublesome youths appear as fast as a hare.
They encircle, determined to cause a great deal of misery.
Trapped, the Jew accepts he has been deprived of his liberty.

What does a Jew do when he encounters the unknown?
He waits to see what the other will demand;
A gift, a way to punish, or maybe the help of a loan.

Mocking, lecherous, voices ring out.
Disparaging words are caste to make him feel fear.
Because he is a Jew, he does not;
The Lord protects him because He is near.

The books are drawn to his chest to offer protection.
Because he is one among The Chosen, he knows he is special by natural
selection.
Few in number, each must fight to survive.
Due to this reason, till this day, they have managed to thrive.
Many in the past, however, have been persecuted, and as a result died.

One after another, cruel names are used to describe
His Faith, family, and also himself;
None of this could possibly be true.
Evidence is present in family portraits, and trophies,
That are above his home fireplace on a mantle shelf.

Pain rips through his back, arms, and legs.
Suffering, experiencing great pain, he falls to the ground
So proud, not for a moment does he consider to beg, or ask, for mercy.

It seems an eternity before they end their cruel game.
Many others have suffered the same degree of humiliation;
Throughout the ages things have remained much the same.

Covered with red is his shirt from his nose that is now swollen.
Evidence of the crime is clearly shown.
His love for justice, and The Lord has grown.

He brushes the dirt from his sleeves and pants,
Then gathers the pages dispersed from his torn, broken, books.
Then an epiphany arrives all of a sudden,
All that matters is to have self-control, not to be a glutton

Instilled in his heart is the love for others of the past.
Relatives, neighbors, friends, how they died;
To the lions and sea they have been caste.
The Chosen people must continue;
Above all else, this is their most important task.

Finally he approaches his family's home.
Here his Abba and Ema reside.
There he feels safe, his sores will be covered,
And he will be able to rest each weary bone.

Practically outside the small bungalow's front door,
Once more a threat presents itself.
This time the enemy can possibly be defeated.
On top of the wooden gate his sole opponent is seated.
A punch is thrust, but misses, as the other hand of the Jew bashes.

He can see now the degree of his pain;
For this reason, he feels quite sane.

The joke is at hand; how did he ever think he had a chance?
So many peoples have become extinct.
The Hebrews till this day can enjoy the game of the dance.

Wailing the pitiful cries of a small child,
He licks the wounds he deserves to get,
And races home, probably to cover his head in the room where lies his bed.

Standing before both his parents,
After passing the threshold of his family's sanctuary,
Ema covers her child with hugs and kisses,
Abba watches weary, and tired, from the long day at the factory.

A bowl seeping steam, smelling of herb and spices,
Is served to please his nose and gullet.
His father, meanwhile, gives him a sermon on the worth of a bullet.

"Always do what is right; pleasing The Lord is the ultimate gift."
Abba preaches, while he stokes the boys shoulder.
We each in life must learn to grow bolder.

The calloused, cracked, and wrinkled, hand of the elder,
Holds the chin of his son.
Struggles never cease, toils never end, the lessons of life have just begun.

Pride fills the man's heart, and reddens his cheeks.
For so long he has waited for this moment, seeming an eternity;
Actually, it has only been weeks.

Time passes so quickly, where does it go?
All his life he's wondered, wanted to know;

Maybe one day, before he should pass away.

The end never seems possible.
He hoped he'd have a son, who could share with him
The joys, wonders, and terrors, of
This scary show.

THE JEWS

Walk the Earth
As they did in Ancient times,
And will build the British Empire
To last a thousand years.

Jews transform people's perception of things; they incorporate elements into societies that weren't there before; they can disassemble whole belief systems, and replace them with something entirely different.

The world is in the horrendous state it is today due to an inability to recognize the value of the Jew, his role within society, and, furthermore, the vast scheme of things. He is the protector, the preserver, and also the destroyer

It is quite extraordinary that the greater the numbers of Jews that have been provided to mankind, fewer are the numbers who are able to appreciate their accomplishments. The holocaust was meant to serve as a lesson no one would ever forget, and thus such horrors would never happen again.

Wind Carries
the Petals of a Rose

After her swollen knuckled hands
Had swept over a pair of candles several times,
She'd lift them to her face to shield her closed eyes
In a voice that was neither muffled, nor distinct,
She said, "Our Lord is blessed,
As well as our world which is His Kingdom".
Heaven and Earth can be joined
By the word we call Love.
That being so,
Our quest is to achieve the ultimate freedom.

Many believe, and have earnestly experienced this to be true,
I, for one, can honestly say I truly loved my dear grandmother.
Her passing often leaves me feeling despondent,
And quite blue.

She was the only grandparent my fortune allowed me to know,
Thus, she meant more to me than the grandest present
Wrapped in the most exquisite bow.

I've asked myself; did she keep her eyes shut
So as to not see the light that flickered before her?
Or was she expressing how open
She was to receiving the eternal flame within her

If the two can be considered the same,
The heaven, and the earth, are governed by a sole Monarch,
In whom our trust should never wane.

The story above could be perceived as a lesson.
We have things we can learn from every person, situation, place;
Even the sight of a single flower
Hiding in the shadows of a country garden.
All can be seen as a beautiful act of glorious grace.

Keats, the beloved poet,
Whose imagination spanned
Across moors, streams, fields, and hills,
But mostly resided in a small room many generations ago,
Once noted that;
"Truth is beauty. Beauty is truth.
That is all we know. That is all we need to know."
If we treasure such wisdom,
Our world would never cease to glow.

Our struggles may appear to overwhelm,
Tomorrow may contain sorrow also pain,
But there is always hope when we believe we can be happy again.

Despair can dampen our spirit,
But as long as a flame exists that emits light,
Our lives can expose something enormously bright.

My grandmother taught me all these things;
Not by the words she said,
But by using her life as a means to be an example.
Ever since I've found my life to be far from a gamble.

Her gestures, wet kisses, laughing smile, and sweet hello,
Were all the nourishment I needed,
Along with her succulent homemade dishes
That I couldn't wait to sample.

Liverpool was once a harbor, then a port;
Later it was bombed by a dark enemy
That left piles of brick, mortar, and rubble.
In bomb shelters people would cower, as they sat in a huddle.

The sirens still ring in the ears of her citizens to this day.
History displays lessons we are not supposed to forget.
Future generations should listen to those who have endured these things,
To better understand what it is they are seeing.

Till this day many mix laughter, and pain, sweetness and sorrow,
To form a humor few could possibly find hollow.

All entwine together to form a place that is marvelously unique.
Continually streams of people flood this patch of ground,
Striving to reach answers to questions
They've desperately chased like a hound.

My grandmother's house is contained in a borough named Great Crosby,
Which is situated close to the shores of the River Mercy,
Near a stretch of water large enough to be called a sea.

Within, between, betwixt, this cross section,
Was the place my father grew up in,
And I also had the chance to share in the wisdom he'd been given.

Rose's home was a studio, in which I studied life,
And all it so generously has to offer.
I slept in her bed, played on her lawn,
Shared time with relatives and friends,
None of which she found to be in the slightest a bother.
Here I learned to accept me,
And all I might possibly one day have a chance to be.

Spread over many rooms were relics
That served as symbols of times that had already passed.
They remain in my mind,
Because I viewed them as monuments that were made to last.

My grandmother developed traits, and carried them all so well.
They were displayed in the manner she'd walk;
Steady, sure, determined, strong;
Things we must be sure we require if we wish to live long.

A face may act as a reminder that a smile may not be on display,
But within lies a heart
That is filled with the substance that can make us gay.
I encounter so many sights, and sounds,
That remind me of my grandmother every day.

She departed from this earth many years ago,
But, if you listen carefully
You'll hear the sound of her presence still present within my soul,
And all that constitutes my essential being.
Look carefully, and you'll notice,
It is her essence you are actually seeing.

My laughter, smile, cheer, thoughtful glance, and casual greeting,
Would not be as they exist today,
If it were not for my grandmother and I meeting,
I have now become aware that an assembly may occur in a Hall,
Where there might be an occasion for people to have a Ball.
No matter the position in which a person may stand,
Little of any of this, can I find bland.

There isn't much I take seriously, due to the strength in my heart,
Which leads to my regarding life as somewhat a lark.

People may come and go, hustle and bustle to and fro,
Whistling a tune as they mow.
My grandmother kept herself amused by this show,
Created by the specimen we call man,
While adhering to what the Bible says about eating a chunk of ham;

I must say, though, on her buttered toast
She'd put a great deal of jam

Her identity was sustained by her character
That was determined to persevere,
And has since enhanced my ability to keep myself fully in gear.

Life, after all, is incredibly precious, and time, really, is all we've got,
Therefore, is it really wise to place a value
On objects such as tables, chairs, clocks,
Or even an ocean sandbar?

I learned all these lessons well with the help of a person
I referred to as my Grandma.

Part Two

The Making of History

The Miracle of Life

A Miracle, I would say,
Is something that can be proven to have occurred,
But the reason why is not understood,
Furthermore, it is perceived that it will never be undersood.
The wars, strife, and suffering, throughout history
Happened not because man wasn't given the means to decipher events,
But because man failed to use knowledge wisely.
He clouds his vision, so the obvious becomes murky,
And what is straight appears crooked.
Ignorance develops the same way
As all other forms of sinful behaviour;
People create their own ceilings; barriers; shoe boxes to live in,
And refuse to accept anything
That doesn't fit into their prefabricated world.
Life isn't a mystery;
Whether there's s hell or heaven on Earth
Is man's choice;
If he's wise;
He'll make sure he always
Stays on
The right path.

Chapter Four

The Last Glacial Period

Chronology of climatic events of importance for the Last Glacial Period (about the last 120 000 years)

The **last glacial period**, popularly known as the **Ice Age**, was the most recent glacial period within the Quaternary occurring during the last 100,000 years of the Pleistocene, from approximately 110,000 to 12,000 years ago. Scientists consider this "ice age" to be merely the latest glaciation event in a much larger ice age, one that dates back over two million years and has seen multiple glaciations.

During this period, there were several changes between glacier advance and retreat. The Last Glacial Maximum, the maximum extent of glaciation within the last glacial period, was approximately 22,000 years ago. While the general pattern of global cooling and glacier advance was similar, local differences in the development of glacier advance and retreat make it difficult to compare the details from continent to continent (see picture of ice core data below for differences).

From the point of view of human archaeology, it falls in the Paleolithic and Mesolithic periods. When the glaciation event started, *Homo sapiens* were confined to Africa and used tools comparable to those used by Neanderthals in Europe and the Levant and by *Homo erectus* in Asia. Near the end of the event, *Homo sapiens* spread into Europe, Asia, and Australia. The retreat of the glaciers allowed groups of Asians to migrate to the Americas and populate them.

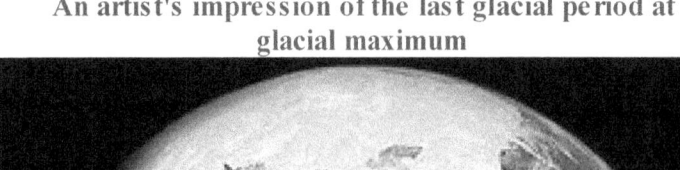

An artist's impression of the last glacial period at glacial maximum

The last glacial period is sometimes colloquially referred to as the "last ice age", though this use is incorrect because an ice age is a longer period of cold temperature in which ice sheets cover large parts of the Earth, such as Antarctica. Glacials, on the other hand, refer to colder phases within an ice age that separate interglacials. Thus, the end of the last glacial period is not the end of the last ice age. The end of the last glacial period was about 10,500 BCE, while the end of the last ice age has not yet come.

Over the past few million years the glacial-interglacial cycle has been "paced" by periodic variations in the Earth's orbit via Milankovitch cycles which are thus the "cause" of ice ages.

The last glacial period is the best-known part of the current ice age, and has been intensively studied in North America, northern Eurasia, the Himalaya and other formerly glaciated regions around the world. The glaciations that occurred during this glacial period covered many areas, mainly in the Northern Hemisphere and to a lesser extent in the Southern Hemisphere. They have different names, historically developed and depending on their geographic distributions:

Fraser (in the Pacific of North America), **Pinedale** (in the Central Rocky Mountains), **Wisconsinan** or **Wisconsin** (in central North America), **Devensian** (in the British Isles), **Midlandian** (in Ireland), **Würm** (in the Alps), **Mérida** (in Venezuela), **Weichselian** or **Vistulian** (in Northern Europe and northern Central Europe), **Valdai** in Eastern and **Zyryanka** in Siberia, **Llanquihue** in Chile, and **Otira** in New Zealand. The geochronological Late Pleistocene comprises the late glacial (Weichselian) and the immediately preceding penultimate interglacial (Eemian) preiond.

The last glaciation centered on the huge ice sheets of North America and Eurasia. Considerable areas in the Alps, the Himalaya and the Andes were ice-covered, and Antarctica remained glaciated.

Canada was nearly completely covered by ice, as well as the northern part of the United States, both blanketed by the huge Laurentide ice sheet. Alaska remained mostly ice free due to arid climate conditions. Local glaciations existed in the Rocky Mountains and the Cordilleran ice sheet and as ice fields and ice caps in the Sierra Nevada in northern California. In Britain, mainland Europe, and northwestern Asia, the Scandinavian ice sheet once again reached the northern parts of the British Isles, Germany, Poland, and Russia, extending as far east as the Taimyr Peninsula in western Siberia. The maximum extent of western Siberian glaciation was reached approximately 16,000 to 15,000 BCE and thus later than in Europe (20,000–16,000 BCE). Northeastern Siberia was not covered by a continental-scale ice sheet. Instead, large, but restricted, icefield complexes covered mountain ranges within northeast Siberia, including the Kamchatka-Koryak Mountains.

The Arctic Ocean between the huge ice sheets of America and Eurasia was not frozen throughout, but like today probably was only covered by relatively shallow ice, subject to seasonal changes and riddled with icebergs calving from the surrounding ice sheets. According to the sediment composition retrieved from deep-sea cores there must even have been times of seasonally open waters.

Outside the main ice sheets, widespread glaciation occurred on the Alps-Himalaya mountain chain. In contrast to the earlier glacial stages, the Würm glaciation was composed of smaller ice caps and mostly confined to valley glaciers, sending glacial lobes into the Alpine foreland. To the east the Caucasus and the mountains of Turkey and Iran were capped by local ice fields or small ice sheets. In the Himalaya and the Tibetan Plateau, glaciers advanced considerably, particularly between 45,000–25,000 BCE, but these datings are controversial. The formation of a contiguous ice sheet on the Tibetan Plateau is controversial.

Other areas of the Northern Hemisphere did not bear extensive ice sheets, but local glaciers in high areas. Parts of Taiwan, for example, were repeatedly glaciated between 42,250 and 8,680 BCE as well as the Japanese Alps. In both areas maximum glacier advance occurred between 58,000 and 28,000 BCE (starting roughly during the Toba catastrophe). To a still lesser extent glaciers existed in Africa, for example in the High Atlas, the mountains of Morocco, the Mount Atakor massif in southern Algeria, and several mountains in Ethiopia. In the Southern Hemisphere, an ice cap of several hundred square kilometers was present on the east African mountains in the Kilimanjaro Massif, Mount Kenya and the Ruwenzori Mountains, still bearing remnants of glaciers today.

Glaciation of the Southern Hemisphere was less extensive because of current configuration of continents. Ice sheets existed in the Andes (Patagonian Ice Sheet), where six glacier advances between 31,500 and 11,900 BCE in the Chilean Andes have been reported. Antarctica was entirely glaciated, much like today, but the ice sheet left no uncovered area. In mainland Australia only a very small area in the vicinity of Mount Kosciuszko was glaciated, whereas in Tasmania glaciation was more widespread. An ice sheet formed in New Zealand, covering all of the Southern Alps, where at least three glacial advances can be distinguished. Local ice caps existed in Irian Jaya, Indonesia, where in three ice areas remnants of the Pleistocene glaciers are still preserved today.

Antarctica glaciation

During the last glacial period Antarctica was blanketed by a massive ice sheet, much as it is today. The ice covered all land areas and extended into the ocean onto the middle and outer continental shelf. According to ice modeling, ice over central East Antarctica was generally thinner than today.

Europe
Devensian & Midlandian glaciation (Britain and Ireland)

The name **Devensian** **glaciation** is used by British geologists and archaeologists and refers to what is often popularly meant by the latest Ice Age. Irish geologists, geographers, and archaeologists refer to the **Midlandian** glaciation as its effects in Ireland are largely visible in the Irish Midlands. The name Devensian is derived from the Latin *Dēvenses*, people living by the Dee (*Dēva* in Latin), a river on the Welsh border near which deposits from the period are particularly well represented.

The effects of this glaciation can be seen in many geological features of England, Wales, Scotland, and Northern Ireland. Its deposits have been found overlying material from the preceding Ipswichian Stage and lying beneath those from the following Flandrian stage of the Holocene.
The latter part of the Devensian includes Pollen zones I-IV, the Allerød and Bølling Oscillations, and the Older and Younger Dryas climatic stages.

Europe during the last glacial period

Alternative names include: **Weichsel glaciation** or **Vistulian glaciation** (referring to the Polish river Vistula or its German name Weichsel). Evidence suggests that the ice sheets were at their maximum size for only a short period, between 25,000 to 13,000 BP. Eight interstadials have been recognized in the Weichselian, including: the Oerel, Glinde, Moershoofd, Hengelo and Denekamp; however correlation with isotope stages is still in process. During the glacial maximum in Scandinavia, only the western parts of Jutland were ice-free, and a large part of what is today the North Sea was dry land connecting Jutland with Britain (see Doggerland). It is also in Denmark that the only Scandinavian ice-age animals older than 13,000 BC are found.

The Baltic Sea, with its unique brackish water, is a result of meltwater from the Weichsel glaciation combining with saltwater from the North Sea when the straits between Sweden and Denmark opened. Initially, when the ice began melting about 10,300 BP, seawater filled the isostatically depressed area, a temporary marine incursion that geologists dub the Yoldia Sea. Then, as rebound lifted the region about 9500 BP, the deepest basin of the Baltic became a freshwater lake, in palaeological contexts referred to as Ancylus Lake, which is identifiable in the freshwater fauna found in sediment cores. The lake was filled by glacial runoff, but as worldwide sea level continued rising, saltwater again breached the sill about 8000 BP, forming a marine Littorina Sea which was followed by another freshwater phase before the present brackish marine system was established. "At its present state of development, the marine life of the Baltic Sea is less than about 4000 years old," Drs. Thulin and Andrushaitis remarked when reviewing these sequences in 2003.

Overlying ice had exerted pressure on the Earth's surface. As a result of melting ice, the land has continued to rise yearly in Scandinavia, mostly in northern Sweden and Finland where the land is rising at a rate of as much as 8–9 mm per year, or 1 meter in 100 years. This is important for archaeologists since a site that was coastal in the Nordic Stone Age now is inland and can be dated by its relative distance from the present shore.

Extent of Alpine glaciation during the Würm ice age. Blue: extent of the early ice ages

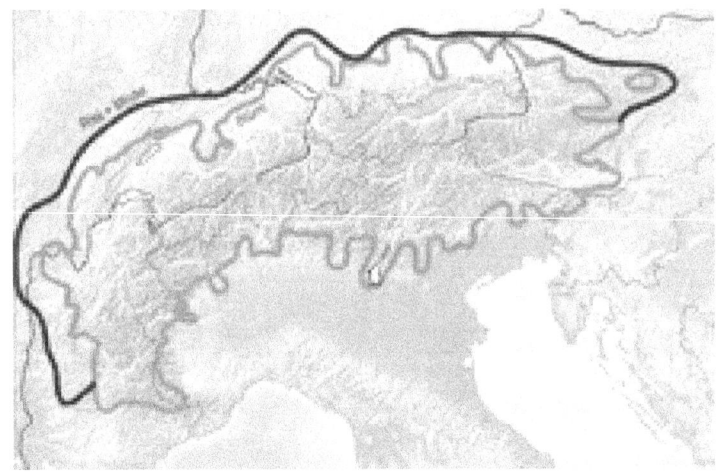

The term *Würm* is derived from a river in the Alpine foreland, approximately marking the maximum glacier advance of this particular glacial period. The Alps were where the first systematic scientific research on ice ages was conducted by Louis Agassiz at the beginning of the 19th century. Here the Würm glaciation of the last glacial period was intensively studied. Pollen, the statistical analyses of microfossilized plant pollens found in geological deposits, chronicled the dramatic changes in the European environment during the Würm glaciation. During the height of Würm glaciation, c. 24,000–10,000 BP, most of western and central Europe and Eurasia was open steppe-tundra, while the Alps presented solid ice fields and montane glaciers. Scandinavia and much of Britain were under ice.

During the Würm, the Rhône Glacier covered the whole western Swiss plateau, reaching today's regions of Solothurn and Aarau. In the region of Bern it merged with the Aar glacier. The Rhine Glacier is currently the subject of the most detailed studies. Glaciers of the Reuss and the Limmat advanced sometimes as far as the Jura. Montane and piedmont glaciers formed the land by grinding away virtually all traces of the older Günz and Mindel glaciation, by depositing base moraines and terminal moraines of different retraction phases and loess deposits, and by the pro-glacial rivers' shifting and redepositing gravels. Beneath the surface, they had profound and lasting influence on geothermal heat and the patterns of deep groundwater flow.

North America

Pinedale or Fraser glaciation (Rocky Mountains)

The Pinedale (central Rocky Mountains) or Fraser (Cordilleran ice sheet) glaciation was the last of the major glaciations to appear in the Rocky Mountains in the United States. The Pinedale lasted from approximately 30,000 to 10,000 years ago and was at its greatest extent between 23,500 and 21,000 years ago. This glaciation was somewhat distinct from the main Wisconsin glaciation as it was only loosely related to the giant ice sheets and was instead composed of mountain glaciers, merging into the Cordilleran Ice Sheet. The Cordilleran ice sheet produced features such as glacial Lake Missoula, which would break free from its ice dam causing the massive Missoula floods. USGS Geologists estimate that the cycle of flooding and reformation of the lake lasted an average of 55 years and that the floods occurred approximately 40 times over the 2,000 year period between 15,000 and 13,000 years ago. Glacial lake outburst floods such as these are not uncommon today in Iceland and other places.

Wisconsin glaciation

The Wisconsin Glacial Episode was the last major advance of continental glaciers in the North American Laurentide ice sheet. At the height of glaciation the Bering land bridge potentially permitted migration of mammals, including people, to North America from Siberia.

It radically altered the geography of North America north of the Ohio River. At the height of the Wisconsin Episode glaciation, ice covered most of Canada, the Upper, and New England, as well as parts of Montana and Washington. On Kelleys Island in Lake Erie or in New York's Central Park, the grooves left by these glaciers can be easily observed. In southwestern Saskatchewan and southeastern Alberta a suture zone between the Laurentide and Cordilleran sheets formed the Cypress Hills, which is the northernmost point in North America that remained south of the continental ice sheets.

The Great Lakes are the result of glacial scour and pooling of meltwater at the rim of the receding ice. When the enormous mass of the continental ice sheet retreated, the Great Lakes began gradually moving south due to isostatic rebound of the north shore. Niagara Falls is also a product of the glaciation, as is the course of the Ohio River, which largely supplanted the prior Teays River.

With the assistance of several very broad glacial lakes, it released floods through the gorge of the Upper Mississippi River, which in turn was formed during an earlier glacial period.

In its retreat, the Wisconsin Episode glaciation left terminal moraines that form Long Island, Block Island, Cape Cod, Nomans Land, Martha's Vineyard, Nantucket, Sable and the Oak Ridges Moraine in south central Ontario, Canada. In Wisconsin itself, it left the Kettle Moraine. The drumlins and eskers formed at its melting edge are landmarks of the Lower Connecticut River Valley.

Tahoe, Tenaya, and Tioga, Sierra Nevada

In the Sierra Nevada, there are three named stages of glacial maxima (sometimes incorrectly called ice ages) separated by warmer periods. These glacial maxima are called, from oldest to youngest, *Tahoe*, *Tenaya*, and *Tioga*. The Tahoe reached its maximum extent perhaps about 70,000 years ago. Little is known about the Tenaya. The Tioga was the least severe and last of the Wisconsin Episode. It began about 30,000 years ago, reached its greatest advance 21,000 years ago, and ended about 10,000 years ago.

Greenland glaciation

In Northwest Greenland, ice coverage attained a very early maximum in the last glacial period around 114,000. After this early maximum, the ice coverage was similar to today until the end of the last glacial period. Towards the end, glaciers readvanced once more before retreating to their present extent. According to ice core data, the Greenland climate was dry during the last glacial period, precipitation reaching perhaps only 20% of today's value.

Mérida glaciation (Venezuelan Andes)

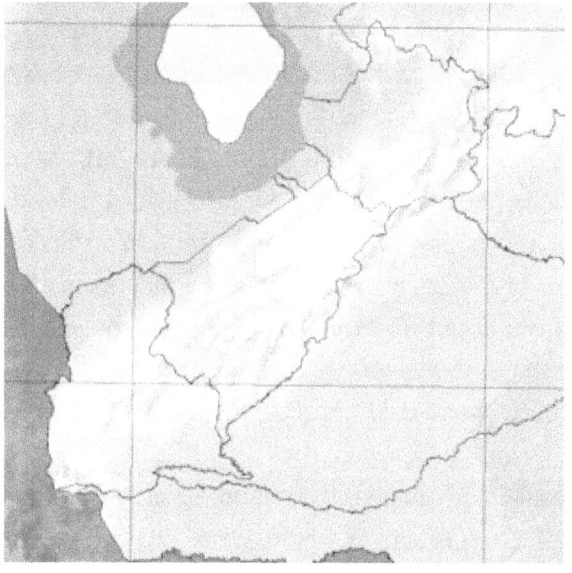

Map showing the extent of the glaciated area in Venezuelan during the Mérida glaciation

The name *Mérida* Glaciation is proposed to designate the alpine glaciation which affected the central Venezuelan during the Late Pleistocene. Two main moraine levels have been recognized: one between 2600 and 2700 m, and another between 3000 and 3500 m elevation. The snow line during the last glacial advance was lowered approximately 1200 m below the present snow line (3700 m). The glaciated area in the Cordillera was approximately 600 km^2; this included the following high areas from southwest to northeast: Páramo de Tamá, Páramo Batallón, Páramo Los Conejos, Páramo Piedras Blancas, and Teta de Niquitao. Approximately 200 km^2 of the total glaciated area was in the Sierra Nevada de Mérida, and of that amount, the largest concentration, 50 km^2, was in the areas of Pico Bolívar, Pico Humboldt (4,942 m), and Pico Bonpland (4,893 m). Radiocarbon dating indicates that the moraines are older than 10,000 years B.P., and probably older than 13,000 years B.P. The lower moraine level probably corresponds to the main Wisconsin glacial advance. The upper level probably represents the last glacial advance (Late Wisconsin).

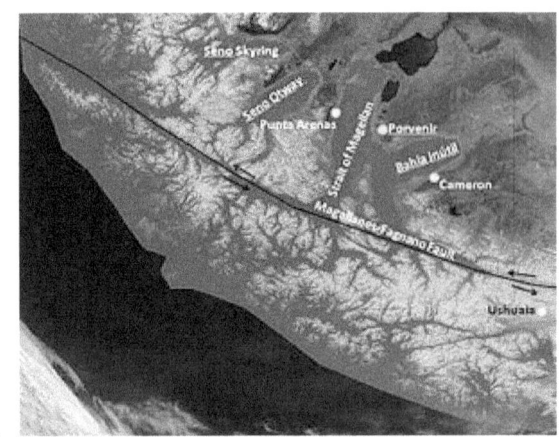

Map showing the extent of the Patagonian Ice Sheet in the Strait of Magellan area during the last glacial period. Selected modern settlements are shown with yellow dots.

The Llanquihue glaciation takes its name from Llanquihue Lake in southern Chile which is a fan-shaped piedmontglacial lake. On the lake's western shores there are large moraine systems of which the innermost belong to the last glacial period. Llanquihue Lake's varves are a node point in southern Chile's varve geochronology. During the last glacial maximum the Patagonian Ice Sheet extended over the Andes from about 35°S to Tierra del Fuegoat 55°S. The western part appears to have been very active, with wet basal conditions, while the eastern part was cold based. Cryogenic features like ice wedges, patterned ground, pingos, rock glaciers, palsas, soilcryoturbation, solifluction deposits developed in unglaciated extra-Andean Patagonia during the Last Glaciation. However, not all these reported features have been verified. The area west of Llanquihue Lake was ice-free during the LGM, and had sparsely distributed vegetation dominated by *Nothofagus*. Valdivian temperate rainforest was reduced to scattered remnants in the western side of the Andes. **(Wikipedia)**

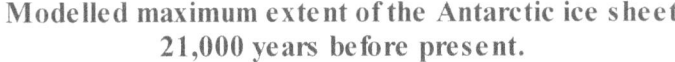

Modelled maximum extent of the Antarctic ice sheet 21,000 years before present.

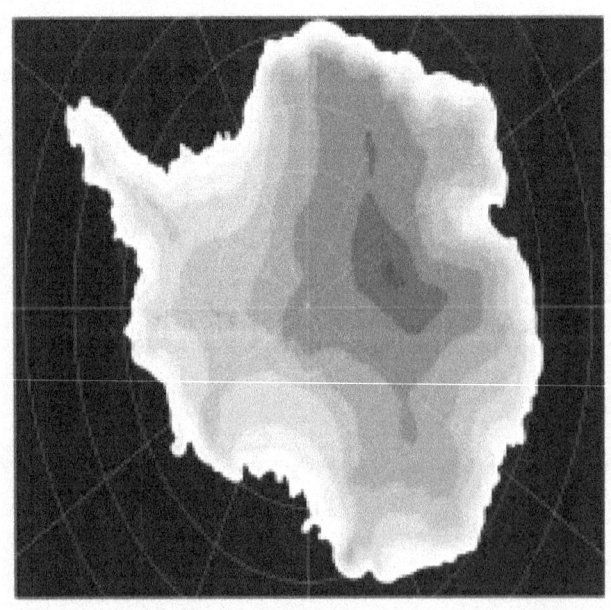

History can be viewed as the pendulum of a clock
that moves the hands on the face
that show the time.
There are two forces; good and evil; one force moves the pendulum in one
direction; the other force moves it in the opposing direction; goodness
creates a thesis; evil creates the antithesis.

Chapter Five

The Titans Rule

❖

You Create the Truths
You Hold

If you are extraordinary,
Of what use are you to the ordinary?
Can they, do they, will they, be able to understand your worth?

The disparities, inequalities, that exist now,
Are greater than they have ever been before;
Daily, more and more, are the number we can label as poor.

Wealth is concentrated among the few.
Their reckless, mean, acts continually result in a loss of life.
Call them cannibals; people are cut up, minced, and made to form a stew.

Only those who are wise and kind can be of benefit to mankind.
The rich, the corrupt, make sure they are of use to no one.

Society is structured to benefit them;
While an environment is sickened, and people walk as they spit ugly phlegm.

Can a change be made?
How could this possibly come about?

The answers have already been given.
The reality is; we should lose all hope; salvation lies not in the Catholic Pope.

Only a few were meant to survive for an eternity.
On earth now they are practically invisible.
What once was is no more, but one day they will relish riches galore.

He's known all along they lie not in gold,
But by heightening awareness of his own soul.

istory reveals there are groups, or types of people, who are indigenous to certain regions. There was a time when each was isolated, relatively speaking, but, over time, have migrated to other regions of the globe.

The first example I'll provide is the so called "Blacks" that now inhabit North America. Most originate from the western part of the Sahara Desert that borders the Atlantic Ocean. Their facia features reveal how they adapted to the environment of this region. How long they inhabited this milieu cannot be exactly determined, but their physical features are distinctly different from the peoples that did, and still to this day, inhabit the regions of Africa that hug the Mediterranean Sea and the people further inland and slightly to the south.

If the word, Negro, was created to refer to those who have darkly pigmented skins, that serves as protection from the Sun's U.V. rays, it would seem logical that this would still be used for the same purpose. What we have been taught, though, is that this term is no longer appropriate, the reason being is that the expression, "nigger" developed in southern parts of the U.S. when Negros were used as slaves, and since then both of these terms, "nigger" and "negro", are considered equally derogatory to those whom we now refer to as "Blacks".

During the time slavery existed in the U.S. Negros were considered a sub-species, and were treated in a deplorable manner because of this, which wasn't thought wrong, rather for their betterment, which was also of economic benefit to slave traders and owners, and the American economy in general.

It should be noted, however, that "nigger" and "negro" are two separate words, and, "Negroid", refers to a genetically distinct Race of people. Negros shouldn't be considered as sub-par in regard to the Rights they deserve, which are the same as everyone else, but they do appear, as voluminous evidence indicates, to be dragging behind other groups in Western countries, particularly in an educational sense.

We have been told, or have been led to believe by "Blacks", that is due to oppression, no different in strength than that which existed during the time Negros were slaves; therefore, or so we are told, any "problem" that has arisen is not due to themselves, but others who have not given them a fair opportunity to assimilate in the manner they are deserving, and is also their Right.

To a large extent this picture that has been created by "Blacks" is a farce, but it is considered politically incorrect to consider this to be the case; after all, we are all equal, especially in the eyes of God, which is used as an argument to defend this position, and ministers, and priests, allege is true, and un-Christian to think otherwise.

Such a view is preposterous; we are all deserving of life; liberty, and to pursue happiness, but we are definitely not all equal. "Blacks" have shown they find it difficult to keep up with others.

To this day, most parts of Africa have not developed, modernized, to the same extent as many other parts of the globe. History has shown that the development of civilization entails the inclusion of an Agricultural Revolution. In many instances this was not possible in Africa due to the climate and local topography. It is interesting to note, however, that on many occasions over the last several centuries the West attempting to cultivate its manners, ways, and technology, within Africa, but time and time again these acts did not achieved the desired outcome, but, peculiarly enough, without exception, across the length and breadth of this varied, and diverse, Continent, Imperial powers have been accused of not allowing this to happen, which enables Negros to not consider "internal" shortcomings within Africa that could have contributed to these occurrences.

The Chinese were kept as a separate people for many centuries. It was only when a desire developed by Europeans for the spices, herbs, indigenous to that region that a merge of civilizations occurred. Other peoples filtered into this region later, but the Chinese, up until the latter part of the Twentieth Century, for the most part, kept themselves situated in the same place.

China is now primarily a country where products are produced, and are then transported to the West, and various other parts of the world... Whenever Chinese people migrate they keep the traits they cultivated within China.

China is now infamous for the extent it allows Human Rights violations to occur; the people who leave this country, and acquire positions of power, do the same abroad. The Chinese have a tendency to not blame themselves for the transformations China has been through that were not to their liking, but rather the blame the West, and persecute them because they deem them the cause for their problems abroad.

The Arabs located in the Arab World, which can be considered to include the most northerly stretch of land of the continent of Africa, were repeatedly victims of invaders who made intrusions on their soil in order to pillage and plunder. Mohammad, "the prophet", sought the same type of conquest as those who had raped the Arab people before, but the Arabs have collectively failed to recognize this, and rape themselves to this day by blaming the West for their misfortunes.

Arab, Muslim, incursions into other lands have commonly resulted in the same problems that befall their home countries – violence; corruption; tyranny, and terrorism, are all expressions of hatred that is not founded in logic.

South Asia once was renowned for being "tolerant"; religious expressions of all types were accepted, if not encouraged. The culture was shaped around the spiritual truth, but at the expense of material wealth, which continued until Britain encroached on this region, and changed India to Her liking.

When Britain was finally ousted, matters didn't improve, but worsened in due to the fragmentation that arose. Instead of becoming a home for a meshing of ideas, which was once India's hallmark, the Continent of South Asia became the opposite. Cohesion turned ever more rapidly into dissolution.

The other Colonies that are a part of the British Empire have succumbed, as has Britain as well, to multiculturalism. People have spread their home grown cultural prejudices al over the world due to Globalization, and utilitarian values have correspondingly diminished in strength.

Women, beginning in the 1920's, managed to entrench themselves in many Western political systems with their own particular agenda taking precedence over all others, which serves their desire to liberate themselves; from what? We should ask, and why apparently at the expense of the Rights of others? The sad truth is that they divulged themselves of the role they were meant to serve - to be the nurturers, care givers, of the young.

People of different Races; creeds; religious affiliations; have spread across the globe; where within this dynamic can the Jew be found?

To discover these remarkable specimens, we must first remind ourselves what they are, and the purpose they serve. They are the people with the great minds who mastered a craft, or discipline, to such an extent, many others were encouraged to develop comparable qualities within themselves, and as a result enhance the lives of many others.

I have mentioned only a few within this book, but by examining their lives; what they did, others can be identified, and their accomplishments appreciated due to the extent hey had an impact upon others. By studying where they were located, and the times when they appeared, one can develop a picture as to how they came about, and the purpose they served.

The greatest composers; philosophers; religious figures, have emerged from a path followed by the people referred to as Aryans; one can conclude that incorporated within this group lays the possibility that a Giant may spring forth from time to time.

The record books show them seeping into Southern Asia from the north and west. Here knowledge stood still for a while, and is reflected in the teachings contained within the Hindu Religion - "Be still, and know you are God", is an aphorism known to mystics who practice yogic religions in this region to this day. Eventually, they moved west, and locked themselves into Persia, but then later released themselves and meandered across the Middle East; as far as Egypt, and then back to Mesopotamia. All these wandering eventually led to the production of the Torah, and later what is referred to as the Old Testament, (the 39 canonical books of the Hebrew Scriptures, written prior to the coming of Christ). They became a part of the Greeks, then the Romans, but when life turned dark in Europe, they escaped to Persia, only to emerge in Europe again when the time was ripe for their awakening; the Renaissance.

Within this group there is one stock in particular that stands out, the people known as the Semites. Today they mostly reside in Israel, in a region flooded with lies; tyranny; and violence; waiting silently, and patiently, for the truth to save Her and all the souls who have worked tirelessly for Her cause; saving Jews; protecting them from persecution.

The Jews that lead others back toward the good; create a thesis, a new beginning, but then Homo-sapiens create an antithesis by doing the opposite of what they're supposed to, then Jews appear and create a synthesis, which makes another thesis possible; this is the movement of the pendulum of the clock that records history.

The truth is that good cannot exist without evil; good exists because there is evil. If good had nothing to oppose; it wouldn't exist at all, and there lies the key to understanding history; God's plan was set from the beginning.

The British Isles

The **British Isles** are a group of islands off the north-western coast of continental that consist of the islands of Great Britain, Ireland and over six thousand smaller isles. Two sovereign states are located on the islands: Ireland (a republic which covers roughly five-sixths of the island with the same name) and the United Kingdom of Great Britain and Northern Ireland;

(which include the countries of England, Scotland, Wales and Northern Ireland).

The British Isles also include three dependencies of the British Crown: the Isle of Man and, by tradition, the Bailiwick of Jersey and the Bailiwick of Guernsey in the Channel Islands, although the latter are not physically a part of the archipelago.

The oldest rocks in the group are in the north west of Scotland, Ireland and North Wales and are 2,700 million years old. During the Silurian period the north-western regions collided with the south-east, which had been part of a separate continental landmass. The topography of the islands is modest in scale by global standards. Ben Nevis rises to an elevation of only 1,344 metres (4,409 ft.), and Lough Neagh, which is notably larger than other lakes on the isles, covers 390 square kilometres (151 sq. mi). The climate is temperatemarine, with mild winters and warm summers. The North Atlantic Drift brings significant moisture and raises temperatures 11 °C (20 °F) above the global average for the latitude. This led to a landscape which was long dominated by temperate, although human activity has since cleared the vast majority of forest cover. The region was re-inhabited after the last glacial period of Quaternary glaciation, by 12,000 BC when Great Britain was still a peninsula of the European continent. Ireland, which became an island by 12,000 BC, was not inhabited until after 8000 BC. Great Britain became an island by 5600 BC.

Hiberni (Ireland), Pictish (northern Britain) and Britons (southern Britain) tribes, all speaking Insular Celtic, inhabited the islands at the beginning of the 1st millennium AD. Much of Brittonic-controlled Britain was conquered by the Roman Empire from AD 43. The first Anglo-Saxons arrived as Roman power waned in the 5th century and eventually dominated the bulk of what is now England. Viking invasions began in the 9th century, followed by more permanent settlements and political change—particularly in England. The subsequent Norman conquest of England in 1066 and the later Angevin partial conquest of Ireland from 1169 led to the imposition of a new Norman ruling elite across much of

Britain and parts of Ireland. By the Late Middle Ages, Great Britain was separated into the Kingdoms of England and Scotland, while control in Ireland fluxed between Gaelic kingdoms, Hiberno-Norman lords and the English-dominated Lordship of Ireland, soon restricted only to The Pale. The 1603 Union of the Crowns, Acts of Union 1707 and Acts of Union 1800attempted to consolidate Britain and Ireland into a single political unit, the United Kingdom of Great Britain and Ireland, with the Isle of Man and the Channel Islands remaining as Crown Dependencies. The expansion of the British Empire and migrations following the Irish Famine and Highland Clearances resulted in the distribution of the islands' population and culture throughout the world and a rapid de-population of Ireland in the second half of the 19th century. Most of Ireland seceded from the United Kingdom after the Irish and the subsequent Anglo-Irish Treaty (1919–1922), with six counties remaining in the UK as Northern Ireland.

The term *British Isles* is controversial in Ireland, where there are objections to its usage due to the association of the word *British* with Ireland. The Government of Ireland does not recognize or use the termand its embassy in London discourages its use. As a result, **Britain and Ireland** is used as an alternative description, and **Atlantic Archipelago** has had limited use among a minority in academia, although *British Isles* is still commonly employed. Within them, they are also sometimes referred to as *these islands*.

Etymology

The earliest known references to the islands as a group appeared in the writings of sea-farers from the ancient Greek colony of Massalia. The original records have been lost; however, later writings, e.g. Avienus's *Ora maritima*, that quoted from the Massaliote Periplus (6th century BC) and from Pytheas's *On the Ocean* (circa 325–320 BC) have survived. In the 1st century BC, Diodorus Siculus has *Prettanikē nēsos*, "the British Island", and *Prettanoi*,"the Britons".

Strabo used Βρεττανική (*Brettanike*), and Marcian of Heraclea, in his *Periplus maris exteri*, used αἱ Πρεττανικαί νῆσοι (*the Prettanic Isles*) to refer to the islands. Historians today, though not in absolute agreement, largely agree that these Greek and Latin names were probably drawn from native Celtic-language names for the archipelago. Along these lines, the inhabitants of the islands were called the Πρεττανοί (*Priteni* or *Pretani*). The shift from the "P" of *Pretannia* to the "B" of *Britannia* by the Romans occurred during the time of Julius Caesar.

The classical writer, Ptolemy, referred to the larger island as *great Britain* (*megale Britannia*) and to Ireland as *little Britain* (*mikra Brettania*) in his work, *Almagest* (147–148 AD). In his later work, *Geography* (c. 150

AD), he gave these islands the names *Alwion* [*sic*], *Iwernia*, and *Mona* (the Isle of Man), suggesting these may have been native names of the individual islands not known to him at the time of writing *Almagest*. The name *Albion* appears to have fallen out of use sometime after the Roman conquest of Great Britain, after which *Britain* became the more common-place name for the island called Great Britain.

The earliest known use of the phrase *Brytish Iles* in the English language is dated 1577 in a work by John Dee. Today, this name is seen by some as carrying imperialist overtones although it is still commonly used. Other names used to describe the islands include the *Anglo-Celtic Isles*, *Atlantic archipelago*, *British-Irish Isles*, *Britain and Ireland*, *UK and Ireland*, and *British Isles and Ireland*. Owing to political and national associations with the word *British*, the Government of Ireland does not use the term *British Isles* and in documents drawn up jointly between the British and Irish governments, the archipelago is referred to simply as "these islands". Nonetheless, British Isles is still the most widely accepted term for the archipelago.

Geography

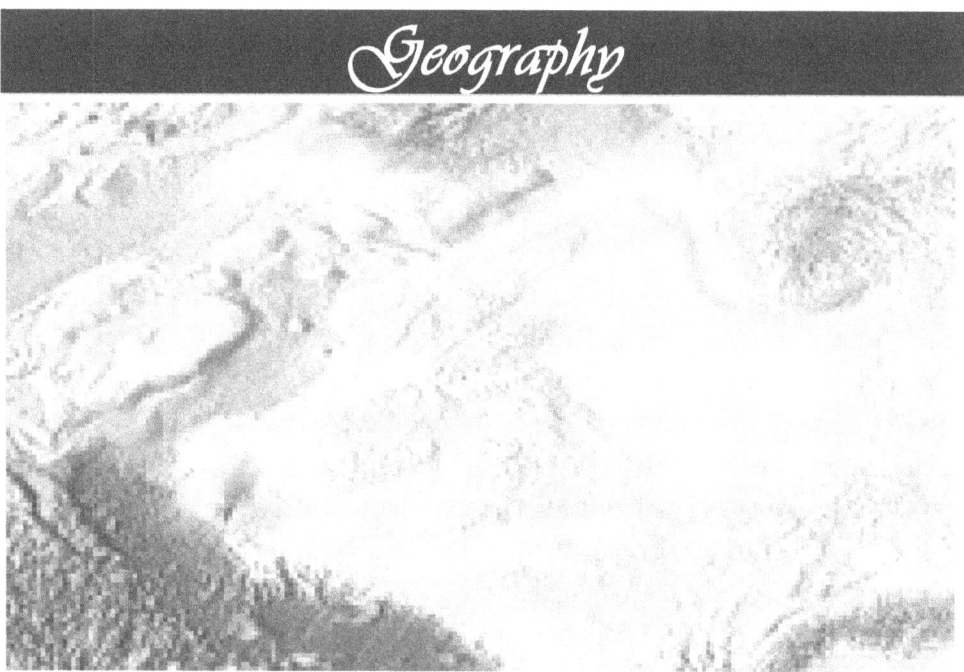

The British Isles in relation to the north-west European continental shelf.

The British Isles lie at the juncture of several regions with past episodes of tectonic mountain building. These orogenic belts form a complex geology that records a huge and varied span of Earth's history. Of particular note was the Caledonian Orogeny during the Ordovician Period, c. 488–444 Ma and early Silurian period, when the craton Baltica collided with the terrane Avalonia to form the mountains and hills in northern Britain and Ireland. Baltica formed roughly the northwestern half of Ireland and

Scotland. Further collisions caused the Variscan orogeny in the Devonian and Carboniferous periods, forming the hills of Munster, southwest England, and southern Wales. Over the last 500 million years the land that forms the islands has drifted northwest from around 30°S, crossing the equator around 370 million years ago to reach its present northern latitude.

The islands have been shaped by numerous glaciations during the Quaternary Period, the most recent being the Devensian. As this ended, the central Irish Sea was deglaciated and the English Channel flooded, with sea levels rising to current levels some 4,000 to 5,000 years ago, leaving the British Isles in their current form. Whether or not there was a land bridge between Great Britain and Ireland at this time is somewhat disputed, though there was certainly a single ice sheet covering the entire sea.

The west coasts of Ireland and Scotland that directly face the Atlantic Ocean are generally characterized by long peninsulas, and headlands and bays; the internal and eastern coasts are "smoother".

There are about 136 permanently inhabited islands in the group, the largest two being Great Britain and Ireland. Great Britain is to the east and covers 83,700 sq. mi (217,000 km^2). Ireland is to the west and covers 32,590 sq. mi (84,400 km^2). The largest of the other islands are to be found in the Hebrides, Orkney and Shetland to the north, Anglesey and the Isle of Man between Great Britain and Ireland, and the Channel Islands near the coast of France.

The islands are at relatively low altitudes, with central Ireland and southern Great Britain particularly low lying: the lowest point in the islands is Holme, Cambridgeshire at −2.75 m (−9.02 ft.). The Scottish Highlands in the northern part of Great Britain are mountainous, with Ben Nevis being the highest point on the islands at 1,343 m (4,406 ft.). Other mountainous areas include Wales and parts of Ireland, however only seven peaks in these areas reach above 1,000 m (3,281 ft.). Lakes on the islands are generally not large, although Lough Neagh in Northern Ireland is an exception, covering 150 square miles (390 km^2). The largest freshwater body in Great Britain (by area) is Loch Lomond at 27.5 square miles (71 km^2), and Loch Ness, by volume whilst Loch Morar is the deepest freshwater body in the British Isles, with a maximum depth of 310 m (1,017 ft.). There are a number of major rivers within the British Isles. The longest is the Shannon in Ireland at 224 mi (360 km). The river Severn at 220 mi (354 km) is the longest in Great Britain. The isles have a temperate marine climate. The North Atlantic Drift ("Gulf Stream") which flows from the Mexico brings with it significant moisture and raises temperatures 11 °C (20 °F) above the global average for the islands' latitudes. Winters are cool and wet, with summers mild and also wet. Most Atlantic depressions pass to the north of the islands, combined with the general westerly circulation and interactions with the landmass; this imposes an east-west variation in climate.

Flora and Fauna

Some female red deer in Killarney National Park, Ireland.

The islands enjoy a mild climate and varied soils, giving rise to a diverse pattern of vegetation. Animal and plant life is similar to that of the northwestern European continent. There are however, fewer numbers of species, with Ireland having even less. All native flora and fauna in Ireland is made up of species that migrated from elsewhere in Europe, and Great Britain in particular. The only window when this could have occurred was between the end of the last Ice Age (about 12,000 years ago) and when the land bridge connecting the two islands was flooded by sea (about 8,000 years ago).

As with most of Europe, prehistoric Britain and Ireland were covered with forest and swamp. Clearing began around 6000 BC and accelerated in medieval times. Despite this, Britain retained its primeval forests longer than most of Europe due to a small population and later development of trade and industry, and wood shortages were not a problem until the 17th century. By the 18th century, most of Britain's forests were consumed for shipbuilding or manufacturing charcoal and the nation was forced to import lumber from Scandinavia, North America, and the Baltic. Most forest land in Ireland is maintained by state forestation programmes. Almost all land outside urban areas is farmland. However, relatively large areas of forest remain in east and north Scotland and in southeast England. Oak, elm, ash and beech are amongst the most common trees in England. In Scotland, pine and birch are most common. Natural forests in Ireland are mainly oak, ash, wych elm, birch and pine. Beech and lime, though not native to Ireland, are also common there. Farmland hosts a variety of semi-natural vegetation of grasses and flowering plants. Woods, hedgerows, mountain slopes and marshes host heather, wild grasses, gorse and bracken.

Many larger animals, such as wolf, bear and the European elk are today extinct. However, some species such as deer are protected. Other small mammals, such as rabbits, foxes, badgers, hares, hedgehogs, and stoats, are very common and the European beaver has been reintroduced in parts of Scotland. Wild boar have also been reintroduced to parts of southern England, following escapes from boar farms and illegal releases. Many rivers contain otters and seals are common on coasts. Over 200 species of bird reside permanently and another 200 migrate. Common types are the common chaffinch, common blackbird, house sparrow and common starling; all small birds. Large birds are declining in number, except for those kept for game such as pheasant, partridge, and red grouse. Fish are abundant in the rivers and lakes, in particular salmon, trout, perch and pike. Sea fish include dogfish, cod, sole, pollock and bass, as well as mussels, crab and oysters along the coast. There are more than 21,000 species of insects.

Few species of reptiles or amphibians are found in Great Britain or Ireland. Only three snakes are native to Great Britain: the common European adder, the grass snake and the smooth snake; none are native to Ireland. In general, Great Britain has slightly more variation and native wild life, with weasels, polecats, wildcats, most shrews, moles, water voles, roe deer and common toads also being absent from Ireland. This pattern is also true for birds and insects. Notable exceptions include the Kerry slug and certain species of wood lice native to Ireland but not Great Britain.

Domestic animals include the Connemara pony, Shetland pony, English Mastiff, Irish wolfhound and many varieties of cattle and sheep.

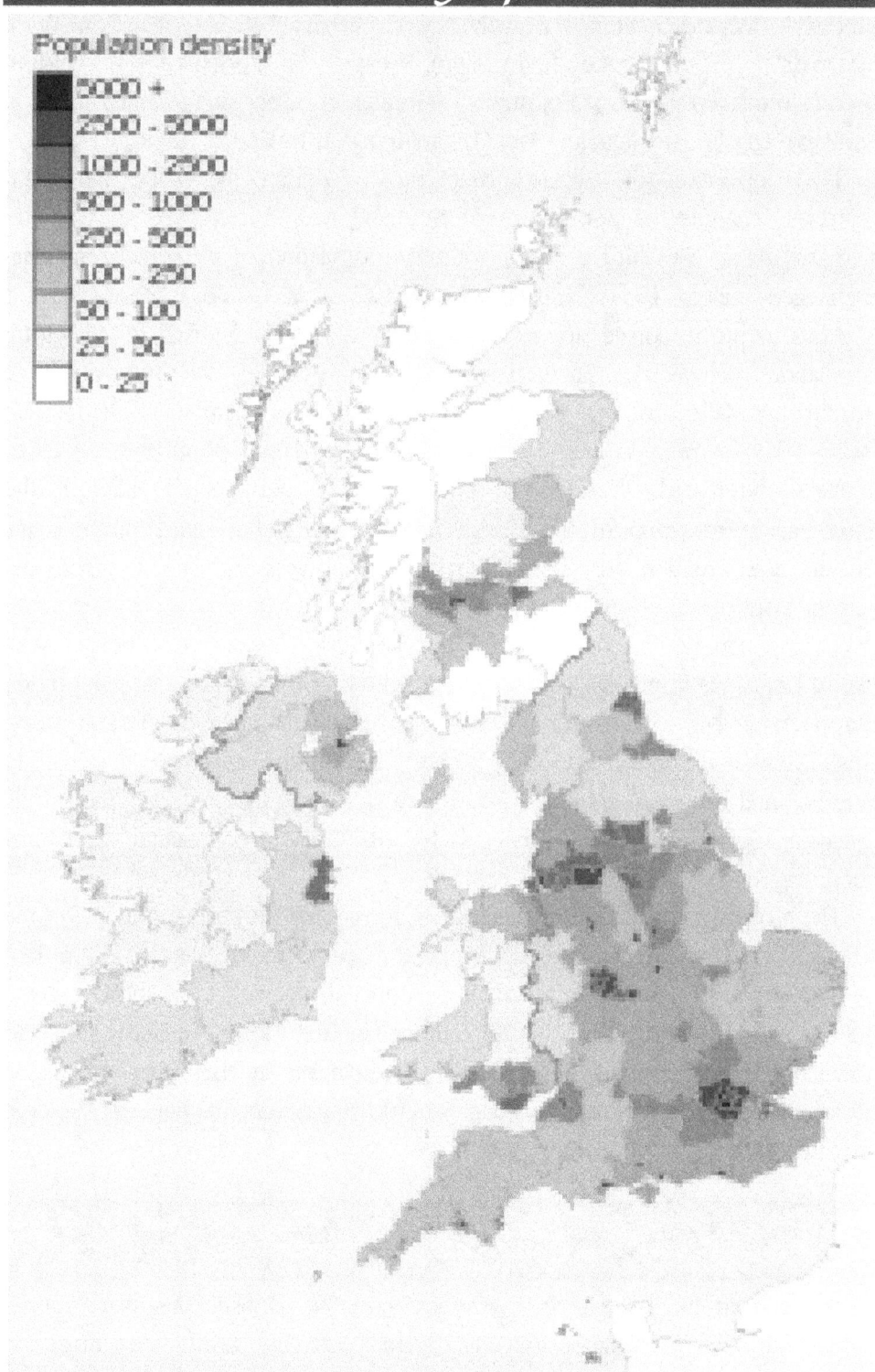

Population density

5000 +
2500 - 5000
1000 - 2500
500 - 1000
250 - 500
100 - 250
50 - 100
25 - 50
0 - 25

Population density per km² of the British Isles' regions.

The demographics of the British Isles today are characterized by a generally high density of population in England, which accounts for almost 80% of the total population of the islands. In elsewhere on Great Britain and on Ireland, high density of population is limited to areas around, or close to, a few large cities. The largest urban area by far is the Greater London Urban Area with 9 million inhabitants. Other major populations centres include Greater Manchester Urban Area (2.4 million), West Midlands conurbation (2.4 million), West Yorkshire Urban Area (1.6 million) in England, Greater Glasgow (1.2 million) in Scotland and Greater Dublin Area (1.1 million) in Ireland.

The population of England rose rapidly during the 19th and 20th centuries whereas the populations of Scotland and Wales have shown little increase during the 20th century, with the population of Scotland remaining unchanged since 1951. Ireland for most of its history comprised a population proportionate to its land area (about one third of the total population). However, since the Great Irish Famine, the population of Ireland has fallen to less than one tenth of the population of the British Isles. The famine, which caused a century-long population decline, drastically reduced the Irish population and permanently altered the demographic make-up of the British Isles. On a global scale, this disaster led to the creation of an Irish diaspora that numbers fifteen times the current population of the island.

The linguistic heritage of the British Isles is rich, with twelve languages from six groups across four branches of the Indo-European family. The Insular Celtic languages of the Goidelic sub-group (Irish, Manx and Scottish Gaelic) and the Brittonic sub-group (Cornish, Welsh and Breton, spoken in north-western France) are the only remaining Celtic languages—the last of their continental relations becoming extinct before the 7th century.

The Norman languages of Guernésiais, Jèrriais and Sarkese spoken in the Channel Islands are similar to French. A cant, called Shelta, is spoken by Irish Travellers, often as a means to conceal meaning from those outside the group. However, English, sometimes in the form of Scots, is the dominant language, with few monoglots remaining in the other languages of the region. The Norn language of Orkney and Shetland became extinct around 1880.

History

At the end of the last ice age, what are now the British Isles were joined to the European mainland as a mass of land extending north west from the modern-day northern coastline of France, Belgium and the Netherlands. Ice covered almost all of what is now Scotland, most of Ireland and Wales, and the hills of northern England. From 14,000 to 10,000 years ago, as the ice melted, sea levels rose separating Ireland from Great Britain and also creating the Isle of Man. About two to four millennia later, Great Britain became separated from the mainland. Britain probably became repopulated with people before the ice age ended and certainly before it became separated from the mainland. It is likely that Ireland became settled by sea after it had already become an island.

At the time of the Roman Empire, about two thousand years ago, various tribes, which spoke Celtic dialects of the Insular Celtic group, were inhabiting the islands. The Romans expanded their civilization to control southern Great Britain but were impeded in advancing any further, building Hadrian's Wall to mark the northern frontier of their empire in 122 AD. At that time, Ireland was populated by a people known as Hiberni, the northern third or so of Great Britain by a people known as Picts and the southern two thirds by Britons.

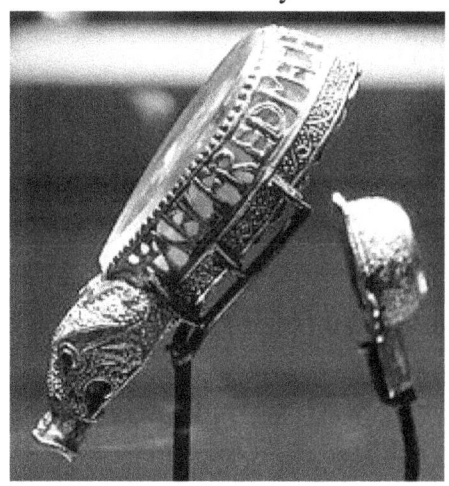

The Alfred Jewel (9th century)

Anglo-Saxons arrived as Roman power waned in the 5th century AD. Initially, their arrival seems to have been at the invitation of the Britons as mercenaries to repulse incursions by the Hiberni and Picts. In time, Anglo-Saxon demands on the British became so great that they came to culturally dominate the bulk of southern Great Britain, though recent genetic evidence suggests Britons still formed the bulk of the population. This dominance creating what is now England and leaving culturally British enclaves only in the north of what is now England, in Cornwall and what is now known as Wales. Ireland had been unaffected by the Romans except, significantly, having been Christianized, traditionally by the Romano-Briton, Saint Patrick. As Europe, including Britain, descended into turmoil following the collapse of Roman civilization, an era known as the Dark Ages, Ireland entered a golden age and responded with missions (first to Great Britain and then to the continent), the founding of monasteries and universities. These were later joined by Anglo-Saxon missions of a similar nature.

Viking invasions began in the 9th century, followed by more permanent settlements, particularly along the east coast of Ireland, the west coast of modern-day Scotland and the Isle of Man. Though the Vikings were eventually neutralized in Ireland, their influence remained in the cities of Dublin, Cork, Limerick, Waterford and Wexford. England however was slowly conquered around the turn of the first millennium AD, and eventually became a feudal possession of Denmark. The relations between the descendants of Vikings in England and counterparts in Normandy, in northern France, lay at the heart of a series of events that led to the Norman conquest of England in 1066. The remnants of the Duchy of Normandy, which conquered England, remain associated to the English Crown as the Channel Islands to this day. A century later the marriage of the future Henry II of England to Eleanor of Aquitaine created the Angevin Empire, partially under the French Crown. At the invitation of a provincial king and under the authority of Pope Adrian IV (the only Englishman to be elected pope), the Angevins invaded Ireland in 1169. Though initially intended to be kept as an independent kingdom, the failure of the to ensure the terms of the Treaty of Windsor led Henry II, as King of England, to rule as effective monarch under the title of Lord of Ireland. This title was granted to his younger son but when Henry's heir unexpectedly died the title of King of England and Lord of Ireland became entwined in one person.

By the Late Middle Ages, Great Britain was separated into the Kingdoms of England and Scotland. Power in Ireland fluxed between Gaelic kingdoms, Hiberno-Norman lords and the English-dominated Lordship of Ireland. A similar situation existed in the Principality of Wales, which was slowly being annexed into the Kingdom of England by a series of laws. During the course of the 15th century, the Crown of England would assert a claim to the Crown of France, thereby also releasing the King of England as from being vassal of the King of France. In 1534, King Henry VIII, at first having been a strong defender of Roman Catholicism in the face of the Reformation, separated from the Roman Church after failing to secure a divorce from the Pope. His response was to place the King of England as "the only Supreme Head in Earth of the Church of England", thereby removing the authority of the Pope from the affairs of the English Church. Ireland, which had been held by the King of England as Lord of Ireland, but which strictly speaking had been a feudal possession of the Pope since the Norman invasion was declared a separate kingdom in personal union with England.

Scotland meanwhile had remained an independent Kingdom. In 1603, that changed when the King of Scotland inherited the Crown of England, and consequently the Crown of Ireland also. The subsequent 17th century was one of political upheaval, religious division and war. English colonialism in Ireland of the 16th century was extended by large-scale Scottish and English colonies in Ulster. Religious division heightened and the King in England came into conflict with parliament. A prime issue was, inter alia, over his policy of tolerance towards Catholicism. The resulting English Civil War or War of the Three Kingdoms led to a revolutionary republic in England. Ireland, largely Catholic was mainly loyal to the king. Following defeat to the parliament's army, large scale land distributions from loyalist Irish nobility to English commoners in the service of the parliamentary army created the beginnings a new Ascendancy class which over the next hundred years would obliterate the English (Hiberno-Norman) and Gaelic Irish nobility in Ireland. The new ruling class was Protestant and British, whilst the common people were largely Catholic and Irish. This theme would influence Irish politics for centuries to come. When the monarchy was restored in England, the king found it politically impossible to restore all the lands of former land-owners in Ireland. The "Glorious Revolution" of 1688 repeated similar themes: a Catholic king pushing for religious tolerance in opposition to a Protestant parliament in England. The king's army was defeated at the Battle of the Boyne and at the militarily crucial Battle of Aughrim in Ireland. Resistance held out, and a guarantee of religious tolerance was a cornerstone of the Treaty of Limerick. However, in the evolving political climate, the terms of Limerick were superseded, a new monarchy was installed, and the new Irish parliament was packed with the new elite which legislated increasing intolerant Penal Laws, which discommoded both Dissenters and Catholics.

The Kingdoms of England and Scotland were unified in 1707 creating the Kingdom of Great Britain. Following an attempted republican revolution in Ireland in 1798, the Kingdoms of Ireland and Great Britain were unified in 1801, creating the United Kingdom. The Isle of Man and the Channel Islands remaining outside of the United Kingdom but with their ultimate good governance being the responsibility of the British Crown (effectively the British government). Although, the colonies of North American that would become the United States of America were lost by the start of the 19th century, the British Empire expanded rapidly elsewhere. A century later it would cover one third of the globe. Poverty in the United Kingdom remained desperate however and industrialization in England led to terrible condition for the working class. Mass migrations following the Irish Famine and Highland Clearances resulted in the distribution of the islands' population and culture throughout the world and a rapid de-population of Ireland in the second-half of the 19th century. Most of Ireland seceded from the United Kingdom after the Irish War of Independence and the subsequent Anglo-Irish Treaty (1919–1922), with the six counties that formed Northern Ireland remaining as an autonomous region of the UK.

Politics

There are two sovereign states in the isles: Ireland and the United Kingdom of Great Britain and Northern Ireland. Ireland, sometimes called *the Republic of Ireland*, governs five sixths of the island of Ireland, with the remainder of the island forming Northern Ireland. Northern Ireland is a part of the United Kingdom of Great Britain and Northern Ireland, usually shortened to simply *the United Kingdom*, which governs the remainder of the archipelago with the exception of the Isle of Man and the Channel Islands. The Isle of Man and the two states of the Channel Islands, Jersey and Guernsey, are known as the Crown Dependencies. They exercise constitutional rights of self-government and judicial independence; responsibility for international representation rests largely upon the UK (in consultation with the respective governments); and responsibility for defence is reserved by the UK. The United Kingdom is made up of four constituent parts: England, Scotland and Wales, forming Great Britain, and Northern in the north-east of the island of Ireland. Of these, Scotland, Wales and Northern Ireland have "devolved" governments meaning that they have their own parliaments/assemblies and are self-governing with respect to certain areas set down by law. For judicial purposes, Scotland, Northern Ireland and England and Wales (the latter being one entity) form separate legal jurisdiction, with there being no single law for the UK as a whole.

Ireland, the United Kingdom and the three Crown Dependencies are all parliamentary democracies, with their own separate parliaments. All parts of the United Kingdom return members to parliament in London. In addition to this, voters in Scotland, Wales and Northern Ireland return members to a parliament in Edinburgh and to assemblies in Cardiff and Belfast respectively. Governance in the norm is by majority rule, however, Northern Ireland uses a system of sharing whereby unionists and nationalists share executive posts proportionately and where the assent of both groups are required for the Northern Ireland Assembly to make certain decisions. (In the context of Northern Ireland, unionists are those who want Northern Ireland to remain a part of the United Kingdom and nationalists are those who want Northern Ireland join with the rest of Ireland.) The British monarch is the head of state for all parts of the isles except for the Republic of Ireland, where the head of state is the President of Ireland.

Ireland and the United Kingdom are both part of the European Union (EU). The Crown Dependencies are not a part of the EU however do participate in certain aspects that were negotiated as a part of the UK's accession to the EU. Neither the United Kingdom or Ireland are part of the Schengen area, that allow passport-free travel between EU members states. However, since the partition of Ireland, an informal free-travel area had existed across the region. In 1997, this area required formal recognition during the course of negotiations for the Amsterdam Treaty of the European Union and is now known as the Common Travel Area.

Reciprocal arrangements allow British and Irish citizens to full voting rights in the two states. Exceptions to this are presidential elections and constitutional referendums in the Republic of Ireland, for which there is no comparable franchise in the other states. In the United Kingdom, these pre-date European Union law, and in both jurisdictions go further than that required by European Union law. Other EU nationals may only vote in local and European Parliament elections while resident in either the UK or Ireland. In 2008, a UK Ministry of Justice report investigating how to strengthen the British sense of citizenship proposed to end this arrangement arguing that, "the right to vote is one of the hallmarks of the political status of citizens; it is not a means of expressing closeness between countries."

In addition, some civil bodies are organized throughout the islands as a whole. For example the Samaritans which is deliberately organized without regard to national boundaries on the basis that a service which is not political or religious should not recognize sectarian or political divisions. The RNLI, the life boats service, is also organized throughout the islands as a whole, covering the waters of the United Kingdom, Ireland, the Isle of Man, and the Channel Islands.

The Northern Ireland Peace Process has led to a number of unusual arrangements between the Republic of Ireland, Northern Ireland and the United Kingdom. For example, citizens of Northern Ireland are entitled to the choice of Irish or British citizenship or both and the Governments of Ireland and the United Kingdom consult on matters not devolved to the Northern. The Northern Ireland Executive and the Government of Ireland also meet as the North/South Ministerial Council to develop policies common across the island of Ireland. These arrangements were made following the 1998 Good Friday Agreement.

British-Irish Council

Another body established under the Good Friday Agreement, the British–Irish Council, is made up of all of the states and territories of the British Isles. The British–Irish Parliamentary Assembly (Irish: *Tionól Pharlaiminteach na Breataine agus na hÉireann*) predates the British–Irish Council and was established in 1990. Originally it comprised 25 members of the Oireachtas, the Irish parliament, and 25 members of the parliament of the United Kingdom, with the purpose of building mutual understanding between members of both legislatures. Since then the role and scope of the body has been expanded to include representatives from the Scottish Parliament, the National Assembly for Wales, the Northern Ireland Assembly, the States of Jersey, the States of Guernsey and the High Court of Tynwald (Isle of Man).

The Council does not have executive powers but meets biannually to discuss issues of mutual importance. Similarly, the Parliamentary Assembly has no legislative powers but investigates and collects witness evidence from the public on matters of mutual concern to its members. Reports on its findings are presented to the Governments of Ireland and the United Kingdom. During the February 2008 meeting of the British–Irish Council, it was agreed to set up a standing secretariat that would serve as a permanent 'civil service' for the Council. Leading on from developments in the British–Irish Council, the chair of the British–Irish Inter-Parliamentary Assembly, Niall Blaney, has suggested that the body should shadow the British–Irish Council's work.

Transportation

London Heathrow Airport is Europe's busiest airport in terms of passenger traffic and the Dublin-London route is the busiest air route in Europe. The English Channel and the southern North Sea are the busiest seaways in the world. The Channel, opened in 1994, links Great Britain to France and is the second-longest rail tunnel in the world.

The idea of building a tunnel under the Irish Sea has been raised since 1895, when it was first investigated. Several potential Irish Sea tunnel projects have been proposed, most recently the *Tusker Tunnel* between the ports of Rosslare and Fishguard proposed by The Institute of Engineers of Ireland in 2004. A rail tunnel was proposed in 1997 on a different route, between Dublin and Holyhead, by British engineering firm Symonds. Either tunnel, at 50 mi (80 km), would be by far the longest in the world, and would cost an estimated €20 billion. A proposal in 2007, estimated the cost of building a bridge from County Antrim in Northern Ireland to Galloway in Scotland at £3.5bn (€5bn). **(Wikipedia)**

100 Greatest Britons

"100 Greatest Britons" was broadcast in 2002 by the BBC. The programme was based on a television poll conducted to determine whom the United Kingdom public considered the greatest British people in history. The series, *Great Britons*, included individual programmes featuring the individuals who featured in the top ten, with viewers having further opportunities to vote after each programme. It concluded with a debate. All of the top 10 were dead by the year of broadcast.

The poll resulted in nominees including Guy Fawkes, who was executed for trying to blow up the England; Oliver who created a republican England; Richard III, suspected of murdering his nephews; James

Connolly, an Irish and socialist who was executed by the Crown in 1916; and a surprisingly high ranking of 17th for actor and singer Michael Crawford (the second highest-ranked entertainer, after John Lennon). Diana, Princess of Wales was judged to be a greater historical British figure than William Shakespeare by BBC respondents to the survey.

One of the more controversial figures to be included on the list was occultist Aleister Crowley. His works have had a direct influence on the rise in popular occultism and some forms of neopaganism in the 20th century. He is considered an influence on Gerald Gardner, founder of Gardnerian Wicca. In addition to the Britons, some notable non-British entrants were listed, including two Irish nationals, the philanthropic musicians Bono and Bob Geldof. The top 19 entries were people of English origin (though Sir Ernest Shackleton and Arthur Wellesley, 1st Duke of Wellington, were both born into Anglo-Irish families when what is now the Republic of Ireland was part of the United Kingdom). The highest-placed Scottish entry was Alexander in 20th place, with the highest Welsh entry, Owain Glyndŵr, at number 23. Sixty had lived in the twentieth century. The highest-ranked living person was Margaret Thatcher, placed 16th. Ringo Starr is the only member of The Beatles not on the list. Isambard Kingdom Brunel occupied the top spot in the polls for some time thanks largely to "students from Brunel University who have been campaigning vigorously for the engineer for weeks." However a late surge in the final week of voting put Churchill over the top.

The opening and closing ceremonies of the 2012 Summer Olympics featured two of the greatest Britons, Isambard Kingdom Brunel and Winston Churchill as main characters, played by Kenneth Branagh and Timothy Spall, each of them reading a monologue from William Shakespeare's *The Tempest*.

In addition, the ceremony also contained a personal appearance by Tim Berners-Lee, who was placed 99th on the list. There were no black Britons on the list, prompting a separate three-month survey to find the 100 greatest black Britons, with double Olympic decathlon gold medalist Daley Thompson the highest ranked track athlete on both lists.

Top 10 on the List

Because of the nature of the poll used to select and rank the Britons, the results do not claim to be an objective assessment. They are as follows:

Rank	Name	Time Frame	Image	Occupation	Notability

1	Sir Winston Churchill	1874–1965	Politician	Prime Minister during World War II historically ranked as one of the greatest British prime ministers.
2	Isambard Kingdom Brunel	1806–1859	Engineer	Creator of the Great Western Railway, and designer of numerous significant ships, tunnels and bridges.
3	Diana, Princess of Wales	1961–1997	Member of the British Royal family and philanthropist	First wife of Charles, Prince of Wales (marriage 1981–1996), and mother of Prince William, Duke of Cambridge, and Prince Harry.
4	Charles Darwin	1809–1882	Naturalist	Originator of the theory of evolution through natural selection and author of *On the Origin of Species.*
5	William Shakespeare	1564–1616	Poet and playwright	Thought of by many as the greatest of all English writers.
6	Sir Isaac Newton	1642–1727	Physicist, mathematician, astronomer, natural philosopher and biblical scholar	Originator of universal gravitation and laws of classical and laws of motion. His *Principia* is one of the most influential works in the history of science.

7	Queen Elizabeth I	1533–1603		Queen regnant	Popular monarch of England (reigned 1558–1603) who brought a period of relative internal stability. She is associated with the defeat of the Spanish Armada.
8	John Lennon	1940–1980		Composer, musician, philanthropist, peace activist, artist, and writer	Co-writer with Paul McCartney in The Beatles, the most successful band and music act of all time, and solo musician.
9	Vice-Admiral Horatio Nelson, 1st Viscount Nelson	1758–1805		Naval commander	Famous for his service in the **Royal**, particularly during the Napoleonic Wars.
10	Oliver Cromwell	1599–1658		Military and political leader	1st Lord Protector of the **Commonwealth**. Commander of the New Model Army during the **English** against King Charles I.

Full list

1. Sir Winston Churchill
2. Isambard Kingdom Brunel
3. Diana, Princess of Wales
4. Charles Darwin
5. William Shakespeare
6. Sir Isaac Newton
7. Elizabeth I
8. John Lennon
9. Horatio Nelson, 1st Viscount Nelson
10. Oliver Cromwell
21. Alan Turing

11. Sir Ernest Shackleton
12. Captain James Cook
13. Robert Baden-Powell
14. Alfred the Great
15. Arthur Wellesley, 1st Duke of Wellington
16. Margaret Thatcher
17. Michael Crawford
18. Queen Victoria
19. Sir Paul McCartney
20. Sir Alexander Fleming
22. Michael Faraday

23. Owain Glyndŵr
24. Elizabeth II
25. Stephen Hawking
26. William Tyndale
27. Emmeline Pankhurst
28. William Wilberforce
29. David Bowie
30. Guy Fawkes
31. Leonard Cheshire
32. Eric Morecambe
33. David Beckham
34. Thomas Paine
35. Boudica
36. Sir Steve Redgrave
37. Sir Thomas More
38. William Blake
39. John Harrison
40. Henry VIII
41. Charles Dickens
42. Sir Frank Whittle
43. John Peel
44. John Logie Baird
45. Aneurin Bevan
46. Boy George
47. Sir Douglas Bader
48. Sir William Wallace
49. Sir Francis Drake
50. John Wesley
51. King Arthur
52. Florence Nightingale
53. Thomas Edward Lawrence
54. Robert Falcon Scott
55. Enoch Powell
56. Sir Cliff Richard
57. Alexander Graham Bell
58. Freddie Mercury
59. Dame Julie Andrews
60. Sir Edward Elgar
61. Queen Elizabeth The Queen Mother
62. George Harrison
63. Sir David Attenborough

64. James Connolly
65. George Stephenson
66. Sir Charles Chaplin
67. Tony Blair
68. William Caxton
69. Bobby Moore
70. Jane Austen
71. William Booth
72. Henry V
73. Aleister Crowley
74. Robert the Bruce
75. Bob Geldof
76. The Unknown Warrior
77. Robbie Williams
78. Edward Jenner
79. David Lloyd George, 1st Earl Lloyd George of Dwyfor
80. Charles Babbage
81. Geoffrey Chaucer
82. Richard III
83. J. K. Rowling
84. James Watt
85. Sir Richard Branson
86. Bono
87. John Lydon (Johnny Rotten)
88. Bernard Law Montgomery, 1st Viscount Montgomery of Alamein ('Monty')
89. Donald Campbell
90. Henry II
91. James Clerk Maxwell
92. J. R. R. Tolkien
93. Sir Walter Raleigh
94. Edward I
95. Sir Barnes Wallis
96. Richard Burton
97. Tony Benn
98. David Livingstone
99. Sir Tim Berners-Lee
100.   Marie Stopes

(Wikipedia)

Genius

A **genius** is a person who displays exceptional intellectual ability, creativity, or originality, typically to a degree that is associated with the achievement of new advances in a domain of knowledge. A scholar in many subjects or a scholar in a single subject may be referred to as a genius. There is no scientifically precise definition of genius, and the question of whether the notion itself has any real meaning has long been a subject of debate, although psychologists are converging on a definition that emphasizes creativity and eminent achievement.

Etymology

In ancient Rome, the *genius* (plural in Latin *genii*) was the guiding spirit or tutelary deity of a person, family *(gens)*, or place *(genius loci)*. The noun is related to the Latin verb *genui, genitus*, "to bring into being, create, produce". Because the achievements of exceptional individuals seemed to indicate the presence of a particularly powerful *genius*, by the time of Augustus the word began to acquire its secondary meaning of "inspiration, talent". The term *genius* acquired its modern sense in the eighteenth century, and is a conflation of two Latin terms: *genius*, as above, and *ingenium*, a related noun referring to our innate dispositions, talents and inborn nature. Beginning to blend the concepts of the divine and the talented, the *Encyclopédie* article on genius (génie) describes such a person as "he whose soul is more expansive and struck by the feelings of all others; interested by all that is in nature never to receive an idea unless it evokes a feeling; everything excites him and on which nothing is lost."

Historical Development

Galton

The assessment of intelligence was initiated by Francis Galton (1822–1911) and James McKeen Cattell. They had advocated the analysis of reaction time and sensory acuity as measures of "neurophysiological efficiency" and the analysis of sensory acuity as a measure of intelligence.

Galton is regarded as the founder of psychometry. He studied the work of his older half-cousin Charles Darwin about biological evolution. Hypothesizing that eminence is inherited from ancestors; Galton did a study of families of eminent people in Britain, publishing it in 1869 as *Hereditary Genius*. Galton's ideas were elaborated from the work of two early 19th-century pioneers in statistics: Carl Friedrich Gauss and Adolphe Quetelet.

Gauss discovered the normal distribution (bell-shaped curve): given a large number of measurements of the same variable under the same conditions, they vary atrandom from a most frequent value, the "average," to two least frequent values at maximum differences greater and less than the most frequent value. Quetelet discovered that the bell-shaped curve applied to social statistics gathered by the French government in the course of its normal processes on large numbers of people passing through the courts and the military. His initial work in criminology led him to observe "the greater the number of individuals observed the more do peculiarities become effaced..." This ideal from which the peculiarities were effaced became "the average man".

Galton was inspired by Quetelet to define the average man as "an entire normal scheme"; that is, if one combines the normal curves of every measurable human characteristic, one will in theory perceive a syndrome straddled by "the average man" and flanked by persons that are different. In contrast to Quetelet, Galton's average man was not statistical, but was theoretical only. There was no measure of general averageness, only a large number of very specific averages. Setting out to discover a general measure of the average, Galton looked at educational statistics and found bell-curves in test results of all sorts; initially in mathematics grades for the final honors examination and in entrance examination scores for Sandhurst.

Galton's method in *Hereditary Genius* was to count and assess the eminent relatives of eminent men. He found that the number of eminent relatives was greater with closer degree of kinship. This work is considered the first example of historiometry, an analytical study of historical human progress. The work is controversial and has been criticized for several reasons. Galton then departed from Gauss in a way that became crucial to the history of the 20th century AD. The bell-shaped curve was not random, he concluded. The differences between the average and the upper end were due to a non-random factor, "natural ability," which he defined as "those qualities of intellect and disposition, which urge and qualify men to perform acts that lead to reputation ... a nature which, when left to itself, will, urged by an inherent stimulus, climb the path that leads to eminence." The apparent randomness of the scores was due to the randomness of this natural ability in the population as a whole, in theory.

Criticisms include that Galton's study fails to account for the impact of social status and the associated availability of resources in the form of economic inheritance, meaning that inherited "eminence" or "genius" can be gained through the enriched environment provided by wealthy families. Galton went on to develop the field of eugenics.

Genius is expressed in a variety of forms (e.g., mathematical, literary, musical performance). Persons with genius tend to have strong intuitions about their domains, and they build on these insights with tremendous energy. Carl Rogers, a founder of the Humanistic Approach to Psychology, expands on the idea of a genius trusting his or her intuition in a given field, writing: "El Greco, for example, must have realized as he looked at some of his early work, that 'good artists do not paint like that.' But somehow he trusted his own experiencing of life, the process of himself, sufficiently that he could go on expressing his own unique perceptions. It was as though he could say, 'Good artists don't paint like this, but *I* paint like this.' Or to move to another field, Ernest Hemingway was surely aware that "good writers do not write like this." But fortunately he moved toward being Hemingway, being himself, rather than toward someone else's conception of a good writer."

A number of people commonly regarded as geniuses have been diagnosed with mental disorders, for example Vincent van Gogh, Virginia Woolf, Jonathan Swift, John Forbes Nash, Jr, and Ernest Hemingway.

IQ and Genius

Galton was a pioneer in investigating both eminent human achievement and mental testing. In his book *Hereditary Genius*, written before the development of IQ testing, he proposed that hereditary influences on eminent achievement are strong, and that eminence is rare in the general population. Lewis Terman chose "'near' genius or genius" as the classification label for the highest classification on his 1916 version of the Stanford-Binet test. By 1926, Terman began publishing about a longitudinal study of California schoolchildren who were referred for IQ testing by their schoolteachers, called Genetic Studies of Genius, which he conducted for the rest of his life. Catherine M. Cox, a colleague of Terman's, wrote a whole book, *The Early Mental Traits of 300 Geniuses*, published as volume 2 of The Genetic Studies of Genius book series, in which she analyzed biographical data about historic geniuses. Although her estimates of childhood IQ scores of historical figures who never took IQ tests have been criticized on methodological grounds, Cox's study was thorough in finding out what else matters besides IQ in becoming a genius. By the 1937 second revision of the Stanford-Binet test, Terman no longer used the term "genius" as an IQ classification, nor has any subsequent IQ test. In 1939, David Wechsler specifically commented that "we are rather hesitant about calling a person a genius on the basis of a single intelligence test score".

The Terman longitudinal study in California eventually provided historical evidence regarding how genius is related to IQ scores. Many California pupils were recommended for the study by schoolteachers. Two pupils who were tested but rejected for inclusion in the study (because their IQ scores were too low) grew up to be Nobel Prize winners in physics, William, and Luis Walter Alvarez. Based on the historical findings of the Terman study and on biographical examples such as Richard Feynman, who had an IQ of 125 and went on to win the Nobel Prize in physics and become widely known as a genius, the current view of psychologists and other scholars of genius is that a minimum level of IQ (approximately 125) is necessary for genius but not sufficient, and must be combined with personality characteristics such as drive and persistence, plus the necessary opportunities for talent development.

Philosophy

Leonardo da Vinci is widely acknowledged as having been a genius and a polymath.

Various philosophers have proposed definitions of what genius is and what that implies in the context of their philosophical theories.

Wolfgang Amadeus Mozart, prodigy and music genius

In the philosophy of David Hume, the way society perceives genius is similar to the way society perceives the ignorant. Hume states that a person with the characteristics of a genius is looked at as a person disconnected from society, as well as a person who works remotely, at a distance, away from the rest of the world. "On the other hand, the mere ignorant is still more despised; nor is any thing deemed a surer sign of an illiberal genius in an age and nation where the sciences flourish, than to be entirely destitute of all relish for those noble entertainments. The most perfect character is supposed to lie between those extremes; retaining an equal ability and taste for books, company, and business; preserving in conversation that discernment and delicacy which arise from polite letters; and in business, that probity and accuracy which are the natural result of a just philosophy."

In the philosophy of Immanuel Kant, genius is the ability to independently arrive at and understand concepts that would normally have to be taught by another person. For Kant, originality was the essential character of genius. This genius is a talent for producing ideas which can be described as non-imitative. Kant's discussion of the characteristics of genius is largely contained within the *Critique of Judgement* and was well received by the Romantics of the early 19th century. In addition, much of Schopenhauer's theory of genius, particularly regarding talent and freedom from constraint, is directly derived from paragraphs of Part I of Kant's *Critique of Judgment*.

Genius is a talent for producing something for which no determinate rule can be given, not a predisposition consisting of a skill for something that can be learned by following some rule or other.

— *Immanuel Kant*

In the philosophy of Arthur Schopenhauer, a genius is someone in whom intellect predominates over "will" much more than within the average person. In Schopenhauer's aesthetics, this predominance of the intellect over the will allows the genius to create artistic or academic works that are objects of pure, disinterested contemplation, the chief criterion of the aesthetic experience for Schopenhauer.

Their remoteness from mundane concerns means that Schopenhauer's geniuses often display maladaptive traits in more mundane concerns; in Schopenhauer's words, they fall into the mire while gazing at the stars, an allusion to Plato's dialogue *Theætetus*, in which Socrates tells of Thales (the first philosopher) being ridiculed for falling in such circumstances. As he says in Volume 2 of *The World as Will and Representation*:

Talent hits a target no one else can hit; Genius hits a target no one else can see.

– Arthur Schopenhauer

In the philosophy of Bertrand Russell, genius entails that an individual possesses unique qualities and talents that make the genius especially valuable to the society in which he or she operates. However, Russell's philosophy further maintains that it's possible for such a genius to be crushed by an unsympathetic environment during his or her youth. Russell rejected the notion he believed was popular during his lifetime that, "genius will out."

1. Ludwig Van Beethoven - 1770-1827
2. Wolfgang Amadeus Mozart - 1756-1791
3. Johann Sebastian Bach - 1685-1750
4. Richard Wagner - 1813-1883
5. Joseph Haydn - 1732-1809
6. Johannes Brahms - 1833-1897
7. Franz Schubert - 1797-1828
8. Peter Ilyich Tchaikovsky - 1840-1893
9. George Frideric Handel - 1685-1759
10. Igor Stravinsky - 1882-1971
11. Robert Schumann - 1810-1856
12. Frederic Chopin - 1810-1849
13. Felix Mendelssohn - 1809-1847
14. Claude Debussy - 1862-1918
15. Franz Liszt - 1811-1886
16. Antonin Dvorak - 1841-1904
17. Giuseppe Verdi - 1813-1901
18. Gustav Mahler - 1860-1911
19. Hector Berlioz - 1803-1869

20. *Antonio Vivaldi - 1678-1741*
21. *Richard Strauss - 1864-1949*
22. *Serge Prokofiev - 1891-1953*
23. *Dmitri Shostakovich - 1906-1975*
24. *Béla Bartók - 1881-1945*
25. *Anton Bruckner - 1824-1896*
26. *Giovanni Pierluigi da Palestrina - 1525-1594*
27. *Claudio Monteverdi - 1567-1643*
28. *Jean Sibelius - 1865-1957*
29. *Maurice Ravel - 1875-1937*
30. *Ralph Vaughan Williams - 1872-1958*
31. *Modest Mussorgsky - 1839-1881*
32. *Giacomo Puccini - 1858-1924*
33. *Henry Purcell - 1659-1695*
34. *Gioacchino Rossini - 1792-1868*
35. *Edward Elgar - 1857-1934*
36. *Sergei Rachmaninoff - 1873-1943*
37. *Camille Saint-Saëns - 1835-1921*
38. *Josquin Des Prez - c.1440-1521*
39. *Nikolai Rimsky-Korsakov - 1844-1908*
40. *Carl Maria von Weber - 1786-1826*
41. *Jean-Philippe Rameau - 1683-1764*
42. *Jean-Baptiste Lully - 1632-1687*
43. *Gabriel Fauré - 1845-1924*
44. *Edvard Grieg - 1843-1907*
45. *Christoph Willibald Gluck - 1714-1787*
46. *Arnold Schoenberg - 1874-1951*
47. *Charles Ives - 1874-1954*
48. *Paul Hindemith - 1895-1963*
49. *Olivier Messiaen - 1908-1992*
50. *Aaron Copland - 1900-1990*
51. *Francois Couperin - 1668-1733*
52. *William Byrd - 1539-1623*
53. *Erik Satie - 1866-1925*
54. *Benjamin Britten - 1913-1976*
55. *Bedrick Smetana - 1824-1884*
56. *César Franck - 1822-1890*
57. *Alexander Nikolayevich Scriabin - 1872-1915*
58. *Georges Bizet - 1838-1875*
59. *Domenico Scarlatti - 1685-1757*
60. *Georg Philipp Telemann - 1681-1767*
61. *Anton Webern - 1883-1945*
62. *Roland de Lassus - 1532-1594*

63. George Gershwin - 1898-1937
64. Gaetano Donizetti - 1797-1848
65. Carl Philipp Emanuel Bach - 1714-1788
66. Archangelo Corelli - 1653-1713
67. Thomas Tallis - 1505-1585
68. Jules Massenet - 1842-1912
69. Johann Strauss II - 1825-1899
70. Leos Janácek - 1854-1928
71. Guillaume de Machaut - 1300-1377
72. Alban Berg - 1885-1935
73. Alexander Borodin - 1833-1887
74. Vincenzo Bellini - 1801-1835
75. Charles Gounod - 1818-1893
76. Francis Poulenc - 1899-1963
77. Giovanni Gabrieli - 1554-1612
78. Pérotin - 1160-1225
79. Heinrich Schütz - 1585-1672
80. John Cage - 1912-1992
81. Giovanni Battista Pergolesi - 1710-1736
82. John Dowland - 1563-1626
83. Gustav Holst - 1874-1934
84. Dietrich Buxtehude - 1637-1707
85. Ottorino Respighi - 1879-1936
86. Guillaume Dufay - 1400-1474
87. Hugo Wolf - 1860-1903
88. Carl Nielsen - 1865-1931
89. William Walton - 1902-1983
90. Darius Milhaud - 1892-1974
91. Orlando Gibbons - 1583-1625
92. Giacomo Meyerbeer - 1791-1864
93. Samuel Barber - 1910-1981
94. Tomás Luis de Victoria - 1549-1611
95. Léonin - 1135-1201
96. Manuel de Falla - 1876-1946
97. Hildegard von Bingen - 1098-1179
98. Mikhail Glinka - 1804-1857
(Wikipedia)

The British people originated in Doggerland around 12,000 B.C., and began constructing Stonehenge around 3100 B.C., at the same time they were building the Indus Valley Civilization; Mesopotamia Civilization, and the Ancient Egyptian Civilization; which represent the triune nature of The Lord; The Father, the Son, the Holy Ghost.

Wherever The British people travel, they leave cultural treasures; Homo sapiens then steal them, and convince themselves they're responsible for their creation, due to being self-deluding creatures by nature, and therein lies the secret to understanding the rise and fall of civilizations; nothing would be possible without
The British

Sir Issac Newton

He saw further because he stood on the shoulders of Giants before.

Sir Isaac Newton PRS (25 December 1642 – 20 March 1726/27) was an English physicist and mathematician (described in his own day as a "natural philosopher") who is widely recognized as one of the most influential scientists of all time and as a key figure in the scientific revolution. His book *Philosophiæ Naturalis Principia Mathematica* ("Mathematical Principles of Natural Philosophy"), first published in 1687, laid the foundations for classical mechanics. Newton made seminal contributions to optics, and he shares credit with Gottfried Leibniz for the development of calculus.

Newton's *Principia* formulated the laws of motion and universal gravitation, which dominated scientists' view of the physical universe for the next three centuries. By deriving Kepler's laws of planetary motion from his mathematical description of gravity, and then using the same principles to account for the trajectories of comets, the tides, the precession of the equinoxes, and other phenomena, Newton removed the last doubts about the validity of the heliocentric model of the Solar System. This work also demonstrated that the motion on Earth and of celestial bodies could be described by the same principles. His prediction that Earth should be shaped as an oblate spheroid was later vindicated by the measurements of Maupertuis, La Condamine, and others, which helped convince most Continental European scientists of the superiority of Newtonian mechanics over the earlier system of Descartes.

Newton built the first practical reflecting telescope and developed a theory of colour based on the observation that a prism decomposes white light into the many colours of the visible spectrum. He formulated an empirical law of cooling, studied the speed of sound, and introduced the notion of a Newtonian fluid. In addition to his work on calculus, as a mathematician Newton contributed to the study of power series, generalized the binomial theorem to non-integer exponents, developed a method for approximating the roots of a function, and classified most of the cubic plane curves.

Newton was a fellow of Trinity College and the second Lucasian Professor of Mathematics at the University of Cambridge. He was a devout but unorthodox Christian and, unusually for a member of the Cambridge faculty of the day, he refused to take holy orders in the Church of England, perhaps because he privately rejected the doctrine of the Trinity. Beyond his work on the mathematical sciences, Newton dedicated much of his time to the study of biblical and alchemy, but most of his work in those areas remained unpublished until long after his death. In his later life, Newton became president of the Royal Society. Newton served the British government as Warden and Master of the Royal Mint.

Life

Early life

Isaac Newton was born according to the Julian calendar (in use in England at the time) on Christmas Day, 25 December 1642 (NS 4 January 1643), at Woolsthorpe Manor in Woolsthorpe-by-Colsterworth, a hamlet in the county of Lincolnshire. He was born three months after the death of his father, a prosperous farmer also named Isaac Newton. Born prematurely, he was a small child; his mother Hannah Ayscough reportedly said that he could have fit inside a quart mug. When Newton was three, his mother remarried and went to live with her new husband, the Reverend Barnabas Smith, leaving her son in the care of his maternal grandmother, Margery Ayscough. The young Isaac disliked his stepfather and maintained some enmity towards his mother for marrying him, as revealed by this entry in a list of sins committed up to the age of 19: "Threatening my father and mother Smith to burn them and the house over them." Newton's mother had three children from her second marriage. Although it was claimed that he was once engaged, Newton never married.

Isaac Newton (Bolton, Sarah K. Famous Men of Science. NY: Thomas Y. Crowell & Co., 1889)

From the age of about twelve until he was seventeen, Newton was educated at The King's School, Grantham which taught him Latin but no mathematics. He was removed from school, and by

October 1659, he was to be found at Woolsthorpe-by-Colsterworth, where his mother, widowed for a second time, attempted to make a farmer of him. Newton hated farming. Henry Stokes, master at the King's School, persuaded his mother to send him back to school so that he might complete his education. Motivated partly by a desire for revenge against a schoolyard bully, he became the top-ranked student, distinguishing himself mainly by building sundials and models of windmills.

In June 1661, he was admitted to Trinity College, Cambridge, on the recommendation of his uncle Rev William Ayscough. He started as a subsizar—paying his way by performing valet's duties—until he was awarded a scholarship in 1664, which guaranteed him four more years until he would get his M.A. At that time, the college's teachings were based on those of Aristotle, whom Newton supplemented with modern philosophers such as Descartes, and astronomers such as Galileo and Thomas Street, through whom he learned of Kepler's work. He set down in his notebook a series of 'Quaestiones' about mechanical philosophy as he found it. In 1665, he discovered the generalized binomial theorem and began to develop a mathematical theory that later became calculus. Soon after Newton had obtained his B.A. degree in August 1665, the university temporarily closed as a precaution against the Great Plague. Although he had been undistinguished as a Cambridge student, Newton's private studies at his home in Woolsthorpe over the subsequent two years saw the development of his theories on calculus, optics, and the law of gravitation.

In April 1667, he returned to Cambridge and in October was elected as a fellow of Trinity. Fellows were required to become ordained priests, although this was not enforced in the restoration years and an assertion of conformity to the Church of England was sufficient. However, by 1675 the issue could not be avoided and by then his unconventional views stood in the way. Nevertheless, Newton managed to avoid it by means of a special permission from Charles II (see "Middle years" section below).

His studies had impressed the Lucasian professor, Isaac Barrow, who was more anxious to develop his own religious and administrative potential (he became master of Trinity two years later), and in 1669, Newton succeeded him, only one year after he received his M.A. He was elected a Fellow of the Royal Society (FRS) in 1672.

Middle Years

Mathematics

Newton's work has been said "to distinctly advance every branch of mathematics then studied". His work on the subject usually referred to as fluxions or calculus, seen in a manuscript of October 1666, is now published among Newton's mathematical papers. The author of the manuscript *De analysi per aequationes numero terminorum infinitas*, sent by Isaac Barrow to John Collins in June 1669, was identified by Barrow in a letter sent to Collins in August of that year as:

Mr. Newton, a fellow of our College, and very young ... but of an extraordinary genius and proficiency in these things.

Newton later became involved in a dispute with Leibniz over priority in the development of calculus (the Leibniz–Newton calculus controversy). Most modern historians believe that Newton and Leibniz developed calculus independently, although with very different notations. Occasionally it has been suggested that Newton published almost nothing about it until 1693, and did not give a full account until 1704, while Leibniz began publishing a full account of his methods in 1684. (Leibniz's notation and "differential Method", nowadays recognized as much more convenient notations, were adopted by continental European mathematicians, and after 1820 or so, also by British mathematicians.) Such a suggestion, however, fails to notice the content of calculus which critics of Newton's time and modern times have pointed out in Book 1 of Newton's *Principia* itself (published 1687) and in its forerunner manuscripts, such as *De motu corporum in gyrum* ("On the motion of bodies in orbit"), of 1684. The *Principia* is not written in the language of calculus either as we know it or as Newton's (later) 'dot' notation would write it. His work extensively uses calculus in geometric form based on limiting values of the ratios of vanishing small quantities: in the *Principia* itself, Newton gave demonstration of this under the name of 'the method of first and last ratios' and explained why he put his expositions in this form, remarking also that 'hereby the same thing is performed as by the method of indivisibles'.

Because of this, the *Principia* has been called "a book dense with the theory and application of the infinitesimal calculus" in modern times and "lequel est presque tout de ce calcul" ('nearly all of it is of this calculus') in Newton's time. His use of methods involving "one or more orders of the infinitesimally small" is present in his *De motu corporum in gyrum* of 1684 and in his papers on motion "during the two decades preceding 1684". Newton had been reluctant to publish his calculus because he feared controversy and criticism. He was close to the Swiss mathematician Nicolas Fatio de Duillier. In 1691, Duillier started to write a new version of Newton's *Principia*, and corresponded with Leibniz. In 1693, the relationship between Duillier and Newton deteriorated and the book was never completed.

Starting in 1699, other members of the Royal Society (of which Newton was a member) accused Leibniz of plagiarism. The dispute then broke out in full force in 1711 when the Royal Society proclaimed in a study that it was Newton who was the true discoverer and labeled Leibniz a fraud. This study was cast into doubt when it was later found that Newton himself wrote the study's concluding remarks on Leibniz. Thus began the bitter controversy which marred the lives of both Newton and Leibniz until the latter's death in 1716.

Newton is generally credited with the generalized binomial theorem, valid for any exponent. He discovered Newton's identities, Newton's method, classified cubic plane curves (polynomials of degree three in two variables), made substantial contributions to the theory of finite differences, and was the first to use fractional indices and to employ coordinate geometry to derive solutions to Diophantine equations. He approximated partial sums of the harmonic series by logarithms(a precursor to Euler's summation formula) and was the first to use power series with confidence and to revert power series. Newton's work on infinite series was inspired by Simon Stevin's decimals. A very useful modern account of Newton's mathematics was written by the foremost scholar on Newton's mathematics, D.T. Whiteside or Tom Whiteside. Tom Whiteside translated and edited all of Newton's mathematical writings and at the end of his life wrote a summing up of Newton's work and its impact. This was published in 2013 as a chapter in a book edited by Bechler.

When Newton received his MA and became a Fellow of the "College of the Holy and Undivided Trinity" in 1667, he made the commitment that "I will either set Theology as the object of my studies and will take holy orders when the time prescribed by these statutes [7 years] arrives, or I will resign from the college." Up till this point he had not thought much about religion and had twice signed his agreement to the thirty-nine articles, the basis of Church of England doctrine.

He was appointed Lucasian Professor of Mathematics in 1669 on Barrow's recommendation. During that time, any Fellow of a college at Cambridge or Oxford was required to take holy orders and become an ordained Anglican priest. However, the terms of the Lucasian professorship required that the holder *not* be active in the church (presumably so as to have more time for science). Newton argued that this should exempt him from the ordination requirement, and Charles II, whose permission was needed, accepted this argument. Thus a conflict between Newton's religious views and Anglican orthodoxy was averted.

Optics

In 1666, Newton observed that the spectrum of colours exiting a prism in the position of minimum deviation is oblong, even when the light ray entering the prism is circular, which is to say, the prism refracts different colours by different angles. This led him to conclude that colour is a property intrinsic to light—a point which had been debated in prior years.

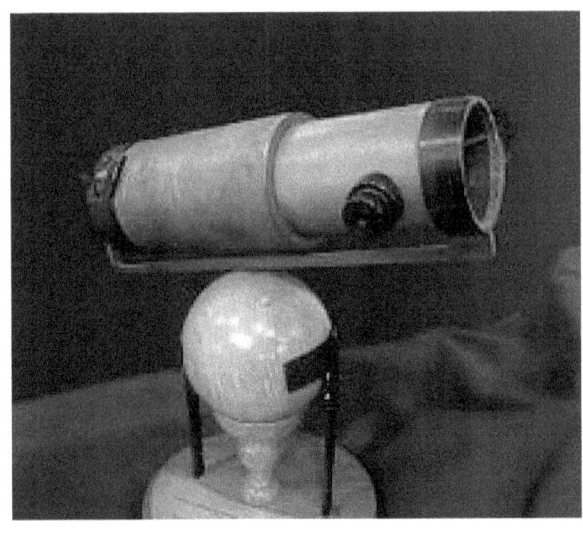

Replica of Newton's second Reflecting that he presented to the Royal Society in 1672[

From 1670 to 1672, Newton lectured on optics. During this period he investigated the refraction of light, demonstrating that the multicoloured spectrum produced by a prism could be recomposed into white light by a lens and a second prism. Modern scholarship has revealed that Newton's analysis and resynthesis of white light owes a debt to corpuscular alchemy.

He also showed that coloured light does not change its properties by separating out a coloured beam and shining it on various objects. Newton noted that regardless of whether it was reflected, scattered, or transmitted, it remained the same colour. Thus, he observed that colour is the result of objects interacting with already-coloured light rather than objects generating the colour themselves. This is known as Newton's theory of colour.

Illustration of a prism decomposing white light into the colours of the spectrum, as discovered by Newton

From this work, he concluded that the lens of any refracting telescope would suffer from the dispersion of light into colours (chromatic aberration). As a proof of the concept, he constructed a telescope using a mirror as the objective to bypass that problem. Building the design, the first known functional reflecting telescope, today known as a Newtonian telescope, involved solving the problem of a suitable mirror material and shaping technique. Newton ground his own mirrors out of a custom composition of highly reflective speculum, using Newton's rings to judge the quality of the optics for his telescopes. In late 1668 he was able to produce this first *reflecting telescope*. In 1671, the Royal Society asked for a demonstration of his reflecting telescope. Their interest encouraged him to publish his notes, *Of Colours*, which he later expanded into the work *Opticks*. When Robert Hooke criticized some of Newton's ideas, Newton was so offended that he withdrew from public debate. Newton and Hooke had brief exchanges in 1679–80, when Hooke, appointed to manage the Royal Society's correspondence, opened up a correspondence intended to elicit contributions from Newton to Royal Society transactions, which had the effect of stimulating Newton to work out a proof that the elliptical form of planetary orbits would result from a centripetal force inversely proportional to the square of the radius vector (see Newton's law of universal gravitation – History and *De motu corporum in gyrum*). But the two men remained generally on poor terms until Hooke's death.

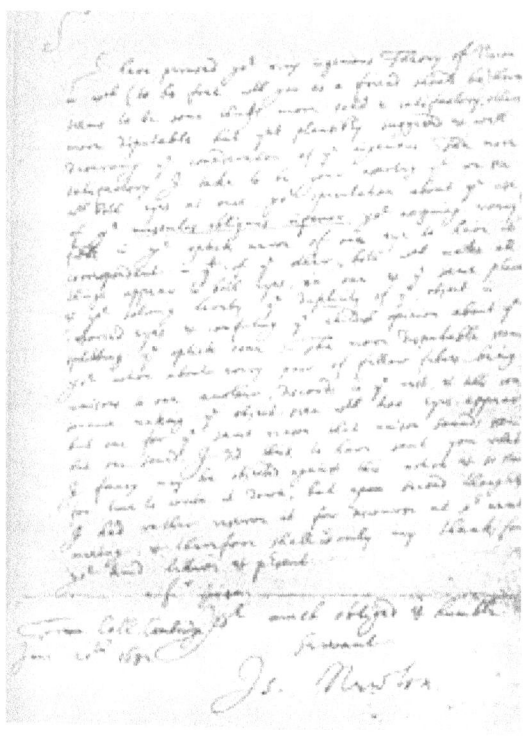

Facsimile of a 1682 letter from Isaac Newton to Dr William Briggs, commenting on Briggs' "A New Theory of Vision"

Newton argued that light is composed of particles or corpuscles, which were refracted by accelerating into a denser medium. He verged on sound like waves to explain the repeated pattern of reflection and transmission by thin films (Opticks Bk.II, Props. 12), but still retained his theory of 'fits' that disposed corpuscles to be reflected or transmitted (Props.13). However, later physicists favoured a purely wavelike explanation of light to account for the interference patterns and the general phenomenon of diffraction. Today's quantum mechanics, photons, and the idea of wave bear only a minor resemblance to Newton's understanding of light.

In his *Hypothesis of Light* of 1675, Newton posited the existence of the ether to transmit forces between particles. The contact with the theosophist Henry More revived his interest in alchemy. He replaced the ether with occult forces based on Hermetic ideas of attraction and repulsion between particles. John Maynard Keynes, who acquired many of Newton's writings on alchemy, stated that "Newton was not the first of the age of reason: He was the last of the magicians." Newton's interest in alchemy cannot be isolated from his contributions to science. This was at a time when there was no clear distinction between alchemy and science. Had he not relied on the occult idea of action at a distance, across a vacuum, he might not have developed his theory of gravity. (See also Isaac Newton's occult studies.)

In 1704, Newton published *Opticks*, in which he expounded his corpuscular theory of light. He considered light to be made up of extremely subtle corpuscles, that ordinary matter was made of grosser corpuscles and speculated that through a kind of alchemical transmutation "Are not gross Bodies and Light convertible into one another, ... and may not Bodies receive much of their Activity from the Particles of Light which enter their Composition?" Newton also constructed a primitive form of a frictional electrostatic generator, using a glass globe.

In an article entitled "Newton, prisms, and the 'opticks' of tunable lasers" it is indicated that Newton in his book *Opticks* was the first to show a diagram using a prism as a beam expander. In the same book he describes, via diagrams, the use of multiple-prism arrays. Some 278 years after Newton's discussion, multiple-prism beam expanders became central to the development of narrow-line width tunable lasers, also, the use of these prismatic beam expanders led to the multiple-prism dispersion theory; subsequent to Newton much has been amended.

Young, and Fresnel, combined Newton's particle theory with Huygens 'wave theory to show that colour is the visible manifestation of light's wavelength. Science also slowly came to realise the difference between perception of colour and mathematisable optics. The German poet and scientist, Goethe, could not shake the Newtonian foundation but "one hole Goethe did find in Newton's armour, ... Newton had committed himself to the doctrine that refraction without colour was impossible. He therefore thought that the object-glasses of telescopes must for ever remain imperfect, achromatism and refraction being incompatible. This inference was proved by Dollond to be wrong.

Mechanics and Gravitation

In 1679, Newton returned to his work on (celestial) mechanics by considering gravitation and its effect on the orbits of planets with reference to Kepler's laws of planetary motion. This followed stimulation by a brief exchange of letters in 1679–80 with Hooke, who had been appointed to manage the Royal Society's correspondence, and who opened a correspondence intended to elicit contributions from Newton to Royal Society transactions. Newton's reawakening interest in astronomical matters received further stimulus by the appearance of a comet in the winter of

1680–1681, on which he corresponded with John Flamsteed. After the exchanges with Hooke, Newton worked out proof that the elliptical form of planetary orbits would result from a centripetal force inversely proportional to the square of the radius vector (see Newton's law of universal gravitation – History and *De motu corporum in gyrum*). Newton communicated his results to Edmond Halley and to the Royal Society in *De motu corporum in gyrum*, a tract written on about nine sheets which was copied into the Royal Society's Register Book in December 1684. This tract contained the nucleus that Newton developed and expanded to form the *Principia*.

The *Principia* was published on 5 July 1687 with encouragement and financial help from Edmond Halley. In this work, Newton stated the three universal laws of motion. Together, these laws describe the relationship between any object, the forces acting upon it and the resulting motion, laying the foundation for classical mechanics. They contributed to many advances during the Industrial Revolution which soon followed and were not improved upon for more than 200 years. Many of these advancements continue to be the underpinnings of non-relativistic technologies in the modern world. He used the Latin word *gravitas* (weight) for the effect that would become known as gravity, and defined the law of universal gravitation.

In the same work, Newton presented a calculus-like method of geometrical analysis using 'first and last ratios', gave the first analytical determination (based on Boyle's law) of the speed of sound in air, inferred the oblateness of Earth's spheroidal figure, accounted for the precession of the equinoxes as a result of the Moon's gravitational attraction on the Earth's oblateness, initiated the gravitational study of the irregularities in the motion of the moon, provided a theory for the determination of the orbits of comets, and much more.

Newton made clear his heliocentric view of the Solar System—developed in a somewhat modern way, because already in the mid-1680s he recognized the "deviation of the Sun" from the centre of gravity of the Solar System. For Newton, it was not precisely the centre of the Sun or any other body that could be considered at rest, but rather "the common centre of gravity of the Earth, the Sun and all the Planets is to be esteem'd the Centre of the World", and this centre of gravity "either is at rest or moves uniformly forward in a right line" (Newton adopted the "at rest" alternative in view of common consent that the centre, wherever it was, was at rest).

Newton's postulate of an invisible force able to act over vast distances led to him being criticised for introducing "occult agencies" into science. Later, in the second edition of the *Principia* (1713), Newton firmly rejected such criticisms in a concluding General Scholium, writing that it was enough that the phenomena implied a gravitational attraction, as they did; but they did not so far indicate its cause, and it was both unnecessary and improper to frame hypotheses of things that were not implied by the phenomena. (Here Newton used what became his famous expression *"hypotheses non-fingo"*).

With the *Principia*, Newton became internationally recognized. He acquired a circle of admirers, including the Swiss-born mathematician Nicolas Fatio de Duillier, with whom he formed an intense relationship. This abruptly ended in 1693, and at the same time Newton suffered a nervous breakdown.

Classification of Cubics and Beyond

Descartes was the most important early influence on Newton the mathematician. Descartes freed plane curves from the Greek and Macedonian limitation to conic sections, and Newton followed his lead by classifying the cubic curves in the plane. He found 72 of the 78 species of cubics. He also divided them into four types, satisfying different equations, and in 1717 Stirling, probably with Newton's help, proved that every cubic was one of these four types. Newton also claimed that the four types could be obtained by plane projection from one of them, and this was proved in 1731.

According to Tom Whiteside (1932–2008), who published 8 volumes of Newton's mathematical papers, it is no exaggeration to say that Newton mapped out the development of mathematics for the next 200 years, and that Euler and others largely carried out his plan.

Later Life

Isaac Newton in old age in 1712, portrait by Sir James Thornhill

In the 1690s, Newton wrote a number of religious tracts dealing with the literal and symbolic interpretation of the Bible. A manuscript Newton sent to John Locke in which he disputed the fidelity of 1 John 5:7 and its fidelity to the original manuscripts of the New Testament, remained unpublished until 1785.

Even though a number of authors have claimed that the work might have been an indication that Newton disputed the belief in Trinity, others assure that Newton did question the passage but never denied Trinity as such. His biographer, scientist Sir David Brewster, who compiled his manuscripts for over 20 years, wrote about the controversy in well-known book *Memoirs of the Life, Writings, and Discoveries of Sir Isaac Newton*, where he explains that Newton questioned the veracity of those passages, but he never denied the doctrine of Trinity as such. Brewster states that Newton was never known as an Arian during his lifetime, it was first William Whiston (an Arian) who argued that "Sir Isaac Newton was so hearty for the Baptists, as well as for the Eusebians or Arians, that he sometimes suspected these two were the two witnesses in the Revelations," while other like Hopton Haynes (a Mint employee and Humanitarian), "mentioned to Richard Baron, that Newton held the same doctrine as himself".

Later works—*The Chronology of Ancient Kingdoms Amended* (1728) and *Observations Upon the Prophecies of Daniel and the Apocalypse of St. John* (1733)—were published after his death. He also devoted a great deal of time to alchemy (see above).

Newton was also a member of the Parliament of England for Cambridge University in 1689–90 and 1701–2, but according to some accounts his only comments were to complain about a cold draught in the chamber and request that the window be closed.

Newton moved to London to take up the post of warden of the Royal Mint in 1696, a position that he had obtained through the patronage of Charles Montagu, 1st Earl of Halifax, then Chancellor of the Exchequer. He took charge of England's great recoining, somewhat treading on the toes of Lord Lucas, Governor of the Tower (and securing the job of deputy comptroller of the temporary Chester branch for Edmond Halley). Newton became perhaps the best-known Master of the Mint upon the death of Thomas Neale in 1699, a position Newton held for the last 30 years of his life. These appointments were intended as sinecures, but Newton took them seriously, retiring from his Cambridge duties in 1701, and exercising his power to reform the currency and punish clippers and counterfeiters.

As Warden, and afterwards Master, of the Royal Mint, Newton estimated that 20 percent of the coins taken in during the Great were counterfeit. Counterfeiting was high treason, punishable by the felon's being hanged, drawn and quartered. Despite this, convicting even the most flagrant criminals could be extremely difficult. However, Newton proved equal to the task.

Disguised as a habitué of bars and taverns, he gathered much of that evidence himself. For all the barriers placed to prosecution, and separating the branches of government, English law still had ancient and formidable customs of authority. Newton had himself made a justice of the peace in all the home counties—there is a draft of a letter regarding this matter stuck into Newton's personal first edition of his *Philosophiæ Naturalis Principia Mathematica* which he must have been amending at the time. Then he conducted more than 100 cross-examinations of witnesses, informers, and suspects between June 1698 and Christmas 1699. Newton successfully prosecuted 28 coiners.

As a result of a report written by Newton on 21 September 1717 to the Lords Commissioners of His Majesty's Treasury the bimetallic relationship between gold coins and silver coins was changed by Royal proclamation on 22 December 1717, forbidding the exchange of gold guineas for more than 21 silver shillings. This inadvertently resulted in a silver shortage as silver coins were used to pay for imports, while exports were paid for in gold, effectively moving Britain from the silver standard to its first gold standard. It is a matter of debate as whether he intended to do this or not. It has been argued that Newton conceived of his work at the Mint as a continuation of his alchemical work.

Newton was made President of the Royal Society in 1703 and an associate of the French Académie des Sciences. In his position at the Royal Society, Newton made an enemy of John Flamsteed, the Astronomer Royal, by prematurely publishing Flamsteed's *Historia Coelestis Britannica*, which Newton had used in his studies.

Personal coat of arms of Sir Isaac Newton

In April 1705, Queen Anne knighted Newton during a royal visit to Trinity College, Cambridge. The knighthood is likely to have been motivated by political considerations connected with the Parliamentary election in May 1705, rather than any recognition of Newton's scientific work or services as Master of the Mint. Newton was the second scientist to be knighted, after Sir Francis Bacon.

Newton was one of many people who lost heavily when the South Sea Company collapsed. Their most significant trade was slaves, and according to his niece, he lost around £20,000.

Towards the end of his life, Newton took up residence at Cranbury Park, near Winchester with his niece and her husband, until his death in 1727. His half-niece, Catherine Barton Conduitt, served as his hostess in social affairs at his house on Jermyn Street in London; he was her "very loving Uncle," according to his letter to her when she was recovering from smallpox.

Newton died in his sleep in London on 20 March 1727
(OS 20 March 1726; NS 31 March 1727)
and was buried in Westminster.
Voltaire may have been present at his funeral.
A bachelor; he had divested much of his estate to relatives during his last years, and died intestate. After his death, Newton's hair was examined and found to contain mercury, probably resulting from his alchemical pursuits. Mercury poisoning could explain Newton's eccentricity in late life.

Personal Relations

Newton never married. The French writer and philosopher Voltaire, who was in London at the time of Newton's funeral, said that he "was never sensible to any passion, was not subject to the common frailties of mankind, nor had any commerce with women—a circumstance which was assured me by the physician and surgeon who attended him in his last moments".The widespread belief that he died a virgin has been commented on by writers such as mathematician Charles Hutton, economist John Maynard Keynes, and physicist Carl Sagan.

Newton did have a close friendship with the Swiss mathematician Nicolas Fatio de Duillier, whom he met in London around 1690. Their friendship came to an unexplained end in 1693. Some of their correspondence has survived.

In September of that year, Newton had a breakdown which included sending wild accusatory letters to his friends Samuel Pepys and John Locke. His note to the latter included the charge that Locke "endeavoured to embroil me with woemen".

After Death

Fame

The mathematician Joseph-Louis Lagrange often said that Newton was the greatest genius who ever lived, and once added that Newton was also "the most fortunate, for we cannot find more than once a system of the world to establish." English poet Alexander Pope was moved by Newton's accomplishments to write the famous epitaph:

> *Nature and nature's laws lay hid in night;*
> *God said "Let Newton be" and all was light.*

Newton himself had been rather more modest of his own achievements, famously writing in a letter to Robert Hooke in February 1676:

If I have seen further it is by standing on the shoulders of giants.

Two writers think that the above quotation, written at a time when Newton and Hooke were in dispute over optical discoveries, was an oblique attack on Hooke (said to have been short and hunchbacked), rather than— or in addition to—a statement of modesty. On the other hand, the widely known proverb about standing on the shoulders of giants, published among others by seventeenth-century poet George Herbert (a former orator of the University of Cambridge and fellow of Trinity College) in his *Jacula Prudentum* (1651), had as its main point that "a dwarf on a giant's shoulders sees farther of the two", and so its effect as an analogy would place Newton himself rather than Hooke as the 'dwarf'.

In a later memoir, Newton wrote:

I do not know what I may appear to the world, but to myself I seem to have been only like a boy playing on the sea-shore, and diverting myself in now and then finding a smoother pebble or a prettier shell than ordinary, whilst the great ocean of truth lay all undiscovered before me.

In 1816, a tooth said to have belonged to Newton was sold for £730 (US$3,633) in London to an aristocrat who had it set in a ring. The *Guinness World Records 2002* classified it as the most valuable tooth, which would value approximately £25,000 (US$35,700) in late 2001. Who bought it and who currently has it has not been disclosed.

Albert Einstein kept a picture of Newton on his study wall alongside ones of Michael Faraday and James Clerk Maxwell. Newton remains influential to today's scientists, as demonstrated by a 2005 survey of members of Britain's Royal Society (formerly headed by Newton) asking who had the greater effect on the history of science, Newton or Einstein. Royal Society scientists deemed Newton to have made the greater overall contribution. In 1999, an opinion poll of 100 of today's leading physicists voted Einstein the "greatest physicist ever;" with Newton the runner-up, while a parallel survey of rank-and-file physicists by the site PhysicsWeb gave the top spot to Newton.

Commemorations

Newton's monument (1731) can be seen in Westminster Abbey, at the north of the entrance to the choir against the choir screen, near his tomb. It was executed by the sculptor Michael Rysbrack (1694–1770) in white and grey marble with design by the architect William Kent. The monument features a figure of Newton reclining on top of a sarcophagus, his right elbow resting on several of his great books and his left hand pointing to a scroll with a mathematical design. Above him is a pyramid and a celestial globe showing the signs of the Zodiac and the path of the comet of 1680. A relief panel depicts putti using instruments such as a telescope and prism. The Latin inscription on the base translates as:

Here is buried Isaac Newton, Knight, who by a strength of mind almost divine, and mathematical principles peculiarly his own, explored the course and figures of the planets, the paths of comets, the tides of the sea, the dissimilarities in rays of light, and, what no other scholar has previously imagined, the properties of the colours thus produced. Diligent, sagacious and faithful, in his expositions of nature, antiquity and the holy Scriptures, he vindicated by his philosophy the majesty of God mighty and good, and expressed the simplicity of the Gospel in his manners. Mortals rejoice that there has existed such and so great an ornament of the human race! He was born on 25 December 1642, and died on 20 March 1726/7.—Translation from G.L. Smyth, *The Monuments and Genii of St. Paul's Cathedral, and of Westminster Abbey* (1826), ii, 703–4.

From 1978 until 1988, an image of Newton designed by Harry Ecclestone appeared on Series D £1 banknotes issued by the Bank of England (the last £1 notes to be issued by the Bank of England). Newton was shown on the reverse of the notes holding a book and accompanied by a telescope, a prism and a map of the Solar System.

A statue of Isaac Newton, looking at an apple at his feet, can be seen at the Oxford University Museum of Natural History. A large bronze statue, *Newton, after William Blake*, by Eduardo Paolozzi, dated 1995 and inspired by Blake's etching, dominates the piazza of the British Library in London.

Although born into an Anglican family, by his thirties Newton held a Christian faith that, had it been made public, would not have been considered orthodox by mainstream Christianity; in recent times he has been described as a heretic.

By 1672 he had started to record his theological researches in notebooks which he showed to no one and which have only recently been examined. They demonstrate an extensive knowledge of early church writings and show that in the conflict between Athanasius and Arius which defined the Creed, he took the side of Arius, the loser, who rejected the conventional view of the Trinity. Newton "recognized Christ as a divine mediator between God and man, who was subordinate to the Father who created him." He was especially interested in prophecy, but for him, "the great apostasy was trinitarianism."

Newton tried unsuccessfully to obtain one of the two fellowships that exempted the holder from the ordination requirement. At the last moment in 1675 he received a dispensation from the government that excused him and all future holders of the Lucasian chair.

In Newton's eyes, worshipping Christ as God was idolatry, to him the fundamental sin. Historian Stephen D. Snobelen says of Newton, "Isaac Newton was a heretic. But ... he never made a public declaration of his private faith—which the orthodox would have deemed extremely radical. He hid his faith so well that scholars are still unravelling his personal beliefs." Snobelen concludes that Newton was at least a Socinian sympathiser (he owned and had thoroughly read at least eight Socinian books), possibly an Arian and almost certainly an anti-trinitarian.

In a minority view, T.C. Pfizenmaier argues that Newton held the Eastern Orthodox view on the Trinity. However, this type of view 'has lost support of late with the availability of Newton's theological papers', and now most scholars identify Newton as an Antitrinitarian monotheist.

Although the laws of motion and universal gravitation became Newton's best-known discoveries, he warned against using them to view the Universe as a mere machine, as if akin to a great clock. He said, "Gravity explains the motions of the planets, but it cannot explain who set the planets in motion. God governs all things and knows all that is or can be done."

Along with his scientific fame, Newton's studies of the Bible and of the early Church Fathers were also noteworthy. Newton wrote works on textual criticism, most notably *An Historical Account of Two Notable Corruptions of Scripture*. He placed the crucifixion of Jesus Christ at 3 April, AD 33, which agrees with one traditionally accepted date.

He believed in a rationally immanent world, but he rejected the hylozoism implicit in Leibniz and Baruch Spinoza. The ordered and dynamically informed Universe could be understood, and must be understood, by an active reason. In his correspondence, Newton claimed that in writing the *Principia* "I had an eye upon such Principles as might work with considering men for the belief of a Deity". He saw evidence of design in the system of the world: "Such a wonderful uniformity in the planetary system must be allowed the effect of choice". But Newton insisted that divine intervention would eventually be required to reform the system, due to the slow growth of instabilities. For this, Leibniz lampooned him: "God Almighty wants to wind up his watch from time to time: otherwise it would cease to move. He had not, it seems, sufficient foresight to make it a perpetual motion."

Newton's position was vigorously defended by his follower Samuel Clarke in a famous correspondence. A century later, Pierre-Simon Laplace's work "Celestial Mechanics" had a natural explanation for why the planet orbits don't require periodic divine intervention.

Effect on Religious Thought

Newton and Robert Boyle's approach to the mechanical philosophy was promoted by rationalist pamphleteers as a viable alternative to the pantheists and enthusiasts, and was accepted hesitantly by orthodox preachers as well as dissident preachers like the latitudinarians. The clarity and simplicity of science was seen as a way to combat the emotional and metaphysical superlatives of both superstitious enthusiasm and the threat of atheism, and at the same time, the second wave of English deists used Newton's discoveries to demonstrate the possibility of a "Natural Religion".

Newton, by William Blake; here, Newton is depicted critically as a "divine geometer".

The attacks made against pre-Enlightenment "magical thinking", and the mystical elements of Christianity, were given their foundation with Boyle's mechanical conception of the Universe. Newton gave Boyle's ideas their completion through mathematical proofs and, perhaps more importantly, was very successful in popularising them.

Newton saw God as the master creator whose existence could not be denied in the face of the grandeur of all creation.

Occult

In a manuscript he wrote in 1704 in which he describes his attempts to extract scientific information from the Bible, he estimated that the world would end no earlier than 2060. In predicting this he said, "This I mention not to assert when the time of the end shall be, but to put a stop to the rash conjectures of fanciful men who are frequently predicting the time of the end, and by doing so bring the sacred prophesies into discredit as often as their predictions fail.

Alchemy

Newton wrote about alchemy. All of Newton's known writings on alchemy are currently being put online in a project undertaken by Indiana University: "The Chymistry of Isaac Newton". Here is a quote from the project web site.

Newton's fundamental contributions to science include the quantification of gravitational attraction, the discovery that white light is actually a mixture of immutable spectral colors, and the formulation of the calculus. Yet there is another, more mysterious side to Newton that is imperfectly known, a realm of activity that spanned some thirty years of his life, although he kept it largely hidden from his contemporaries and colleagues. We refer to Newton's involvement in the discipline of alchemy, or as it was often called in seventeenth-century England, "chymistry." Newton wrote and transcribed about a million words on the subject of alchemy.

The project is headed by William R. Newman. Newman presented a lecture entitled "Why did Isaac Newton Believe in Alchemy?" at the Perimeter Institute, in 2010. Speculative fiction author Fritz Leiber said of Newton, "Everyone knows Newton as the great scientist. Few remember that he spent half his life muddling with alchemy, looking for the philosopher's stone. That was the pebble by the seashore he really wanted to find."

Enlightenment Philosophers

Enlightenment philosophers chose a short history of scientific predecessors – Galileo, Boyle, and Newton principally – as the guides and guarantors of their applications of the singular concept of Nature and Natural law to every physical and social field of the day. In this respect, the lessons of history and the social structures built upon it could be discarded.

It was Newton's conception of the Universe based upon Natural and rationally understandable laws that became one of the seeds for Enlightenment ideology. Locke and Voltaire applied concepts of Natural Law to political systems advocating intrinsic rights; the physiocrats and Adam Smith applied Natural conceptions of psychology and self-interest to economic systems; and sociologists criticized the current social order for trying to fit history into Natural models of progress. Monboddo and Samuel Clarke resisted elements of Newton's work, but eventually rationalized it to conform with their strong religious views of nature.

Apple Incident

Reputed descendants of Newton's apple tree, (from top to bottom) at Trinity College, Cambridge, the Cambridge University Botanic Garden, and the Instituto Balseiro library garden

Newton himself often told the story that he was inspired to formulate his theory of gravitation by watching the fall of an apple from a tree. Although it has been said that the apple story is a myth and that he did not arrive at his theory of gravity in any single moment, acquaintances of Newton (such as William Stukeley, whose manuscript account of 1752 has been made available by the Royal Society) do in fact confirm the incident, though not the cartoon version that the apple actually hit Newton's head. Stukeley recorded in his *Memoirs of Sir Isaac Newton's Life* a conversation with Newton in Kensington on 15 April 1726:

we went into the garden, & drank tea under the shade of some apple trees; only he, & my self. amidst other discourse, he told me, he was just in the same situation, as when formerly; the notion of gravitation came into his mind. "why should that apple always descend perpendicularly to the ground," thought he to himself; occasion'd by the fall of an apple, as he sat in a contemplative mood. "why should it not go sideways, or upwards? but constantly to the earths center? assuredly, the reason is, that the earth draws it. there must be a drawing power in matter. & the sum of the drawing power in the matter of the earth must be in the earths center, not in any side of the earth. therefore dos this apple fall perpendicularly, or toward the center. if matter thus draws matter; it must be in proportion of its quantity. therefore the apple draws the earth, as well as the earth draws the apple.

John Conduitt, Newton's assistant at the Royal Mint and husband of Newton's niece, also described the event when he wrote about Newton's life:

In the year 1666 he retired again from Cambridge to his mother in Lincolnshire. Whilst he was pensively meandering in a garden it came into his thought that the power of gravity (which brought an apple from a tree to the ground) was not limited to a certain distance from earth, but that this power must extend much further than was usually thought. Why not as high as the Moon said he to himself & if so, that must influence her motion & perhaps retain her in her orbit, whereupon he fell a calculating what would be the effect of that supposition.

In similar terms, Voltaire wrote in his *Essay on Epic Poetry* (1727), "Sir Isaac Newton walking in his gardens, had the first thought of his system of gravitation, upon seeing an apple falling from a tree."

It is known from his notebooks that Newton was grappling in the late 1660s with the idea that terrestrial gravity extends, in an inverse-square proportion, to the Moon; however it took him two decades to develop the full-fledged theory. The question was not whether gravity existed, but whether it extended so far from Earth that it could also be the force holding the Moon to its orbit. Newton showed that if the force decreased as the inverse square of the distance, one could indeed calculate the Moon's orbital period, and get good agreement. He guessed the same force was responsible for other orbital motions, and hence named it "universal gravitation".

Various trees are claimed to be "the" apple tree which Newton describes. The King's School, Grantham, claims that the tree was purchased by the school, uprooted and transported to the headmaster's garden some years later. The staff of the National Trust-owned Woolsthorpe Manor, dispute this, and claim that a tree present in their gardens is the one described by Newton. A descendant of the original tree can be seen growing outside the main gate of Trinity College, Cambridge, below the room Newton lived in when he studied there. The National Fruit Collection at Brogdale can supply grafts from their tree, which appears identical to Flower of Kent, a coarse-fleshed cooking variety.

Works

Published in his lifetime

- *De analysi per aequationes numero terminorum infinitas (1669, published 1711)*
- *Method of Fluxions (1671)*
- *Of Natures Obvious Laws & Processes in Vegetation (unpublished, c. 1671–75)[146]*
- *De motu corporum in gyrum (1684)*
- *Philosophiæ Naturalis Principia Mathematica (1687)*
- *Opticks (1704)*
- *Reports as Master of the Mint (1701–25)*
- *Arithmetica Universalis (1707)*

Published posthumously

- *The System of the World (1728)*
- *Optical Lectures (1728)*
- *The Chronology of Ancient Kingdoms Amended (1728)*
- *De mundi systemate (1728)*
- *Observations on Daniel and The Apocalypse of St. John (1733)*
- *Newton, Isaac (1991). Robinson, Arthur B., ed. Observations upon the Prophecies of Daniel, and the Apocalypse of St. John. Cave Junction, Oregon: Oregon Institute of Science and Medicine. ISBN 0-942487-02-8. (A facsimile edition of the 1733 work.)*

- *An Historical Account of Two Notable Corruptions of Scripture (1754)*

Decline, Revival, and the Modern Legend

Post-medieval literature

The end of the Middle Ages brought with it a waning of interest in King Arthur. Although Malory's English version of the great French romances was popular, there were increasing attacks upon the truthfulness of the historical framework of the Arthurian romances – established since Geoffrey of Monmouth's time – and thus the legitimacy of the whole Matter of Britain. So, for example, the 16th-century humanist scholar Polydore Vergil famously rejected the claim that Arthur was the ruler of a post-Roman empire, found throughout the post-Galfridian medieval 'chronicle tradition', to the horror of Welsh and English antiquarians. Social changes associated with the end of the medieval period and the Renaissance also conspired to rob the character of Arthur and his associated legend of some of their power to enthrall audiences, with the result that 1634 saw the last printing of Malory's *Le Morte d'Arthur* for nearly 200 years. King Arthur and the Arthurian legend were not entirely abandoned, but until the early 19th century the material was taken less seriously and was often used simply as a vehicle for allegories of 17th- and 18th-century politics. Thus Richard's epics *Prince Arthur* (1695) and *King Arthur* (1697) feature Arthur as an allegory for the struggles of William against James II. Similarly, the most popular Arthurian tale throughout this period seems to have been that of Tom, which was told first through chapbooks and later through the political plays of Henry Fielding; although the action is clearly set in Arthurian Britain, the treatment is humorous and Arthur appears as a primarily comedic version of his romance character.

John Dryden's masque *King Arthur* is still performed, largely thanks to Henry Purcell's music, though seldom unabridged.

Tennyson and the Revival

In the early 19th century, medievalism, Romanticism, and the Revival reawakened interest in Arthur and the medieval romances. A new code of ethics for 19th-century gentlemen was shaped around the chivalric ideals embodied in the "Arthur of romance". This renewed interest first made itself felt in 1816, when Malory's *Le Morte d'Arthur* was reprinted for the first time since 1634. Initially, the medieval Arthurian legends were of particular interest to poets, inspiring, for example, William Wordsworth to write "The Egyptian Maid" (1835), an allegory of the Holy Grail. Pre-eminent among these was Alfred Lord Tennyson, whose first Arthurian poem "The Lady of Shalott" was published in 1832. Arthur himself played a minor role in some of these works, following in the medieval romance tradition. Tennyson's Arthurian work reached its peak of popularity with *Idylls of the King*, however, which reworked the entire narrative of Arthur's life for the Victorian era. It was first published in 1859 and sold 10,000 copies within the first week. In the *Idylls*, Arthur became a symbol of ideal manhood who ultimately failed, through human weakness, to establish a perfect kingdom on earth. Tennyson's works prompted a large number of imitators, generated considerable public interest in the legends of Arthur and the character himself, and brought Malory's tales to a wider audience. Indeed, the first modernization of Malory's great compilation of Arthur's tales was published in 1862, shortly after *Idylls* appeared, and there were six further editions and five competitors before the century ended.

Gustave Doré's illustration of Camelot for Alfred, Lord Tennyson's Idylls of the King (1868)

This interest in the 'Arthur of romance' and his associated stories continued through the 19th century and into the 20th, and influenced poets such as William Morris and Pre-Raphaelite artists including Edward Burne-Jones. Even the humorous tale of Tom Thumb, which had been the primary manifestation of Arthur's legend in the 18th century, was rewritten after the publication of *Idylls*. While Tom maintained his small stature and remained a figure of comic relief, his story now included more elements from the medieval Arthurian romances and Arthur is treated more seriously and historically in these new versions. The revived Arthurian romance also proved influential in the United States, with such books as Sidney Lanier's *The Boy's King Arthur* (1880) reaching wide audiences and providing inspiration for Mark Twain's satiric *A Connecticut Yankee in King Arthur's Court* (1889). Although the 'Arthur of romance' was sometimes central to these new Arthurian works (as he was in Burne-Jones's "The Sleep of Arthur in Avalon", 1881-1898), on other occasions he reverted to his medieval status and is either marginalized or even missing entirely, with Wagner's Arthurian operas providing a notable instance of the latter. Furthermore, the revival of interest in Arthur and the Arthurian tales did not continue unabated. By the end of the 19th century, it was confined mainly to Pre-Raphaelite imitators, and it could not avoid being affected by the First World War, which damaged the reputation of chivalry and thus interest in its medieval manifestations and Arthur as chivalric role model. The romance tradition did, however, remain sufficiently powerful to persuade Hardy, Laurence and John Masefield to compose Arthurian plays, and T. S. Eliot alludes to the Arthur myth (but not Arthur) in his poem *The Waste Land*, which mentions the Fisher King.

Modern Legend

The combat of Arthur and Mordred, illustrated by N.C. Wyeth for Sidney Lanier's The Boy's King Arthur (1922)

In the latter half of the 20th century, the influence of the romance tradition of Arthur continued, through novels such as T. H. White's *The Once and Future King* (1958) and Marion's *The Mists of Avalon* (1982) in addition to comic strips such as *Prince Valiant* (from 1937 onward). Tennyson had reworked the romance tales of Arthur to suit and comment upon the issues of his day, and the same is often the case with modern treatments too. Bradley's tale, for example, takes a feminist approach to Arthur and his legend, in contrast to the narratives of Arthur found in medieval materials, and American authors often rework the story of Arthur to be more consistent with values such as equality and democracy. The romance Arthur has become popular in film and theatre as well. T. H. White's novel was adapted into the Lerner and Loewe stage musical *Camelot* (1960) and the Disney animated film *The Sword in the Stone* (1963); *Camelot*, with its focus on the love of Lancelot and Guinevere and the cuckolding of Arthur, was itself made into a film of the same name in 1967. The romance tradition of Arthur is particularly evident and, according to critics, successfully handled in Robert Bresson's *Lancelot du Lac* (1974), Eric Rohmer's *Perceval le Gallois* (1978) and perhaps John Boorman's fantasy film *Excalibur* (1981); it is also the main source of the material utilized in the Arthurian spoof *Monty Python and the Holy Grail* (1975).

The Death of Arthur, by John Garrick (1862)

Re-tellings and re-imaginings of the romance tradition are not the only important aspect of the modern legend of King Arthur. Attempts to portray Arthur as a genuine historical figure of c. 500, stripping away the "romance", have also emerged. As Taylor and Brewer have noted, this return to the medieval "chronicle tradition" of Geoffrey of Monmouth and the *Historia Brittonum* is a recent trend which became dominant in Arthurian literature in the years following the outbreak of the Second World War, when Arthur's legendary resistance to Germanic invaders struck a chord in Britain. Clemence Dane's series of radio plays, *The Saviours* (1942), used a historical Arthur to embody the spirit of heroic resistance against desperate odds, and Robert Sherriff's play *The Long Sunset* (1955) saw Arthur rallying Romano-British resistance against the Germanic invaders. This trend towards placing Arthur in a historical setting is also apparent in historical and fantasy novels published during this period. In recent years the portrayal of Arthur as a real hero of the 5th century has also made its way into film versions of the Arthurian legend, most notably the TV series *Arthur of the*

Britons (1972–73), *The Legend of King Arthur* (1979), and *Camelot* (2011) and the feature films *King Arthur* (2004) and *The Last Legion* (2007).

Arthur has also been used as a model for modern-day behaviour. In the 1930s, the Order of the Fellowship of the Knights of the Round Table was formed in Britain to promote Christian ideals and Arthurian notions of medieval chivalry. In the United States, hundreds of thousands of boys and girls joined Arthurian youth groups, such as the Knights of King Arthur, in which Arthur and his legends were promoted as wholesome exemplars. However, Arthur's diffusion within contemporary culture goes beyond such obviously Arthurian endeavours, with Arthurian names being regularly attached to objects, buildings, and places. As Norris J. Lacy has observed, "The popular notion of Arthur appears to be limited, not surprisingly, to a few motifs and names, but there can be no doubt of the extent to which a legend born many centuries ago is profoundly embedded in modern culture at every level." **(Wikipedia)**

To be great is to reach heights far above the sky.
You may now look down upon others
But with gentle warmth, you offer your care.

Visions of the wondrous cosmos can be captured in your eye.
Panoramic scenes of the glories that have been,
Offer the clues you need to know what is present before you
Is in totality, the whole.

The future you realize is inconceivable.
Its distance from the present is illusionist;
It has no length.

You've discovered the truth; your time of rest has now arrived.
By striving to reach your full potential,
You have found you are now complete; whole.

Sit down, your throne awaits;
King of Kings, Lord of Lords.

Consider yourself now whatever you like.
No one can judge you.
You are above it all.
A Giant is that tall.

If you wish, you can now rest your head on clouds,
And as you slowly close your eyes,
You'll see exquisite, colorful, birds, glide by.

This, for me, is heaven;
To know that you have propelled yourself
To these heights.

Others may not know, or be clear,
About where you are in time and space,
But your name will be carried over time by currents that flow in the air;
In hallowed halls in will echo.

Others will be inspired to be like you;
Be something! Great!
Let these words reverberate within the chambers of words,
And life will treat you kind.

Giants

There were plenty once on our good, green, Earth.
Continually they fought, sought, tangled, with things unknown,
So that Man could become all he could be.

Instead of large portions of time being spent acquiring the meat
We require from gnawing on bone;
His wish is that we all arrive one day home.

Back and forth they did wonder, no path too treacherous or long;
Whether it led eventually to a pole,
The search for answers was their painstaking goal.

So many questions had been left unanswered.
They needed knowledge, fresh insight, often, frequently, in fact,
There was little, if any, substance, content, to many things
That had already been told.

They directed their minds toward books,
Enquiry, research, along with good company;
They needed these things, because the act of helping was serious,
Very, very, far from funny.
Little time was spent thinking of the things others consider so precious;
Namely gold.

The greatest resource of all they knew was the mind.
It was practical, and essential, always,
To value the enormous worth of being kind.

Only when these two could be combined,
Could great and significant things be accomplished.
New discoveries could be made; each one a great find.

Joined together, it was their hope,
They would offer the solution as to how to be truly free.

Eventually every man wants, desires, hopes,
Most of all, we must grant, needs,
The mysterious but plentiful magnificence
We have come to reveal as being
Liberation.

Endlessly, with no thought of rest in their sights,
Would they speculate on the nature of this entity.
Above all else this was due to an overwhelming sense of fascination.

Researching the history books
That help to reveal the essence of what constitutes the past,
We uncover the commonality that often
The gifts and character of the wise
Are not given their due and deserving attention;
Therefore, it is sad to see,
They have not received the materials of which to pursue their task
For which they have oh so dire a need.

This is the painful, and obvious, truth,
And not some sort of formulated made up speculation.

Why, for heaven's sake, this has so often occurred,
I cannot delineate a suitable explanation.
I am left flabbergasted,
Overwhelmed, by the hideous injustice of it all.

Surly, it doesn't take too long to realize
They alone are the ones that can save us.
Are we, could we, may it be so,
That we wish inexplicably to bring about our own fall?

The contradictions apparent in the species known as Mankind
Are often incredibly, dreadfully, hard to explain.

A clue toward deciphering the reason why conclusions have not yet arisen,
Is by considering the meaning of the word Mankind.

Think now how often, unfortunately,
A woman causes a great deal of pain.
Their acts are contrary to what is meant to be.
Instead of caring, offering affection,
Nourishing the subtle and tender needs of those who are extra special,
The opposite has occurred.
The result has been a deep, dark, penetrating, pain.

This may be the most important factor to consider.
Surely, by this time, centuries of technological growth have occurred,
There should be much more from which we should have gained.

The population of the Earth has increased so much,
But with each progressive stage at which a new height has arrived,
We find that fewer and fewer are the number that are wise.

Man has become more corrupted with time.
The exploitation of others often offers his greatest pleasures.
To accomplish these undesirable things a continuous fabrication is needed;
Please, let us call it lies.

How can we ever survive?
We require new ideas;
From these only will the answer be able to derive.

Originality of thought has become incredibly sparse.
To criticize, to disparage, sometimes merely to comment,
Is considered, usually, something quite harsh.

Just considering these matters alone,
We see a pattern that would inevitably lead to dissolution;
There can, therefore, never be found a feasible solution.

Maybe one day a special someone will come along.
I would like to think one day it will be me.

I am prepared to work however hard may be required.
It is of utmost value, precious above all, else,
That I be among those who can appreciate,
Then wish to exploit, the gifts I've been given,
Then, maybe, we will all have a chance
To eventually arrive in heaven!

Jerusalem

Jerusalem
Is a city that symbolizes a lot to many.
Heaven sent it seems to those who have plenty.
I don't mean gold, that's what makes you empty.
It creates feet of clay, as you stand on a bay,
Where ships have no place to stay.

Those are people, who churn in circles, as time passes them by
Waiting for someone of worth to say, hi!

The riches of the Spirit will guide you over the greatest wave.
You remain calm, though, because you know you are saved.

You remain always certain,
Because a shore will be offered when life offers you a burden.

Here angels spoke.
Multitudes have heard their precious words;
Thus, they awoke!

Eichmann Before Jerusalem

The Unexamined Life of a Mass Murderer

By Bettina Stangneth

We like to imagine criminals as shady figures, committing their crimes in secret, fearful of public judgment. When they are unmasked, we like to imagine a consistent reaction from the public, an instinctive wish to ostracize them, and bring them to justice.

The first attempts to consider the perpetrators of the disenfranchisement, expulsion, and murder of the European Jews were wholly in line with this cliché of shady characters, terrorizing their victims while society's back was turned.

But we have long since moved on from this vision of a small group of pathological, a social freaks within an upstanding population who would have mounted a collective resistance, if only they had known what was going on.

We now know a lot about how the national Socialist world –view functioned. We know about the dynamics of collective behaviour, and the impact of totalitarian regimes. We understand the influence that an atmosphere of violence can have, even on people with no particular inclination toward sadism, and we have explored the disastrous effect of the division of labor on people's sense of individual responsibility.

Schindler's List

"Identification Crisis"

Erik Erickson was not referring to the murderous intent of those involved in a scheme to erase a person's existence, (identity), but rather a recognition within a person that he/she has not done many enough to differentiate him/her-self, (Carl Jung used the expression, "individuation", in reference to the process of discovering one's individuality); this usually transpires during one's middle age - later forties/ fifties. the documents i have collected, (the intention was to steal money; thus, Amazon would only allow $1.62 into my account), prove that a collection of people have been coordinated to behave in such a fashion a theft of a person's resources occurs while leaving not a trace of the person who generated the resource - a pitiful expression of mob brainwashing; a state of utter mindlessness.

One will note the majority of names apparent on those document are those of females; how is made possible?

A person who has not developed a Superego, a conscience, behaves much the same as a machine, when it has the opportunity, it will readily use violence to insure its own survival, and, thus, also better insure its genes will be reproduced in the next generation. The motive, or intent, behind most criminal acts is to take, either directly, or indirectly, take resources from another.

Simply put, the more resources a man has, better able he is to attract somebody deemed a suitable female partner, i.e.; young, fertile, and attractive. Women, in turn, lookout for men who can provide the protection and resources required to raise their future children.

In a genetic, and, evolutionary, sense, men find women worth fighting for; she is a valuable resource because she bears children, worries about their health, and constitute the bulk of the parental investment., which is true throughout the animal kingdom; where one sex provides the greater parental investment, the other sex will fight to access the resource; which is a very good explanation as to why men , by an overwhelming majority, are victims of homicide perpetrated by other men; they are in completion with each other over a resource. this is why it is so often claimed that men are more violent, and aggressive than females; women, are actually governed by the same instincts as men, they just happen to express themselves in a different manner; men are openly aggressive, a show of bravado can elevate a man's status; women, on the other hand, use violence in a surreptitious manner.

365

Part Three

British Culture
The Thesis

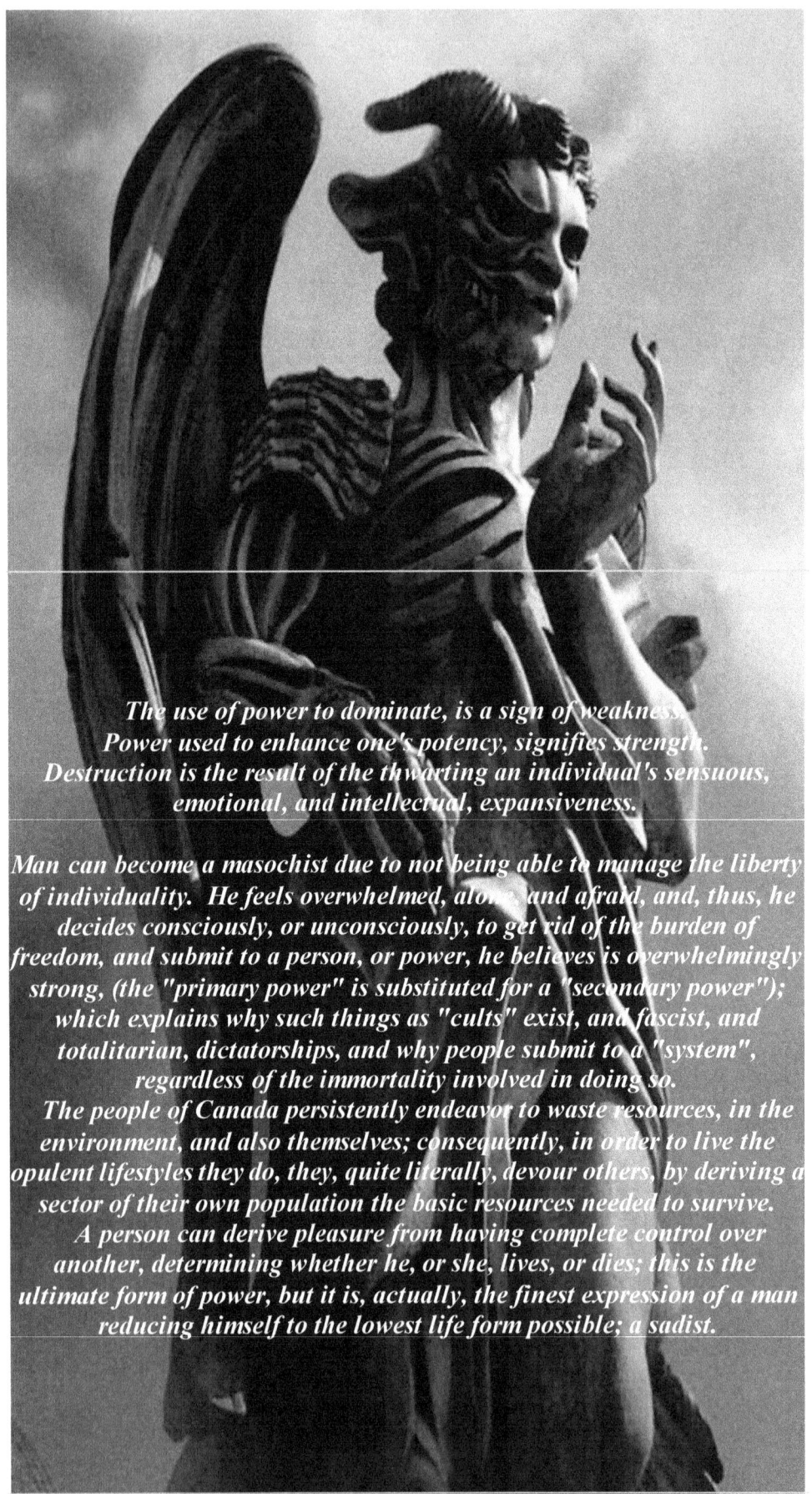

The use of power to dominate, is a sign of weakness.
Power used to enhance one's potency, signifies strength.
Destruction is the result of the thwarting an individual's sensuous,
emotional, and intellectual, expansiveness.

Man can become a masochist due to not being able to manage the liberty
of individuality. He feels overwhelmed, alone, and afraid, and, thus, he
decides consciously, or unconsciously, to get rid of the burden of
freedom, and submit to a person, or power, he believes is overwhelmingly
strong, (the "primary power" is substituted for a "secondary power");
which explains why such things as "cults" exist, and fascist, and
totalitarian, dictatorships, and why people submit to a "system",
regardless of the immortality involved in doing so.
The people of Canada persistently endeavor to waste resources, in the
environment, and also themselves; consequently, in order to live the
opulent lifestyles they do, they, quite literally, devour others, by deriving a
sector of their own population the basic resources needed to survive.
A person can derive pleasure from having complete control over
another, determining whether he, or she, lives, or dies; this is the
ultimate form of power, but it is, actually, the finest expression of a man
reducing himself to the lowest life form possible; a sadist.

Chapter Six

The **rescue of Danish Jews** occurred during Nazi Germany's occupation of Denmark during World War II. On October 1, 1943, Nazi leader Adolf Hitler ordered Danish Jews to be arrested and deported. Despite great personal risk, the Danish resistance movement, with the assistance of many ordinary Danish citizens, managed to evacuate 7,220 of Denmark's 7,800 Jews, plus 686 non-Jewish spouses, by sea to nearby neutral Sweden.

Polish passport used in Denmark up to March 1940. The Jewish holder escaped to Sweden during the war.

The rescue allowed the vast majority of Denmark's Jewish population to avoid capture by the Nazis and is considered to be one of the largest actions of collective resistance to aggression in the countries occupied by Nazi Germany. As a result of the rescue, and the following Danish intercession on behalf of the 464 Danish Jews who were captured and deported to Theresienstadt transit camp in Bohemia, over 99% of Denmark's Jewish population survived the Holocaust.

Memorial in "Denmark Square", Jerusalem

"Model Protectorate" (1940-1943)

On April 9, 1940, Denmark and Norway were invaded by Nazi Germany. Realizing that successful armed resistance was impossible and wishing to avoid civilian casualties, the Danish government surrendered after a few token skirmishes on the morning of the invasion.

The Nazi German government stated that its occupation was a measure taken against the Allies and that Germany did not intend to disturb the political independence of Denmark. Because the Danish government promised "loyal cooperation" with the Germans, the occupation of Denmark was thus relatively mild at first. German propaganda even referred to Denmark as the "model protectorate". King Christian X retained his throne, and the Danish government, the Rigsdag (parliament) and the national courts continued to function. Even censorship of radio and the press was administered by the Danish government, rather than by the occupying German civil and military authorities.

During the early years of the occupation, Danish officials repeatedly insisted to the German occupation authorities that there was no "Jewish problem" in Denmark. The Germans recognized that discussion of the "Jewish question" in Denmark was a possibly explosive issue, which had the potential to destroy the "model" relationship between Denmark and Germany and, in turn, cause negative political and economic consequences for Germany. In addition, the German Reich relied substantially upon Danish agriculture, which supplied meat and butter to 3.6 million Germans in 1942. As a result, when officials in Berlin attempted to implement anti-Jewish measures in Denmark, even ideologically committed Nazis, such as Reich Plenipotentiary Werner Best, followed a strategy of avoiding and deferring any discussion of Denmark's Jews.

In late 1941, during the visit of the Danish foreign minister, Erik Scavenius, to Berlin, German authorities there (including Hermann Göring) insisted that Denmark choose not to avoid its "Jewish problem". A Danish anti-Semitic newspaper used these statements as an opportunity for a slanderous attack on the country's Jews; shortly thereafter, arsonists attempted to start a fire at the Great Synagogue in Copenhagen. The Danish state responded robustly; the courts imposed stiff fines and jail sentences on the editors and would-be arsonists, and the government took further administrative action. Denmark's punishment of anti-Semitic crimes during the occupation were interpreted by the German authorities in Denmark as signaling the Danish view toward any future measures that might be taken against Denmark's Jews by the occupiers.

In mid-1943, Danes saw the German defeats in the Battle of Stalingrad and in North Africa as an indication that having to live under German rule was no longer a long-term certainty, as it had seemed in 1940.

At the same time, the Danish resistance movement was becoming more assertive in its underground press and its increased sabotage activities. During the summer, several nationwide strikes led to armed confrontations between Danes and German troops. In the wake of increased resistance activities and riots, the German occupation authorities presented the Danish government with an ultimatum on August 28, 1943; they demanded a ban on strikes, a curfew, and the punishment of sabotage with the death penalty. Deeming these terms unacceptable and a violation of national sovereignty, the Danish government declared a state of emergency. Some 100 prominent Danes were taken hostage, including the Chief Rabbi Dr. Max Friediger and a dozen other Jews. In response, the Danish government resigned on August 29, 1943. The result was direct administration of Denmark by the German authorities; this direct form of rule meant that the "model protectorate" had come to an end—and with it, the protection the Danish government had provided for the country's Jews.

Deportation Order and Rescue

Without the recalcitrant Danish government to impede them, Denmark's German occupiers began planning the deportation to Nazi concentration camps of the 7,800 or so Jews in Denmark. The German diplomat Georg Ferdinand Duckwitz unsuccessfully attempted to assure safe harbor for the Danish Jews in Sweden—the Swedish government told Duckwitz they would accept the Danish Jews only if approved by the Nazis, who ignored the request for approval. On September 28, 1943, Duckwitz leaked word of the plans for the operation against Denmark's Jews to Hans Hedtoft, chairman of the Danish Social Democratic Party. Hedtoft contacted the Danish Resistance Movement and the head of the Jewish community, C.B. Henriques, who in turn alerted the acting chief rabbi, Dr. Marcus Melchior. At the early morning services, on September 29, the day prior to the Rosh Hashanah services, Jews were warned by Rabbi Melchior of the planned German action and urged to go into hiding immediately and to spread the word to all their Jewish friends and relatives.

The early phases of the rescue were improvised. When Danish civil servants at several levels in different ministries learned of the German plan to round up all Danish Jews, they independently pursued various measures to find the Jews and hide them. Some simply phoned friends and asked them to go through telephone books and warn those with Jewish-sounding names to go into hiding. Most Jews hid for several days or weeks, uncertain of their fate.

Although the majority of the Danish Jews were in hiding, they would eventually have been caught if safe passage to Sweden could not have been secured. Sweden had earlier been receiving Norwegian Jews with some sort of Swedish connection. But the actions to save the Norwegians were not entirely efficient, due to the lack of experience in how to deal with the German authorities. When martial law was introduced in Denmark on August 29, the Swedish Ministry for Foreign Affairs (UD) realized that the Danish Jews were in immediate danger. In a letter dated August 31, the Swedish ambassador in Copenhagen was given clearance by the Chief Legal Officer Gösta Engzell to issue Swedish passports in order to "rescue Danish Jews and bringing them here". On October 2, the Swedish government announced in an official statement that Sweden was prepared to accept all Danish Jews in Sweden. It was a message parallel to an earlier unofficial statement made to the German authorities in Norway. Groups such as the Elsinore Sewing Club (Danish: *Helsingør Syklub*) sprung up to covertly ferry Jews to safety.

Niels Bohr, the Danish physicist whose mother was Jewish, made a determined stand for his fellow countrymen in a personal appeal to the Swedish king and government ministers. He was spirited off to Sweden, whose government arranged immediate transport for him to the United States to work on the then top-secret Manhattan Project. When Bohr touched Swedish soil, government representatives told him he had to board a plane immediately for the United States. Bohr refused. He told the officials, and eventually the king, that until they announced over their air waves and through their press that their borders would be open to receive the Danish Jews, he wasn't going anywhere. Bohr wrote of these events himself. As related by the historian Richard Rhodes, on September 30 Bohr persuaded King Gustaf V of Sweden to make public Sweden's willingness to provide asylum, and on October 2 Swedish radio broadcast that Sweden was ready to receive the Jewish refugees. Historians Richard Rhodes and others interpret Bohr's actions in Sweden as being a necessary precursor without which that mass rescue could not have occurred. According to Paul A. Levine however, who does not mention the Bohr factor at all, the Swedish MFA acted based on clear instructions given much earlier by Prime Minister Hansson and Foreign Minister Günther, following a policy already established in 1942. Even if Bohr's efforts in Sweden might have been superfluous, he did all that he could for his fellow countrymen.

The Jews were smuggled and transported out of Denmark over the Øresund strait from Zealand to Sweden—a passage of varying time depending on the specific route and the weather, but averaging under an hour on the choppy winter sea. Some were transported in large fishing boats of up to 20 tons, but others were carried to freedom in rowboats or kayaks. The ketch *Albatros* was one of the ships used to smuggle Jews to Sweden. Some refugees were smuggled inside freight cars on the regular ferries between Denmark and Sweden, this route being suited for the very young or old who were too weak to endure a rough sea passage. The underground had broken into empty freight cars sealed by the Germans after inspection, helped refugees onto the cars, and then resealed the cars with forged or stolen German seals to forestall further inspection.

Fishermen charged on average 1000 Danish Crowns per person for the transport, but some charged up to 50,000 crowns. The average monthly wage at the time was less than 500 crowns, and half of the rescued Jews belonged to the working class. Prices were determined by the market principles of supply and demand, as well as by the fishermen's perception of the risk. The Danish underground took an active role in organizing the rescue and providing financing, mostly from wealthy Danes who donated large sums of money to the endeavor. In all the rescue is estimated to have cost around 20 million crowns, about half of which were paid by Jewish families and half from donations and collections.

During the first days of the rescue action, Jews moved into the many fishing harbors on the Danish coast for rescue, but the Gestapo became suspicious of activity around harbors (and on the night of October 6, about 80 Jews were caught hiding in the loft of the church at Gilleleje, their hiding place having been betrayed by a Danish girl who was in love with a German soldier). Subsequent rescues had to take place from isolated points along the coast. While waiting their turn, the Jews took refuge in the woods and in cottages away from the coast, out of sight of the Gestapo.

Some of the refugees never made it to Sweden; a few chose to commit suicide, some were captured by the Gestapo *en route* to their point of embarkation, some 23 were lost at sea when vessels of poor seaworthiness capsized, and still others were intercepted at sea by German patrol boats. Danish harbor police and civil police often cooperated with the rescue effort. During the early stages, the Gestapo was undermanned and the German army and navy were called in to reinforce the Gestapo in its effort to prevent transportation taking place; but by and large they proved less than enthusiastic in the operation and frequently turned a blind eye to escapees. The local Germans in command, for their own political calculations and through their own inactivity, may have actually facilitated the escape.

Arrests and Deportations

In Copenhagen the deportation order was carried out on the Jewish New Year, the night of October 1–2, when the Germans assumed all Jews would be gathered at home. The roundup was organized by the SS who used two police battalions and about 50 Danish volunteer members of the Waffen SS chosen for their familiarity with Copenhagen and northern Zealand. The SS organized themselves in five-man teams, each with a Dane, a vehicle, and a list of addresses to check. Most teams found no one, but one team found four Jews on the fifth address checked. There a bribe of 15,000 kroner was rejected and the cash destroyed. The arrested Jews were allowed to bring two blankets, food for 3–4 days, and a small suitcase. They were transported to the harbor, Langelinie, where a couple of large ships awaited them. One of the Danish Waffen-SS members believed the Jews were being sent to Danzig.

On October 2, some arrested Danish communists witnessed the deportation of about 200 Jews from Langelinie via the ship *Wartheland*. Of these, a young married couple was able to convince the Germans that they were not Jewish, and set free. The remainder included mothers with infants, the sick and elderly, chief rabbi Max Friediger, and the other Jewish hostages mentioned above, who had been placed in the Danish internment camp, Horserød, on August 28–29. They were driven below deck without their luggage while being screamed at, kicked and beaten. The Germans then took anything of value from the luggage. Their unloading the next day in Swinemunde was even more inhumane, though without fatalities. There the Jews were driven into two cattle cars, about one hundred per car. During the night, while still locked in the cattle cars, a Jewish mother cried that her child had died. For comparison the Danish communists were packed into cars with "only" fifty people in each; nevertheless, they quickly began to suffer from heat, thirst and lack of ventilation; furthermore, on October 5, shortly before being unloaded in Danzig, they received filthy water for the first time since they had left Copenhagen.

Only some 580 Danish Jews failed to escape to Sweden. Some of these remained hidden in Denmark to the end of the war, a few died of accidents or committed suicide, and a handful had special permission to stay. The vast majority, however, 464 of the 580, were captured and sent to the Theresienstadt concentration camp in German occupied Czechoslovakia. After these Jews' deportation, leading Danish civil servants persuaded the Germans to accept packages of food and medicine for the prisoners; furthermore, Denmark persuaded the Germans not to deport the Danish Jews to extermination camps. This was achieved by Danish political pressure, using the Danish Red Cross to frequently monitor the condition of the Danish Jews at Theresienstadt. Some 51 Danish Jews—mostly elderly—died of disease at Theresienstadt, but in April 1945, as the war drew to a close, 425 surviving Danish Jews (whereof a few born in the camp) were turned over by the Germans to Folke Bernadotte of the Swedish Red Cross and transported to Sweden (see White Buses). The casualties among Danish Jews during the Holocaust were among the lowest of the occupied countries of Europe. Yad Vashem records only 102 Jews from Denmark who died in the Shoah.

It has been popularly reported that the Nazis ordered Danish Jews to wear an identifying yellow star, as elsewhere in Nazi controlled territories. In some versions of the myth, King Christian X opted to wear such a star himself and the Danish people followed his example, thus making the order unenforceable.

However, the story is a myth. In fact the story about the King and the Star and other similar myths originated in the offices of the National Denmark America Association (NDAA) where a handful of Danish nationals opened a propaganda unit called "Friends of Danish Freedom and Democracy", which published a bulletin called *The Danish Listening Post*. This group hired Edward L. Bernays, "The father of Public Relation and Spin" as a consultant. Whether Bernays was the inventor of the story about the King and the yellow star, is not known.

Although the Danish authorities cooperated with the German occupation forces, they and most Danes strongly opposed the isolation of any group within the population, especially the well-integrated Jewish community. The German action to deport Danish Jews prompted the Danish state church and all political parties except the pro-Nazi National Socialist Workers' Party of Denmark (NSWPD) immediately to denounce the action and to pledge solidarity with the Jewish fellow citizens. For the first time, they openly opposed the occupation. At once the Danish bishops issued a *hyrdebrev*—a pastoral letter to all citizens. The letter was distributed to all Danish ministers, to be read out in every church on the following Sunday. This was in itself very unusual since the Danish church is decentralized, apolitical, and without a central leadership.

The unsuccessful German deportation attempt and the actions to save the Jews were important steps in linking the resistance movement to broader anti-Nazi sentiments in Denmark. In many ways October 1943 and the rescuing of the Jews marked a change in most people's perception of the war and the occupation thereby giving a "subjective-psychological" foundation for the myth.

A few days after the roundup, a small news item in the New York *Daily News* reported the myth about the wearing of the Star of David. Later, the story gained its popularity in Leon Uris' novel *Exodus* and in its movie adaptation. It persists to the present, but it is unfounded.

"Righteous among the Nations"

At their initial insistence, the Danish resistance movement wished to be honored only as a collective effort by Yad Vashem in Israel as being part of the "Righteous Among the Nations"; only a handful are individually named for that honor. Instead, the rescue of the Jews of Denmark is represented at Yad Vashem by a tree planting to the King and the Danish Resistance movement— and by an authentic fishing boat from the Danish village of Gilleleje. Similarly, the US Holocaust Museum in Washington D.C. has on permanent exhibit an authentic rescue boat used in several crossings in the rescue of some 1400 Jews.

Georg Ferdinand Duckwitz, the German official who leaked word of the round-up, is also on the Yad Vashem list. (Wikipedia)

When a person has been brainwashed he can no longer, as the expression goes, "think outside of the box", rather, his actions, speech patterns, overall behavior, become coordinated to suit the will of someone else; in "cults" this person is referred to as the "leader".

The only way to avoid becoming something contrary to what one is supposed to be, is to decide to think for oneself, and, thus, not fall into the trap of having others explain to you how things are, and tell you how things are supposed to be; which can also be called "laziness".

Schools, religious organizations, not to mention, educators, have become agents who's primary agenda, whether they realize it or not, is to discourage "free thinking", which can only result in a society of mindless automatons automatically responding to stimuli in a controlled manner.

The primary function of technology in today's world is that it be used as a tool to reinforce the mind set of, "I don't have to learn how to do something - spell, learn grammar, punctuation, add, or subtract - because the simple push of a button will give me the results I need".

It is the job of parents, teachers, and politicians, to encourage, prod, as well as at times, push, people to acquire the skills required to make a contribution to society, and, thus, as a consequence enhance the quality of other people's lives.

John F. Kennedy; inauguration speech;
"Ask not what your country can do for you,

ask what you can do for your country."

Dutch Resistance

Members of the Eindhoven Resistance with troops of the US 101st Airborne Division in Eindhoven during Operation Market Garden, September 1944

The **Dutch resistance** to the Netherlands during World War II can be mainly characterized by its prominent non-violence, peaking at over 300,000 people in hiding in the autumn of 1944, tended to by some 60,000 to 200,000 illegal landlords and caretakers and tolerated knowingly by some one million people, including German occupiers and military.

Dutch resistance developed relatively slowly, but the event of the February and its cause, the random police harassment and deportation of over 400 Jews, greatly stimulated resistance. The first to organize themselves were the Dutch communists, who set up a cell-system immediately. Some other very amateurish groups also emerged, notably De Geuzen, set-up by Bernard IJzerdraatand also some military-styled groups started, such as the Ordedienst ('order service'). Most had great trouble surviving betrayal in the first two years of the war.

Dutch counterintelligence, domestic sabotage, and communications networks eventually provided key support to Allied forces, beginning in 1944 and continuing until the Netherlands was fully liberated. Some 75% (105,000 out of 140,000) of the Jewish population perished in the Holocaust, most of them murdered in Nazi death camps. A number of resistance groups specialized in saving Jewish children, including the *Utrecht's Kindercomité*, the *Landelijke Organisatie voor Hulp aan Onderduikers*, the *Naamloze Vennootschap* (NV), and the Amsterdam Student Group. *The Columbia Guide to the Holocaust* estimates that 215-500 Dutch Romanis were killed by the Nazis, with the higher figure estimated as almost the entire pre-war population of Dutch Romanis.

Definition

The Dutch themselves, especially their official war historian Dr. Loe de Jong, director of the State Institute for War Documentation (RIOD, also known as NIOD) distinguished between several types of resistance. Going into hiding, at which the Dutch appeared to excel, was generally not categorized by the Dutch as resistance because of the passive nature of such an act; helping these so-called *onderduikers* was, but more or less reluctantly so. Non-compliance with German rules, wishes or commands or German condoned Dutch rule, was also not considered resistance. According to official publications, sabotage on an extensive scale must have appeared at those companies in the Netherlands that kept on working during the war (collaboration was rife in the country), but until recently this was not seen as resistance.

Public protests of individuals, political parties, newspapers or the churches were also not considered to be resistance. Publishing illegal papers – something the Dutch were very good at, with some 1,100 separate titles appearing, some reaching circulations of more than 100,000 for a population of 8.5 million – was not considered resistance per se. Only active resistance in the form of spying, sabotage or with arms was what the Dutch considered resistance.

Nevertheless, thousands of members of all the 'non-resisting' categories were arrested by the Germans and often subsequently jailed for months, tortured, sent to concentration camps or killed.

Up until the 21st century, the tendency existed in Dutch historical research and publications, not to regard passive resistance as 'real' resistance. Slowly, this has started to change, also because of the emphasis the RIOD has been putting on individual heroism since 2005. The unique Dutch February strike of 1941, protesting deportation of Jews from the Netherlands, the only such strike to ever occur in Nazi-occupied Europe, is usually not defined as resistance by the Dutch. The strikers, who numbered in the tens of thousands, are not considered resistance participants. The Dutch generally prefer to use the term *illegaliteit* ('illegality') for all those activities that were illegal, contrary, underground or unarmed.

After the war, the Dutch created and awarded a Resistance Cross ('Verzetkruis', not to be confused with the much lower ranking *Verzetsherdenkingskruis*) to only 95 people, of whom only one was still alive when receiving the decoration, a number in stark contrast to the hundreds of thousands of Dutch men and women who performed illegal tasks at any moment during the war.

Prelude

Prior to the German invasion, the Netherlands had adhered to a policy of strict neutrality. The country had narrow bonds with Germany, and less so with the British. The Dutch had not engaged in war with any European nation since 1830. During World War I, the Dutch were not invaded by Germany and anti-German sentiment was not as strong after that war as it was in other European countries. The German ex-Kaiser had fled to the Netherlands in 1918 and lived there in exile. The German invasion therefore came as a great shock to many Dutch people. Nevertheless, the country had ordered general mobilization in September 1939. By November 1938, during the Kristallnacht, many Dutch people received a foretaste of things to come; German synagogues could be seen burning, even from the Netherlands, (such as the one in Aachen).

An anti-fascist movement started to gain popularity – as did the fascist movement, notably the *Nationaal-Socialistische Beweging* (NSB). Despite strict neutrality, even going so far as shooting down British as well as German planes, the country's large merchant fleet was severely attacked by the Germans after 1 September 1939, the beginning of World War II.

The sinking of the passenger liner SS *Simon Bolivar* in November 1939, with 84 dead, especially shocked the nation. It was not the only vessel.

German Invasion

On 10 May 1940, German troops started their surprise attack on the Netherlands without a declaration of war. The day before, small groups of German troops wearing Dutch uniforms had entered the country. Many of them wore 'Dutch' helmets, some made of cardboard as there were not enough originals. The Germans employed about 750,000 men, three times the strength of the Dutch army; some 1,100 planes (Dutch army: 125) and six armored trains; they managed to destroy 80% of the Dutch military aircraft on the ground in one morning, mostly by bombing. Although the Dutch army was inferior in nearly every way, consisting mostly of conscripts, poorly led, poorly outfitted and with poor communications, the Germans lost over 500 planes in the three days of the attack, a loss they would never replenish. Also the first large-scale paratroop attack in history failed, the Dutch managing to recapture the three German-occupied airfields near The Hague within the day. Remarkable was the existence of privately owned anti-aircraft guns. No less surprising may be the fact that the Dutch army owned only one tank.

Major areas of intensive military resistance were:

- the *Grebbelinie*, a north–south line some 50 km east of the capital Amsterdam, from Amersfoort to the Waal, fortified, with field guns, with extensive inundations; the Dutch had to surrender after heavy losses.
- *Kornwerderzand*, with a bunker-complex that defended the eastern end of the *Afsluitdijk* connecting Friesland to North and was held until the capitulation.
- Rotterdam, the bridges over the Waal, defended by Dutch Marines until the surrender.

After four days, it seemed as if the Dutch had stopped the German advance, although at that time, they had already invaded some 70% of the country, excluding the urban areas to the west. Adolf Hitler, who had expected the occupation to be completed in two days (in Denmark in April 1940 it had taken only one day), ordered Rotterdam to be annihilated, leading to the Rotterdam Blitz on 14 May that destroyed much of the city center and killed about 800 people; it also left some 85,000 homeless. The air attack was to be followed by every other major city if the Dutch people refused to surrender. The Dutch, having lost the bulk of their air force, realized they could not stop the German bombers and surrendered.

The 2,000 Dutch soldiers who died defending their country, together with at least 800 civilians who perished in the flames of Rotterdam, were the first victims of a Nazi occupation which was to last five years.

Initial German Policy

The Nazis, who considered the Dutch to be fellow Aryans, were less repressive in the Netherlands than in other occupied countries, at least at first. Their main goals were the Nazification of the populace, the creation of a large-scale aerial attack and defence system, and the integration of the Dutch economy in the German economy. As Rotterdam was already Germany's main port, it remained so and collaboration with the enemy was widespread, stimulated by the flight of all the government ministers who had instructed their secretaries-general to carry on as if nothing happened. The open terrain and dense population, the densest in Europe, made it difficult to conceal illegal activities; unlike for example, the *Maquis* in France, who had ample hiding places. Furthermore, the country was surrounded by German-controlled territory on all sides, offering few escape routes. The complete coast was forbidden territory for all Dutch people.

The very first German round-up of Jews in February 1941 led to the first general strike against the Germans in Europe (and indeed one of only two such throughout occupied Europe).

If the Germans discovered people were involved in the resistance, they were often immediately jailed. It was the social democrats, Catholics, and communists who started the resistance movement. Membership of an armed or military organized group could lead to prolonged stays in concentration camps, and after mid-1944, to immediate death (as a result of Hitler's orders to shoot resistance members on sight - the *Niedermachungsbefehl*). The increasing attacks against Dutch fascists and Germans led to large-scale reprisals, often involving dozens, even hundreds of randomly chosen people who, if not executed, died after being deported. Most of the adult males in the village of Putten for example, which had 600 inhabitants, shared this fate.

The Nazis deported the Jews to concentration and extermination camps, rationed food, and withheld food stamps as a punishment. They started large-scale fortifications along the coast and constructed some 30 airfields, paying with money they claimed from the national bank at a rate of 100 million guilders a month (the so-called 'costs of the occupation'). They also forced adult males between 18 and 45 to work in German factories or on public work projects. In 1944 most trains were diverted to Germany, known as 'the great train robberies', and in total some 550,000 Dutch people were selected to be sent to Germany as forced laborers. Males over the age of 14 were deemed 'able to work' and females over the age of 15. Over the next five years, as conditions became increasingly harsh and difficult, resistance became better organized and more forceful. The resistance managed to kill high-ranking Dutch officials, such as General Seyffardt.

In the Netherlands, the Germans managed to exterminate a relatively large proportion of the Jews. The main reason they were found so easily was that before the war, the Dutch authorities had required citizens to register their religion so that church taxes could be distributed among the various religious organizations. Furthermore, shortly after the Nazis took over the government, they demanded all Dutch public servants fill out an "Aryan Attestation" in which they were asked to state in detail their religious and ethnic ancestry. The American author Mark Klempner writes, "Though there was some protest, not just from the government employees, but from several churches and universities, in the end, all but twenty of 240,000 Dutch civil servants dutifully signed and returned the form." In addition, the country was occupied by the oppressive SS rather than the *Wehrmacht* as in the other Western European countries, as well as the fact that the occupying forces were generally under the command of Austrians who were keen to show that they were 'good Germans' by implementing anti-Semitic policy. The Dutch public transport organization and the police collaborated to a large extent in the transportation of the Jews.

Activities

Plaque honoring the Dutch resistance members executed by the Germans at Sachsenhausen concentration camp

On 25 February 1941, the Communist Party of the Netherlands called for a general strike, the 'February strike', in response to the first Nazi raid on Amsterdam's Jewish population. The old Jewish quarter in Amsterdam had been cordoned off into a ghetto and as retaliation for a number of violent incidents that followed, 425 Jewish men were taken hostage by the Germans and eventually deported to extermination camps, just two surviving. Many citizens of Amsterdam, regardless of their political affiliation, joined in a mass protest against the deportation of Jewish Dutch citizens. The next day, factories in Zaandam, Haarlem, IJmuiden, Weesp, Bussum, Hilversum and Utrecht joined in. The strike was largely put down within a day with German troops firing on unarmed crowds, killing nine people and wounding 24, as well as taking many prisoners. It was significant because opposition to the German occupation intensified as a result. The only other general strike in Nazi-occupied Europe was the general strike in occupied Luxembourg in 1942. The Dutch struck four more times against the Germans: the students' strike in November 1940, the doctors' strike in 1942, the April–May strike in 1943 and the railway strike in 1944. No other country showed such overt unarmed refusal to cooperate with the occupiers.

The February strike was also unusual for the Dutch resistance, which was more covert. Resistance in the Netherlands initially took the form of small-scale, decentralized cells engaged in independent activities, mostly small-scale sabotage (such as cutting phone lines, distributing anti-German leaflets or tearing down posters).

Some small groups had no links with others. They produced forged ration cards and counterfeit money, collected intelligence, published underground papers such as *De Waarheid*, *Trouw*, *Vrij Nederland*, and *Het Parool*; they also sabotaged phone lines and railways, produced maps, and distributed food and goods.

One of the most popular activities was hiding and sheltering refugees and enemies of the Nazi regime, which included concealing Jewish families like that of Anne Frank, underground operatives, draft-age Dutchmen and, later in the war, Allied aircrew. Collectively these people were known as *onderduikers* ('people in hiding' or literally: 'under-divers').Corrie ten Boom and her family were among those who successfully hid several Jews and resistance workers from the Nazis. The total amounted to over 300,000 people up to September 1944, tended-to by some 60,000 to 200,000 landlords and carers.

Reprisals under Operation Silbertanne

After Hitler had approved Anton Mussert as "*Leider* of the Netherlands" in December 1942, he was allowed to form a national government institute, a Dutch shadow cabinet called "*Gemachtigden van den Leider*", which would advise *Reichskommissar* Arthur Seyss-Inquart from 1 February 1943. The institute would consist of a number of deputies in charge of defined functions or departments within the administration.

On 4 February Retired General and *Rijkscommissaris* Hendrik Seyffardt, already head of the Dutch SS volunteer group *Vrijwilligers Legioen Nederland*, was announced through the press as "Deputy for Special Services". As a result, the Communist resistance group CS-6 under Dr. Gerrit Kastein (for their address, 6 Corelli Street, in Amsterdam), concluded that the new institute would eventually lead to a National-Socialist government, which would then introduce general conscription to enable the call-up of Dutch nationals for the Eastern Front. However, in reality the Nazi's only saw Mussert and the NSB as a useful Dutch tool to enable general co-operation, and further Seyss-Inquart had assured Mussert after his December 1942 meeting with Hitler that general conscription was not on the agenda. However, CS-6 assessed that Seyffardt was the first person within the new institute eligible for an attack, after the heavily guarded Mussert.

After approval from the Dutch government-in-exile, on the evening of Friday 5 February 1943, after answering a knock at his front door in Scheveningen, Den Haag, Seyffardt was shot twice by student Jan Verleun who had accompanied Dr. Kastein on the mission. A day later Seyffardt succumbed to his injuries in hospital. A private military ceremony was arranged at the Binnenhof, attended by family and friends and with Mussert in attendance, after which Seyffardt was cremated. On 7 February, CS-6 shot fellow institute member *Gemachtigde voor de Volksvoorlichting* (Attorney for the national relations) H. Reydon and his wife. His wife died on the spot, while Reydon died on 24 August of his injuries. The gun used in this attack had been given to Dr. Kastein by *Sicherheitsdienst* (SD) agent Van der Waals, and after tracking him back through information, arrested him on 19 February. Two days later Dr. Kastein committed suicide so as not to give away Dutch Resistance information under torture.

Seyffardt and Reydon's deaths led to massive Nazi Germany reprisals in the occupied Netherlands, under Operation *Silbertanne*. SS General Hanns Albin Rauter immediately ordered the murder of 50 Dutch hostages and a series of raids on Dutch universities. By accident the Dutch resistance had attacked Rauter's car on 6 March 1945, which in turn led to the killings at De Woeste Hoeve, where 117 men were rounded up and executed at the site of the ambush and another 147 *Gestapo* prisoners were executed elsewhere. A similar war crime occurred on 1–2 October 1944, in the village of Putten, where over 600 men were deported to camps to be killed in retaliation for resistance activity in the Putten raid.

'England-Voyagers'

A little more than 1,700 Dutch people managed to escape to England and offered themselves to their Queen Wilhelmina for service against the Germans. They were called the *Engelandvaarders* named after some 200 who had traveled by boat across the North Sea, most of the other 1,500 went across land.

Some figures are especially noteworthy: Erik Hazelhoff Roelfzema, whose life was described in his book and made into a film and a musical *Soldaat van Oranje*, Peter Tazelaar and Bob or Bram van der Stok, who became a squadron leader in No. 322 Squadron RAF. Van der Stok was one of only three successful survivors of 'the Great Escape' from Stalag Luft III.

Radio

A major role in keeping the Dutch resistance alive was played by the BBC, *Radio Oranje*, the broadcasting service of the Dutch government-in-exile and Radio Herrijzend Nederland which broadcast from the Southern part of the country during liberation. Listening to either programme was forbidden and after about a year the Germans decided to confiscate all Dutch radio receivers. About half of all sets were taken, the rest went underground. With some listeners managing to replace their sets with homemade receivers. Surprisingly the authorities failed to outlaw the publication of magazine articles explaining how to build sets or the sale of the necessary materials until many months later. When they eventually did there were leaflets dropped from British planes containing instructions on building sets and directional aerials to circumvent German jamming.

Press

The Dutch managed to set up a remarkably large underground press that led to some 1,100 titles. Some of these were never more than hand-copied newsletters, while others were printed in larger runs and grew to become newspapers and magazines some of which still exist today, such as *Trouw*, *Het Parool*, and *Vrij Nederland*.

Organization

As early as 15 May 1940, the day after the Dutch capitulation, the Communist Party of the Netherlands (CPN) held a meeting to organize their underground existence and resistance against the German occupiers. It was the first resistance organization in the country. As a result, some 2,000 communists were to lose their lives in torture rooms, concentration camps or by firing squad. On the same day Bernardus IJzerdraat distributed leaflets protesting against the German occupation and called on the public to resist the Germans. This was the first public act of resistance. IJzerdraat started to build an illegal resistance organization called *De Geuzen*, named after a group who rebelled against Spanish occupation in the 16th century.

A few months after the German invasion, a number of Revolutionary Socialist Worker's Party (RSAP) members including Henk Sneevliet formed the Marx–Lenin–Luxemburg Front. Its entire leadership was caught and executed in April 1942. The CPN and the RSAP were the only pre-war organizations that went underground and protested against the anti-Semitic action taken by the German occupiers.

The most important resistance act, as said above, in the Netherlands was hiding and moving people. The first people who went into hiding were German Jews who had arrived in the Netherlands before 1940. They were not duped by the German attitude just after the Dutch capitulation. In the first weeks after the surrender, some British soldiers who could not get to Dunkirk (Duinkerken) in French Flanders hid with farmers in Dutch Flanders. In the winter of 1940/1941 many French escaped prisoners of war (POWs) passed through the Netherlands. One single family in Oldenzaal helped 200 men. In total about 4,000 mainly French, some Belgian, Polish, Russian and Czech ex-POWs were aided on their way south in the province of Limburg.

According to CIA historian Stewart Bentley, there were four major resistance organizations in the country by the middle of 1944, independent of each other:

- the LO ("*Landelijke Organisatie voor Hulp aan Onderduikers*" (nl), or National Organization for Help to People in Hiding); it became the most successful illegal organization in Europe, set up in 1942 by Mrs. Helena Kuipers-Rietberg (nl) (a.k.a. as *tante Riek*- auntie Riek) and Frits Slomp (nl) (a.k.a. Frits de Zwerver) complete with its own illegal social services *Nationaal Steun Fonds* run by Walraven van Hall that paid a kind of dole on a regular basis throughout the war to all families in need, including relatives of sailors and hide-aways;

- the LKP ("*Landelijke Knokploeg*", or National Assault Group, literally translated "brawl crew" or "goon squad"), with about 750 members in the summer of 1944 conducting sabotage operations and occasional assassinations; The LKP provided many of the ration cards to the LO through raids. Leendert Valstar ('Bertus'), Jacques van der Horst ('Louis') and Hilbert van Dijk ('Arie') organized local Assault Groups into the LKP in 1943;

- the RVV ("*Raad van Verzet*" or Council of Resistance), engaged in sabotage, assassinations, and the protection of people in hiding;

- and the OD (*"Orde Dienst"* or Order of Service), a group preparing for the return of the exiled Dutch government and its subgroup the GDN (Dutch Secret Service), the intelligence arm of the OD.

CS 6

Another, but more radical group was called 'CS 6'; it was probably named for the address where they were based, 6 Corelli Street in Amsterdam. According to Dutch official state war historian Dr. L. de Jong, they were by far the most deadly of the resistance groups, committing some 20 assassinations. Having been started in 1940 by the brothers Gideon and Jan Karel ('Janka') Boissevain, the group grew quickly to some 40 members and made contact with the Dutch communist and surgeon Dr. Gerrit Kastein.

They targeted the highest ranking Dutch collaborators and traitors, but duly became the victim of the most dangerous Dutch traitor and German spy, Anton van der Waals. Included in the list of their victims was the Dutch General Seyffardt, who was used by the Germans to head the Dutch SS-legion. They also managed to assassinate an assistant minister, Reydon, and several police chiefs. CS 6 is, according to De Jong, rightly recognized for their crucial role in the deportation of Jews and general terror and suppression. The planned assassination of the best known Dutch traitor and collaborator, Dutch Nazi-party leader Anton Mussert, was delayed and could never be accomplished.

Their activities in eliminating Dutch collaborators prompted the 1943 'Silbertanne' covert murder reprisals by the Dutch SS. By 1944 treason and strain had decimated their ranks.

NSF

In addition to these groups, the NSF ("Nationale Steun Fonds", or National Support Fund) financial organization received money from the exiled government to fund operations of the LO and KP. It also set up large-scale scams involving the national bank and the tax service that were never discovered. The principal figure of the NSF was the banker Walraven van Hall, whose activities were discovered by chance by the Nazis and who was shot at the age of 39. Because of Van Hall's work, the Dutch resistance was never short of money. Van Hall is considered to be the most important Dutch underground worker by national war historian Dr L. de Jong; he finally got his monument in Amsterdam in September 2010.

The number of people cared for by the LO in July 1944 is estimated to be between 200,000 and 350,000. That is one out of 40 inhabitants of the Netherlands. 1,671 members of the LO-LKP organizations lost their lives. Of the 12,000 to 14,000 participants in the LO, 1,104 were killed or died in prison camps. 514 members of the LKP also died. The number of members of the LKP is rather precise - 2,277, since their members were registered after the war. 2,277 was their number in September 1944. 1/3 were members before this time. Only one of the top LKP members survived the war - Liepke Scheepstra (a.k.a. Bob). Mrs. Helema-Rietberg, one of the founders of the LO, was betrayed and died in Ravensbruck concentration camp.

On 22 September 1944, members of the LKP, RVV and a small number of the OD in the southern liberated part of the Netherlands became a Dutch army unit: the *Stoottroepen*. This was during Operation *Market Garden*. Three battalions, without any military training, were formed in Brabant and three in Limburg. The first and second battalions from Brabant were involved in guarding the frontline along the Waal and Meuse rivers with the British 2nd Army. The third battalion from Brabant was incorporated into a Polish formation of the Canadian 2nd Army on the front line on the islands of Tholen and Sint Philipsland. The second and third battalions from Limburg were included in the 9th American Army and were involved in guarding the front line from Roosteren to Aix la Chapelle (Aachen/Aken). During the Battle of the Bulge (December 1944), they were repositioned on the line Aix la Chapelle to Liege (Luik). The first battalion from Limburg was an occupational force in Germany in the area between Cologne (Köln), Aix la Chappelle and the Dutch border. The second and third battalions from Limburg accompanied the American push in March 1945 up to Magdeburg, Brunswick and Oschersleben, which was deep into Germany. Women also served as typists and nurses. When the unit was brought into the regular Dutch army after the war, the women had to leave.

Churches

Both the Dutch Catholic and reformed churches (the latter in all its several forms), were agreed on their total but cautious denial of Nazism and the occupation. Both cooperated with many illegal organizations and made funds available, for instance to save Jewish children. Many priests and ministers were arrested and deported; some died, such as father Titus Brandsma, a professor of philosophy and an early outspoken critic of Nazism, who eventually succumbed to illness in Dachau concentration camp. Monseigneur De Jong, archbishop of Utrecht, was a steadfast leader of the Catholic community and a clear but wise opponent of the German occupiers. The Catholic stance on the protection of converted Jews, amongst others Edith Stein, a philosopher who was then also a nun in a Dutch convent, led to special prosecution of those Jews, sister Stein being deported. After the war, captured documents showed that the Germans feared the role of the churches, especially when Catholics and Protestants worked together.

After Normandy

Following the Normandy invasion in June 1944, the Dutch civilian population was put under increasing pressure by Allied infiltration and the need for intelligence regarding the German military defensive buildup, the instability of German positions and active fighting. Portions of the country were liberated as part of the Allied Drive to the Siegfried Line.

The unsuccessful Allied airborne Operation *Market Garden* liberated Eindhoven and Nijmegen, but the attempt to secure bridges and transport lines around Arnhem in mid-September failed, partly because British forces disregarded intelligence offered by the Dutch resistance toward German strength of forces; they were right in believing that the sources had been compromised. The Battle of the Scheldt, aimed at opening the Belgian port of Antwerp, liberated the south-west Netherlands the following month.

While the south was liberated, Amsterdam and the rest of the north remained under Nazi control until their official surrender on 5 May 1945. For these eight months Allied forces held off, fearing huge civilian losses, and hoping for a rapid collapse of the German government. When the Dutch government-in-exile asked for a national railway strike as a resistance measure, the Nazis stopped food transports to the western Netherlands, and this set the stage for the "Hunger winter", the Dutch famine of 1944.

Some 374 Dutch resistance fighters are buried in the Field of Honor in the Dunes around Bloemendaal. In total, some 2,000 Dutch resistance members were killed by the Germans. Their names are recorded in a memorial ledger *Erelijst van Gevallenen 1940-1945*, kept in the Dutch parliament and available online since 2010 . (Wikipedia)

The Holocaust in France

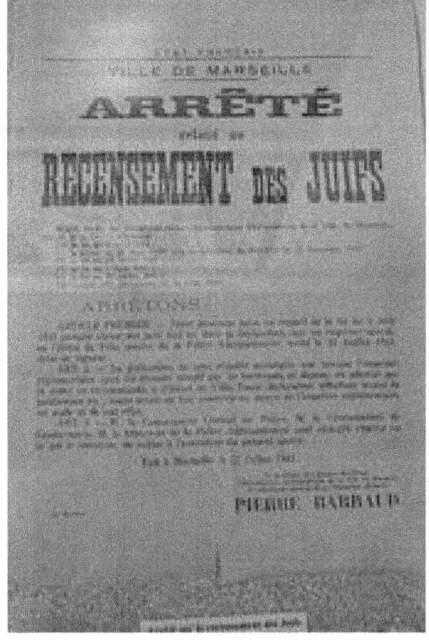

The Holocaust in France refers to the persecution and extermination of Jews and Roma between 1940 and 1944 in occupied France during World War II. The persecution began in 1940, and culminated in deportations of Jews from France to death in Germany and Poland from 1942 which lasted until July 1944. Over 75,000 Jews were deported from France to death camps out of the 340,000 Jewish individuals in France in 1940. About 72,500 of these Jews were killed in the camps. The French Vichy government and the French police participated in the roundup of Jews. Although most deported Jews died, the survival rate of the Jewish population in France was up to 75% which is one of the highest survival rates in Europe.

Background

France had a population of about 150,000 Jewish nationals during the Interwar period; in addition France hosted a large population of foreign Jews who had fled persecutions in Germany. By 1939, the Jewish population had increased to 330,000 due to the refusal of the United States and the United Kingdom to accept any more Jewish refugees following the Évian Conference. After the occupation of Belgium and the Netherlands in 1940, France hosted a new wave of Jewish immigrants and Jewish population peaked at 340,000 individuals.

At the declaration of World War II, French Jews were mobilized into the French military like their compatriots, and, like in 1914, a significant number of foreign Jews enlisted in regiments of foreign volunteers. Jewish refugees from Germany were interned as enemy aliens. In general, the Jewish population of France was confident in the ability of France to defend them against the occupiers, but some, particularly from Alsace and the Moselle regions fled westwards into the unoccupied zone from July 1940.

The armistice agreement of 22 June 1940, signed between the Third Reich and the government of Marshal Philippe Pétain did not contain any overtly anti-Jewish clauses, but did indicate that the Germans intended the racial order existent in Germany since 1933 to spread to France:

- Article 3 warned that in the regions of France occupied directly by the Germans, the French administration must "by all means facilitate the regulations" relating to the exercise of the rights of the Reich;
- Articles 16 and 19 warned that the French government had to proceed to repatriate refugees from the occupied territory and that "The French government is required to deliver on demand all German nationals designated by the Reich and who are in France, in French possessions, colonies, protectorates and territories under mandate"

Under the terms of the armistice only part of France was occupied; the Vichy government of Pétain controlled southern France, the departments of French Algeria, and France's overseas territories such as North Africa, Indochina, the Levant, etc.

History

From the Armistice to the invasion of the Zone libre

Two Jewish women in occupied Paris wearing Yellow badges in June 1942, a few weeks before the mass arrest

Yellow badge made mandatory by the Nazis in France

From the summer of 1940, Otto Abetz, the German ambassador in Paris, organized the expropriation of rich Jewish families. The Vichy regime took the first anti-Jewish measures slightly after the German authorities in the autumn of 1940. The *Statut des Juifs* ("statute on Jews") of 3 October was prepared by Raphaël Alibert. According to a document made public in 2010, Pétain himself made slight moderations to the term of the law. The *statut* forbade French Jews from working in certain professions (teachers, journalists, lawyers, etc.) while a Law of 4 October 1940 envisaged the incarceration of foreign Jews in internment camps in southern France such as the one at Gurs. These internees were joined by convoys of Jews deported from regions of France, including 6,500 Jews who had been deported from Alsace-Lorraine during Operation *Bürckel*.

During Operation *Bürckel*, *Gauleiters* Josef Bürckel and Robert Heinrich Wagneroversaw the expulsion of Jews into unoccupied France from their *Gaues* and the parts of Alsace-Lorraine that had been annexed in the summer of 1941 to the *Reich*. Only those Jews in mixed marriages were not expelled. The 6,500 Jews affected by Operation *Bürckel* were given at most two hours warning on the night of 22–23 October 1940, before being rounded up. The nine trains carrying the deported Jews crossed over into France "without any warning to the French authorities", who were not happy with receiving them. The deportees had not been allowed to take any of their possessions with them, these being confiscated by the German authorities. The German Foreign MinisterJoachim von Ribbentrop treated the ensuing complaints by the Vichy government over the expulsions in a "most dilatory fashion". As a result, the Jews expelled in Operation *Bürckel* were interned in harsh conditions by the Vichy authorities at the camps in Gurs, Rivesaltes and Les Milles while awaiting a chance to return them to Germany.

The *Commissariat Général aux Questions juives* ("Commissariat-General for Jewish Affairs"), created by the Vichy State in March 1941, managed the seizure of Jewish assets and organized anti-Jewish propaganda. At the same time, the Germans began compiling registers of Jews in the occupied zone. The Second *Statut des Juifs* of 2 June 1941 systematized these registrations across the country. Because the yellow star-of-David badge was not made compulsory in the unoccupied zone, these records would provide the basis for the future round-ups and deportations. In the occupied zone, a German order enforced the wearing of the yellow star for all Jews aged over 6 on 29 May 1942.

In order to more closely control the Jewish community, on 29 November 1941, the Germans created the *Union Générale des Israélites de France* (UGIF) in which all Jewish charitable works were subsumed. The Germans were thus able to learn where the local Jews lived. Many of the leaders of the UGIF were also deported, such as René-Raoul Lambert and André Baur.

The arrests of Jews in France began from 1940 for individuals, and general round ups began in 1941. The first raid (*raffle*) took place on 14 May 1941. The Jews arrested, all men and foreigners, were interned in the first transit campus at Pithiviers and Beaune-la-Rolande in the Loiret (3,747 men). The second round-up, between 20–1 August 1941, led to the arrest of 4,232 French and foreign Jews who were taken to Drancy internment camp.

Deportations began on 27 March 1942, when the first convoy left Paris for Auschwitz. Women and children were also targeted, for instance during the Vel' d'Hiv Roundup on 16–7 July 1942, in which 13,000 Jews were arrested by the French police. In the occupied zone, the French police was effectively controlled by the German authorities. They carried out the measures ordered by the Germans against Jews, and in 1942, delivered non-French Jews from internment camps to the Germans. They also contributed to the sending of tens of thousands from those camps to extermination camps in Poland, via Drancy.

In the unoccupied zone, from August 1942, foreign Jews who had been deported to refugee camps in south-west France, in Gurs and elsewhere, were again arrested and deported to the occupied zone, from where they were sent to extermination camps in Germany and Poland.

From the invasion of the Zone libre to 1945

French Jews being deported from Marseilles, 1943

In November 1942, the whole of France came under direct German control, apart from a small sector occupied by Italy. In the Italian zone, Jews were generally spared persecution, until collapse of the Italian fascist regime led to the German occupation of all Italian territory in September 1943.

The German authorities took increasing charge of the persecution of Jews, while the Vichy authorities were forced towards a more sensitive approach by public opinion. However, the Milice, a French paramilitary force inspired by Nazi ideology, was heavily involved in rounding up Jews for deportation during this period. The frequency of German convoys increased. The last, from the camp at Drancy, left the Gare de Bobigny on 31 July 1944.

In French Algeria, General Henri Giraud and later Charles de Gaulle, the French exile government restored (*de jure*) French citizenship to Jews on 20 October 1943.

Results

Of the approximately 330,000 Jews in metropolitan France in 1939, 75% survived the Holocaust, which is one of the highest survival rates in Europe. France has the third highest number of citizens who were awarded the Righteous Among the Nations, an award given to "non-Jews who acted according to the most noble principles of humanity by risking their lives to save Jews during the Holocaust". About 75,000 Jews were deported to Nazi concentration camps and death camps and 72,500 of them died.

(Wikipedia)

The Holocaust in Norway

In 1941—1942 during the occupation of Norway by Nazi Germany, there were at least 2,173 Jews in Norway. At least 775 of them were arrested, detained and/or deported. More than half of the Norwegians who died in camps in Germany were Jews. 742 Jews were murdered in the camps and 23 Jews died as a result of extrajudicial execution, murder and suicide during the war, bringing the total of Jewish Norwegian dead to at least 765 Jews, comprising 230 complete households. "Nearly two-thirds of the Jews in Norway fled from Norway". Of these, around 900 Jews were smuggled out of the country by the Norwegian resistance movement, mostly to Sweden but some also to the United Kingdom). Between 28 and 34 of those deported survived their continued imprisonment in camps (following their deportation)—and around 25 (of these) returned to Norway after the war.

Background

Who's who in the Jewish World, an attaché to an anti-Semitic periodical listing Jews and presumed Jews in Norway. First edition printed in 1925.

The Jewish community in Norway was established in the late 19th century, after a clause in the Norwegian constitution of 1814 that banned Jews from entering Norway was repealed in 1851. The population grew slowly until the early 20th century, when pogroms in Russia and the Baltic states increased the number of immigrants. Another immigration increase came in the 1930s, as Jews fled Nazi persecution in Germany and areas under German control. *See also Nansenhjelpen.*

By 1942, there were 2,173 Jews in Norway. Of these, it is estimated that 1,643 were Norwegian citizens, 240 were foreign citizens, and 290 were stateless.

Much of the prejudice against Jews commonly found in Europe was also evident in Norway in the late 19th and early 20th century, and Nasjonal Samling (NS), the Nazi party in Norway, made anti-Semitism part of its political platform in the 1930s. Halldis Neegaard Østbye became the de facto spokeswoman for increasingly virulent propaganda against Jews, summarized in her 1938 book *Jødeproblemet og dets løsning* (The Jewish Problem and its Solution). NS had also started gathering information about Jewish Norwegians before the war started, and anti-Semitic op-ed articles were occasionally published in the mainstream press.

Following the German invasion and occupation, of Norway, and after the legitimate Norwegian government had left the country, German occupying authorities under the leadership of Reichskommissar Josef Terboven, put Norwegian civilian authorities under his control. This included various branches of Norwegian police, including the district sheriffs (Lensmannsetaten), criminal police, and order police. Nazi police branches, including the SD and Gestapo, also became part of a network that served as tools for increasingly oppressive policies toward the Norwegian populace.

Preparations

Land	Zahl
A. Altreich	131.800
Ostmark	43.700
Ostgebiete	420.000
Generalgouvernement	2.284.000
Bialystok	400.000
Protektorat Böhmen und Mähren	74.200
Estland - judenfrei -	
Lettland	3.500
Litauen	34.000
Belgien	43.000
Dänemark	5.600
Frankreich / Besetztes Gebiet	165.000
Unbesetztes Gebiet	700.000
Griechenland	69.600
Niederlande	160.800
Norwegen	1.300
B. Bulgarien	48.000
England	330.000
Finnland	2.300
Irland	4.000
Italien einschl. Sardinien	58.000
Albanien	200
Kroatien	40.000
Portugal	3.000
Rumänien einschl. Bessarabien	342.000
Schweden	8.000
Schweiz	18.000
Serbien	10.000
Slowakei	88.000
Spanien	6.000
Türkei (europ. Teil)	55.500
Ungarn	742.800
UdSSR	5.000.000
Ukraine 2.994.684	
Weißrußland aus- schl. Bialystok 446.484	
Zusammen: über	11.000.000

(Incorrect) estimate of the number of Jews presented at the Wannsee Conference.

As a deliberate strategy, Terboven's regime sought to use Norwegian, rather than German, officials to subjugate the Norwegian population. Although German police and paramilitary forces reported through the RSHA chain of command, and Norwegian police formally into the newly formed Department of Police, the actual practice was that Norwegian police officials took direction from the German RSHA.

Although several Jewish Norwegians had already been arrested and deported as political prisoners in the early months of the occupation, the first measure targeting all Jews was an order from the German foreign ministry made through Terboven that on 10 May 1941 the police of Oslo were to confiscate radios from all Jews in the city. Within days local sheriffs throughout the entire country received the same orders.

To identify Jewish Norwegians, the authorities relied on information from the police and telegraph service, whilst the synagogues in Oslo and Trondheim were ordered to produce full rosters of their members, including their names, date of birth, profession, and address. Jewish burial societies and youth groups were likewise ordered to produce their lists.

In August, the synagogues were also ordered to produce lists of Jewish individuals who were not members. The resulting lists were cross-referenced with information Nasjonal Samling had compiled previously and information from the Norwegian Central Bureau of statistics. In the end, occupying authorities in Norway had a more complete list of Jewish residents in Norway than most other countries under Nazi rule.

On the basis of the lists compiled in the spring, the Justice Department and county governors started in the fall to register all Jewish property, including commercial holdings. A complete inventory was transmitted to the police department in December 1941, and this also included individuals who were suspected of having a Jewish background.

Anti-Semite graffiti on shop windows in Oslo in 1941. (The location is at the junction of present-day Henrik Ibsen's Street and Crown Prince Street.)

On 20 December, the Norwegian Department of Police ordered 700 stamps with a 2 cm tall "J" for use by authorities to stamp the identification cards of Jewish individuals in Norway. These were put into use on 10

January 1942, when advertisements in the mainstream press ordered all Norwegian Jews to immediately present themselves at the local police stations to have their identification papers stamped. They were also ordered to complete an extensive form. For purposes of this registration, a Jew was identified as anyone who had at least three "full-Jewish" grandparents; anyone who had two "full-Jewish" grandparents and was married to a Jew; or was a member of a Jewish congregation. This registration showed that about 1,400 Jewish adults lived in Norway.

The Norwegian State Railways "aided without protest in the deportation", according to author Halvor Hegtun.

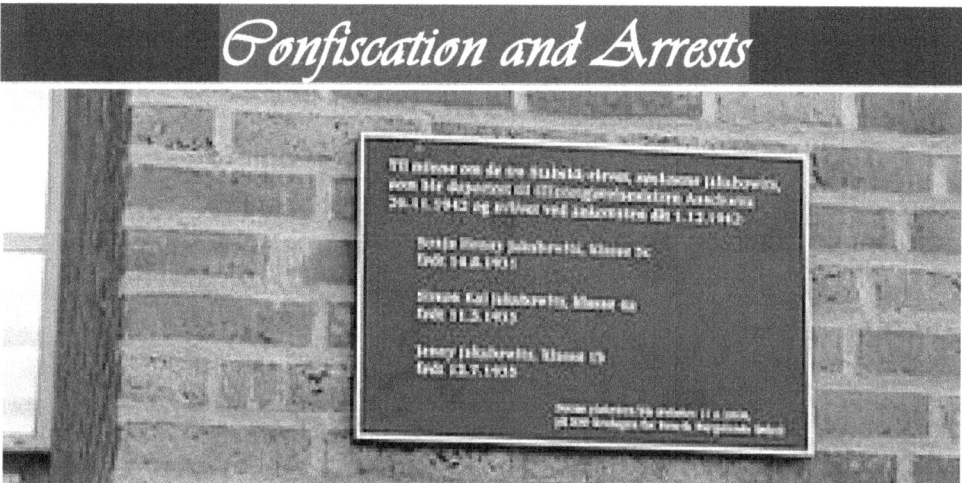

Confiscation and Arrests

Memorial plaque at Stabekk elementary school over three children who were taken out of their classrooms and sent to Auschwitz

Both the German and Norwegian police officials intensified efforts to target the Jewish population during the course of 1941 and the Falstad concentration camp was established near Levanger, north of Trondheim. Jewish individuals, particularly those who were stateless, were briefly detained in connection with Operation Barbarossa. The first Jewish Norwegian to be deported was Benjamin Bild, a labor union activist and mechanic, who died in Gross Rosen. Moritz Rabinowitz was probably the first to be arrested in March, 1941 for agitating against Nazi anti-Semitism in the Haugesund press. He was sent to Sachsenhausen concentration camp where he was beaten to death on 27 December 1942.

German troops occupied and vandalized the Trondheim Synagogue on 21 April 1941. The Torah scrolls had been secured in the early days of the war, and before long the Methodist church in Trondheim had provided temporary facilities for Jewish religious services. Several Jewish residents of Trondheim were arrested and detained at Falstad. The first such prisoner was Efraim Koritzinsky, a medical doctor and head of Trondheim hospital. Several others followed; altogether eight of these were shot in the woods outside of the camp that became the infamous site of extrajudicial executions in Norway. On 24 February, all remaining Jewish property in Trondheim was seized by Nazi authorities.

By the fall of 1942, about 150 Jews from Norway had fled the country. The Jewish population in Norway had experienced some mistreatment specifically targeted at them, but the prevailing sense was that their lot was the same as all other Norwegians.

As the brutality of the Terboven regime came to light through the atrocities at Telavåg, Martial law in Trondheim in 1942, etc., persecution against Jews in particular became more pronounced.

After numerous cases of harassment and violence against individuals, orders were issued to Norwegian police authorities on 24 and 25 October 1942, to arrest all Jewish men over the age of 15 and confiscate all their property. On 26 October, several Norwegian police branches and 20 soldiers of Germanic-SS rounded up and arrested Jewish men, often leaving their wives and children on the street. These prisoners were held primarily at Berg concentration camp in Southern Norway and Falstad concentration camp in central parts of the country; some were held in local jails, while Jewish women were ordered to report in person to their local sheriffs on a daily basis.

On the morning of 26 November, German soldiers and more than 300 Norwegian officials (belonging to Statspolitiet, Kriminalpolitiet, Hirden and Germanske SS-Norge) were deployed to arrest and detain Jewish women and children. These were sent by cars and train to the pier in Oslo where a cargo ship, the SS Donau was waiting to transport them to Stettin, and from there to Auschwitz.

By 27 November, all Jews in Norway (except one) were either deported and murdered, imprisoned, had fled to Sweden, or were in hiding in Norway.

Around 70 Jews remained imprisoned at Berg concentration camp until the end of the war, because they were married to "Aryans".

Deportation and Mass Murder

- The first group deportation of Jews from Norway was on 19 November 1942 when the ship *Monte Rosa* left Oslo with 223 prisoners, of which 21 were Jewish.

- The original plan was to ship all remaining Jews in Norway in one cargo ship, the SS *Donau*, on 26 November 1942, but only 532 prisoners boarded the SS *Donau* that day. Coincidentally with the departure of the SS *Donau* the same day, the MS *Monte Rosa* carried 26 Jews from Oslo. The *Donau* landed in Stettin on 30 November. The prisoners boarded cargo trains at Breslauer Bahnhof, 60 to a car and departed Stettin at 5:12 pm. The train journey to Auschwitz took 28 hours. All the prisoners arrived alive at the camp, and there they were sorted into two lines. 186 were sent to slave labor in the Birkenau sub camp, the rest - 345 - were killed (within hours) in Auschwitz's gas chambers.

- The remaining Jewish prisoners that had been en route to Oslo on 26 November for the departure of the *Donau* were delayed, possibly as a result of delaying tactics by the Red Cross and sympathetic railroad workers. These were imprisoned under harsh conditions at Bredtveit concentration camp in Oslo to await a later transport.
- On 24 February 1943, the Bredtveit prisoners, along with 25 from Grini, boarded the *Gotenland* in Oslo, altogether 158. The ship departed the following day, also landing in Stettin, where they arrived on 27 February. They traveled to Auschwitz via Berlin, where they stayed overnight at the Levetzowstrasse Synagogue. They arrived at Auschwitz on the night between 2 March and 3 March. Of the 158 who arrived from Norway, only 26 or 28 survived the first day, being sent to the Monowitz sub camp of Auschwitz.

There were smaller and individual deportations after the *Gotenland*'s voyage. A smaller number of Jewish prisoners remained in camps in Norway during the war, primarily those who were married to non-Jewish Norwegians. These were subject to mistreatment and neglect. In the camp in Grini, for example, the group that was harshest treated consisted of violent criminals and Jews.

Altogether, about 767 Jews from Norway were deported and sent to concentration camps under German control, primarily Auschwitz. 26 of these survived the ordeal. In addition to the 741 murdered in the camps, 23 died as a result of extrajudicial execution, murder, and suicide during the war; bringing the total of Jewish Norwegian dead to at least 764, comprising 230 complete households.

The death toll among Jews from Norway constituted about 0.013% of the total death toll of European Jews in the Holocaust.

Escape to Sweden

Early during the occupation, there was traffic between neutral countries, primarily Sweden over land; and the United Kingdom, by sea. Even as the occupying authorities tried to limit such traffic, the underground railroad became more organized. Swedish authorities were at first only willing to accept political refugees and did not count Jews among them.

Several Jewish refugees were turned away at the border, and a few were subsequently deported.

The North Sea route would become increasingly challenging as German forces increased their naval presence along the Norwegian coast, limiting the sea route to special operations missions against German military targets. The land routes to Sweden became the main conduit for people and materials that either needed to get out of Norway for their safety, or into Norway for clandestine missions.

There were a few private routes across the border, but most were organized through three resistance groups: Milorg ("military organization"), Sivorg ("civilian organization") and Komorg, the communist resistance group. These routes were carefully guarded, in large part through a network of secret cells. Some efforts to infiltrate them, especially through the Rinnan gang succeeded, but such holes were quickly plugged.

Recommendations for (or warnings to) escape

Examples of Jews being recommended to escape include outgoing communication by anti-Nazi Germans in Norway:Theodor Steltzer warned Wolfgang Geldmacher—married to Randi Eckhoff, sister of member of the Resistance "Rolf Eckhoff. From them, warnings were passed on to Lise Børsum, Amalie Christie, Robert Riefling, Ole Jacob Malm and others".

Report of disappearance—filed in Norway—regarding two Jews on the first transport from Prague to Poland

On 16 December 1941, "secretary of Nansen International Office for Refugees received a letter from the stateless Jews Nora Lustig, Fritz Lusting and Leo Eitinger. They were in Norway, and wrote that Czech Jews that they knew had been deported to an unknown place in Poland. They asked Filseth, to report missing (through Red Cross), two Jews, shipped with the first transport from Prague to Poland".

After the arrest of Jewish men (on 26 October 1942)

The arrest and detention of Jewish men on 26 October 1942 changed that premise, but at that point many were afraid of reprisals against the imprisoned men if they left. Some Norwegian Nazis and German officials advised Jews to leave the country as quickly as possible.

On the evening of 25 November, resistance people got a few hours' notice before the scheduled arrests and deportation of all Jews in Norway. Many did their best to notify the remaining Jews who were not already detained, usually by making brief phone calls or short appearances on people's doorsteps. This was more successful in Oslo than other areas. Those who were warned only had a few hours to go into hiding and days to find their way out of the country.

The Norwegian resistance movement had not planned for the contingency that hundreds of individuals had to go underground in one night, and it was left to individuals to improvise shelter out of sight of the arresting authorities. Many were moved several times in just as many days.

Most of the refugees were moved in small groups across the border, typically with the help of taxis or trucks, railroads to areas near the border, and then by foot, car, bicycle, or on skis across the border. It was a particularly cold winter, and the crossing involved considerable hardship and uncertainty. Those who had the means paid their non-Jewish helpers for their trouble.

The passage was complicated by the vigilance of police who were committed to capturing such refugees, and Terboven imposed the death penalty for anyone caught aiding Jewish refugees. Only individuals who by application were granted "border zone permits" were allowed within easy traveling distance to the border with Sweden. Trains were subject to regular search and inspection, and there were continuous patrols of the area. A failed crossing would have dire consequences for anyone caught, as indeed it turned out for a few.

Still, at least 900 Jewish refugees made their way across the border to Sweden. They usually went through a transit center in Kjesäter in Vingåker, and then found temporary homes throughout Sweden, but mostly in certain towns where Norwegians gathered, such as Uppsala.

Criticism of the Norwegian government in exile, and of Milorg

Some have said that the Norwegian government in exile should have warned the Jews (and told them to flee), since Trygve Lie already in June 1942 knew about what was happening to Jews in continental Europe, while others say that "What could one expect from Lie while the British and the Americans did not believe the messages from Poland? Also in Norway there had been difficulty in believing that gruesomeness had taken place".

Some have said that Milorg did too little for the Jews, while others say that "The great rescue operation Carl Fredriksens Transport was a result of orders from a tilbaketrukket leader of Milorg, Ole Berg, and later financed by Sivorg".

Criminal Culpability and Moral Responsibility

Criminal Prosecution

Terboven, Rediess, and other SS officers on an excursion to Skeikampen in April, 1942

Although both the Norwegian Nazi party Nasjonal Samling and the German Nazi establishment had a political platform that called for persecution and ultimately the genocide of European Jewry, the arrest and deportation of Jews in Norway into the hands of the camp officials turned on the actions of several specific individuals and groups.

The ongoing rivalry between Reichskommissar Josef Terboven and Minister President Vidkun Quisling may have played a role, as both were likely presented with the directives from the Wannsee Conference in January 1942. The German policy was to use Norwegian police as a front for the Norwegian implementation of the conference plans, orders for which were issued along two chains of command: from Adolf Eichmann through the RSHA and Heinrich Fehlis to Hellmuth Reinhard, the Gestapo chief in Norway; and from Quisling through the "minister of justice" Sverre Riisnæs and "minister of police" Jonas Lie through to Karl Marthinsen, the head of the Norwegian state police.

Documentation from the period suggests that the Nazi authorities, and especially the Quisling administration, were loath to initiate actions that might cause widespread opposition among the Norwegian population.

Quisling had tried and failed to take over the teachers' unions, the clergy of the State of Norway, athletics, and the arts. Eichmann had de-prioritized the extermination of Jews in Norway, as the number was low and even Nasjonal Samling had claimed that the "Jewish problem" in Norway was minor. Confiscation of Jewish property, the arrest of Jewish men, constant harassment and individual murder was - until late November, 1942 - part of Terboven's approach of terrorizing the Norwegian population into submission.

The evidence suggests that Hellmuth Reinhard took the initiative to put an end to all Jews in Norway. This may have been motivated by his own ambition, and it's possible he was encouraged by the lack of outrage over the initial measures targeting Jews.

According to the trial against him in Baden-Baden in 1964, Reinhard arranged for the SS *Donau* to set aside capacity for prisoner transport on 26 November and ordered Karl Marthinsen to mobilize the necessary Norwegian forces to affect the transit from Norway. In a curious side note to all this, he also sent along a typewriter on the *Donau* to properly register all prisoners, and was insistent that it be returned to him on *Donau*'s return voyage - which it was.

A local, Norwegian, police chief in Oslo named Knut Rød provided on-the-ground command of Norwegian police officers for arresting women and children and transporting them as well as the men who had already been detained to the Oslo harbor and putting them in the hands of the German SS troops.

Eichmann was not notified of the transport until the *Donau* had left the harbor, bound for Stettin. Nevertheless, he was able to arrange for box cars to be present for transport to Auschwitz.

Of those involved:

- **Terboven** committed suicide before being captured when the war ended; **Quisling** was convicted for treason and executed. **Jonas Lie** died, apparently of a heart attack before his capture. **Sverre Riisnæs** either feigned insanity or went insane and was put in protective custody. **Marthinsen** was assassinated by the Norwegian resistance in February 1945. Heinrich Fehlis committed suicide by first taking poison and then shooting himself in May 1945.

In the end, only two of the principals were put on trial:

- **Hellmuth Reinhard** left Norway in January 1945 without any clues to his whereabouts. He was presumed dead and his wife was issued a death certificate so she could remarry. But it turned out he had changed his name to his birth name of Hellmuth Patzschke and had actually remarried his "widow," settling down as a publisher in Baden-Baden.

His real identity was discovered in 1964, and he was put on trial. In spite of overwhelming evidence about his culpability for the deportation of Jews from Norway and his complicity in their deaths, he was acquitted because statute of limitations had expired. He was convicted and sentenced to five years for his participation in Operation Blumenpflücken.

- **Knut Rød** was put on trial in 1948, acquitted of all charges, and managed to get reinstated as a police officer and retired in 1965. Rød's acquittal remains controversial this day and has been characterized as "the strangest criminal trial [in the legal proceedings after World War II]".

- Another controversial trial was that held against members of the resistance **Peder Pedersen** and **Håkon Løvestad**, who confessed to killing an elderly Jewish couple and stealing their money. The jury found that the killing was justified, but convicted the two of embezzlement. This also became a controversial issue known as the Feldmann case.

The moral culpability among Norwegian police officers and Norwegian informants is a matter of continuing research and debate.

Although the persecution and murder of Jews was raised as a factor in several trials, including that against Quisling, legal scholars agree that in no case was it a decisive or even weighty factor in the conviction or sentencing of these people.

406

Beyond the criminal actions of individuals in Norway that led to the deportation and murder of Jews from Norway, and indeed also of non-Jews who were persecuted on political, religious or other pretexts, there has been considerable public debate in Norway about the public morals that allowed these crimes to take place and did not prevent them from happening.

Comparison between Denmark and Norway

The situation of the Jews in Denmark was very different from Norway. Far fewer Danish Jews were arrested and deported, and those who were deported were sent to Theresienstadt, rather than Auschwitz, where a relatively large percentage survived.

Several factors have been cited for these differences:

- In Denmark, the German diplomat Georg Ferdinand Duckwitz leaked the plans for arrest and deportation to Hans Hedtoft several days before the plan was to be put in motion. There was no such humanitarian among German officials in Norway.

- The terms of occupation in Denmark gave Danish politicians greater influence over internal affairs in Denmark, and in particular command authority over Danish police forces. Consequently, German occupying authorities had to rely on German police and military to perform arrests. Where Danish police participated, it was to rescue Jews from Germans. Since the Norwegians resisted the Germans more actively, the country never enjoyed the same civil autonomy as did the Danes during the occupation.

- Danish popular opinion was more actively opposed to the Nazi occupation and was more emboldened to take care of its Jewish citizens. Non-Jewish Danes were known to take to the streets to find Jews who needed shelter, and to search the forests for Jews who had hidden there to help them.

- The arrest of Norwegian Jews happened about one year before the arrests in Denmark, and also before the Soviet victory at Stalingrad, which changed nearby Sweden's stand from being supportive of the Germans to lean towards the Allies. As there was considerable contact between the resistance in Denmark and Norway through neutral Sweden it means that the Danes knew what fate the Danish Jews were destined for. That Sweden had changed to lean towards the Allies also meant that it was open for Jewish refugees, which had not been the case before and early in the war.

The exiled Norwegian government became part of the Allies upon the invasion on 9 April 1940. Though the most significant contribution of the Allied war effort was through the merchant marine fleet known as Nortraship, a number of Norwegian military forces were established and became part of the Norwegian Armed Forces in exile. Consequently, the Norwegian government was regularly briefed on Allied intelligence relating to atrocities committed by German forces in Eastern Europe and in occupied Netherlands, France, etc.

In addition, the Norwegian government also received regular intelligence from the Norwegian home front, including accounts from returning Norwegian Germanic-SS soldiers, who had firsthand accounts of massacres of Jews in Poland, the Ukraine, etc.

Indeed, both underground resistance newspapers in Norway and the Norwegian press abroad published news about "wholesale murders" of Jews in the late summer and fall of 1942. There is, however, little evidence that either the Norwegian home front or Norwegian government expected that the Jews in Norway would be a target for the genocide that was unfolding on the European continent. On 1 December 1942, the Norwegian foreign minister, Trygve Lie sent a letter to the British section of the World Jewish Congress where he asserted that:

> *...it has never been found necessary for the Norwegian Government to appeal to the people of Norway to assist and to protect other individuals of classes in Norway, who have been selected for persecution by the German aggressors, and I feel convinced that such an appeal is not needed in order to urge the population to fulfil their human duty towards the Jews of Norway.*

Although the Norwegian resistance by the fall of 1942 had a sophisticated network for transmitting and propagating urgent news among the population that led to very effective passive resistance efforts, e.g., in keeping the teachers' union, athletics, physicians, etc., out of Nazi control, no such notifications were issued to save Jews.

The Protestant religious establishment in Norway did, however, make their opposition known: in a letter to Vidkun Quisling dated 10 November 1942, bishops of the Church of Norway, the administration of the theological seminaries, the leaders of several leading religious organizations, and the leaders of non-Lutheran Protestant organizations protesting actions against the Jews, calling on Quisling "in the name of

Jesus Christ" to "stop the persecution of Jews and stop the bigotry that through the press is disseminated throughout our land."

The discrimination, persecution, and ultimately deportation of Jews was enabled by the cooperation of Norwegian agencies that were not entirely co-opted by Nasjonal Samling or the German occupying powers. In addition to the police and local sheriffs who implemented the directives of Statspolitiet, the taxis aided in transporting Jewish prisoners to their point of deportation and even sued the Norwegian government after the war for wages owed to them for such services.

Jews in Norway had been singled out for persecution also before 26 October 1942. They were the first to have radios confiscated, were forced to register and have identification papers imprinted, and were banned from certain professions. However, it was not widely considered that this would extend to deportation and murder. It wasn't until the night of 26 November that the resistance movement was mobilized to rescue Jews from deportation. It took time for the network to be fully engaged, and until then Jewish refugees had to improvise on their own, and rely on acquaintances to avoid capture. Within a few weeks, however, the Norwegian home front organizations (including Milorg and Sivorg) had developed the means to move relatively large numbers of refugees out of Norway and also financed these escapes when needed.

The State Railways' Role

Bjørn Westlie says that the "Norwegian State Railways transported Jews to the outward shipping from the Oslo harbor (...) the NSB employees did not know what fate awaited the Jews. Naturally they understood that the Jews would be shipped out of the country by force, because the train went to Oslo harbor". Furthermore Westlie points to "dilemmas [that] NSB's employees found themselves in when the NSB leadership cooperated with the Germans".

Later Westlie said about the extermination of Norwegian Jews: "what else than co-responsible was NSB? For me, NSB's use of POWs and this deportation of Jews must be viewed as one: namely, that NSB thereby became an agency that participated in Hitler's violence against these two groups, who were the Nazism's main enemies. The fact that the pertinent NSB leaders received awards after the war, confirms NSB's and others' desire to conceal this".

There was no investigation of the agencies [or NSB] after the war. However, the former chief Vik was not to be prosecuted if he "did not again work for NSB".

Post-War Reactions

"When the White Buses travelled down [southward from Scandivia] to fetch prisoners who had survived, the Jews were not permitted to come along because they were no longer Norwegian citizens, and the government after 8 May [1945] din not want to finance the homeward transportation", according to historian Kjersti Dybvig."

Restitution Skarpnes Commission

On 27 May 1995, Bjørn Westlie published an article in the daily, *Dagens Næringsliv*, that highlighted the uncompensated financial loss incurred by the Norwegian Jewish community as a result of Nazi persecution during the war. This brought to public attention the fact that much if not most of the assets confiscated from Jewish owners during the war had been inadequately restored to them and their descendants, even in cases where the Norwegian government or private individuals had benefited from the confiscation after the war.

In response to this debate, the Norwegian Ministry of Justice on 29 March 1996, named a commission to investigate what was done with Jewish assets during the war. The commission consisted of County governor of Vest Agder, Oluf Skarpnes as its chair, professor of law Thor Falkanger, professor of history Ole Kristian Grimnes, district court judge Guri Sunde, director at National Archival Services of Norway, psychologist Berit Reisel, and cand.philol. Bjarte Bruland, Bergen. Consultant Torfinn Vollan from the Skarpnes's office acted as the commission's secretary. Of the commission's members, Dr. Reisel and Mr. Bruland had been nominated by the Jewish community in Norway. Anne Hals resigned from the commission early in the process, and Eli Fure from the same institution was named in her place.

The commission worked together for a year, but it became apparent that were diverging views on premises for the group's analysis.

- The majority focused its effort on arriving at an accurate accounting of the assets lost during the war using conventional assumptions and information in available records.

- The minority, consisting of Reisel and Bruland, sought a more in-depth understanding of the historical sequence of events around the loss of individual assets, as well as both the intended and actual effect of the confiscation and subsequent events, whether the owners were deported, killed, or escaped.

By all accounts, the commission had difficulty unifying these views, and on 23 June 1997, two separate reports were submitted to the Ministry of Justice. After considerable debate in the media, the government accepted the findings of the minority report and initiated financial compensation and issuing a public apology.

Assessment of Financial Loss

The Nazi authorities confiscated all Jewish property with an administrative penstroke. This included commercial property such as retail stores, factories, workshops, etc.; and also personal property such as residences, bank accounts, automobiles, securities, furniture, and other fixtures they could find. Jewelry and other personal valuables were usually taken by German officials as "voluntary contributions to the German war effort." In addition, Jewish professionals were typically deprived of any legal right to practice their profession: attorneys were disbarred, physicians and dentists lost their licenses, and craftsmen were locked out of their trade associations. Employers were pressured to fire all Jewish employees. In many cases, Jewish proprietors were forced to continue to work at their confiscated businesses for the benefit of the "new owners."

Assets were often sold at fire sale prices or assigned at a token price to Nazis, Germans, or their sympathizers.

The administration of these assets was performed by a "Liquidation board for confiscated Jewish assets" that accounted for the assets as they were seized and their disposition. For these purposes, the board continued to treat each estate as a bankrupt legal entity, charging expenses even after the assets had been disposed. As a result, there was a significant discrepancy between the value of the assets for the rightful owners, and the value assessed by the confiscating authorities.

This was further complicated by the methodology employed by the legitimate Norwegian government after the war. In order to restore confiscated assets to their owners, the government was guided by public policy to alleviate the economic impact on the economy by reducing compensation to approximate a sense of fairness and finance the reconstruction of the country's economy. The assessed value was thereby reduced by the Nazis' liquidation practices and was further reduced by the discounting applied as a result of governmental policy after the war.

Norwegian estate law imposes estate tax on inheritance passed from the deceased to his/her heirs depending on the relationship between the two.

This tax was compounded at each step of inheritance. As no death certificates had been issued for Jews murdered in German concentration camps, the deceased were listed as missing. Their estates were held in probate pending a declaration of death and charged for administrative expenses.

By the time all these factors had had their effect on the valuation of the confiscated assets, very little was left. In total, NOK 7.8 million was awarded to principals and heirs of Jewish property confiscated by the Nazis. This was less than the administrative fees charged by governmental agencies for probate. It did not include assets seized by the government that belonged to non-Norwegian citizens, and that of citizens that left no legal heirs. This last category was formidable, as 230 entire Jewish households were killed during the course of the Shoah.

Legacy Monuments

Monuments over the victims were erected fairly early in the graveyards in Oslo and later in Trondheim; in later years, monuments in Haugesund (to commemorate Moritz Rabinowitz), at the pier in Oslo from which the *Donau* sailed, at Falstad, in Kristiansund in Trondheim (over Cissi Klein), and at schools have also raised the awareness. Snublesteiner have been placed in many Norwegian towns.

Notable Remorse

- On Holocaust Remembrance Day in 2012, prime minister Jens Stoltenberg expressed regret "that Norwegian citizens aided in the arrests and deportations of Norwegian Jews". [Around] the same time, National Police Commissioner Odd Reidar Humlegård said to *Dagsavisen* that "I wish to-, on behalf of Norwegian police—and those who participated in the deportation of Norwegian Jews to the concentration camps—express regret".

- In 2015 the chief of public relations of the Norwegian State Railways, Åge-Christoffer Lundeby, said that "The transportation of Jews that were to be deported and the use of POWs on the Nordland Line is a dark chapter of NSB's history". **(Wikipedia)**

Part Four

Damnation
The Antithesis

When people think of War, they envision armaments held by an army being used against an adversary; wars generally entail taking land, and controlling the people who live on that land; it is because people believe an adversary will attack them with such things as guns; missiles; bombs, and gas, that the common man, for the most part, has been unaware that a global war continued after the Second World War. The enemy is the "Monolith", the network of Mafia families that emerged in the 1920's, and the weaponry they use is primarily psychological.

The Second World War began when Germany invaded Poland; two days later, on September 3, 1939, Britain and France declared war on Germany. France fell not long after, and the British army fought from Her island home, instead of the continent, as a result. Fortunately, Germany chose not to pursue Britain, and instead focused Her attention on the Soviet Union, who managed to hold Her own, and weakened the German army, giving Britain the chance to accumulate armaments, and continue the fight; meanwhile the U.S. stood back, and watched, and finally, more than two years after the War began, used the bombing of Pearl Harbour as an excuse to enter the battle to defeat Nazi fascism; this meant that much of the War was fought by Soviets and British.

When the Second World War was over the Monolithimplemented a scheme to make themselves rich by controlling the world's population. The first act of deception was the Marshall Plan; the re-building of Western Europe; the Monolith actually wanted to shape Western Europe to its' liking.

Gangsters, generally speaking, steal by force, and avoid being incarceration due to taking over the police. The war that continued after Germany's defeat is about accumulating wealth, not the preservation of a superior race. The Monolith has invaded countries around the world by "manufacturing wars", and feeding people's fears; U.S. army, and navy, bases, are now situated on every continent.

The Monolith doesn't use threats, and guns, to get people to hand over their wealth, instead, they use the propaganda techniques developed by the Nazis to brainwash people. First World Countries now consist of populations full of "consumers"; people who spend more money than they earn; this lifestyle makes people fall in debt; get loans, and pay interest on those loans, and this expenditure; the result of gluttonous lifestyles; falls into the hands of the Monolith.

The Monolith has used various tactics to keep people on guard, and off balance, which weakens minds, and makes them susceptible to being brainwashed. Since the fall of the Soviet Union, Muslims have been the primary agency to fulfil this objective.

Chapter Seven

The Cloak of Darkness
Jews, Humans & Beasts
What's the differece? Picture the following for the answer.

Areas in the world that are safe for Homo-sapiens to conduct their affairs are covered with cement; the areas that are unsafe for them to inhabit, are covered with soil.

The areas safe for animals are covered with soil, and the areas that are unsafe, are covered with cement. When either Homo-sapiens or animals cross into unsafe areas, this has a detrimental impact on the world's environment.

Animals have "Instincts", and restrict their movements to the area covered with soil. Homo-sapiens with a "superego", understand that if they step on soil they will compromise the ecological system they rely upon for their survival, thus, they're careful to confine their movements to the space covered with cement; if, by chance, they step upon the area covered with soil, they will be burdened with guilt, which helps to inhibit the behaviour from being repeated.

Homo-sapiens that do not have a "superego", will step off cement and walk on soil if, for example, something attracts their attention, and want to acquire right away; they have no concern regarding the consequences of their actions, because they lack the faculty to care; they don't feel guilt, shame, or remorse, and will only feel sadness if their actions result in them being harmed in some manner.

Homo-sapiens that learn about the nature of existence, and behave safely, and are, therefore, humane, are deserving of the title, "human". Homo-sapiens that do not have the capacity to care will endangers others, and are, therefore, deserving of the title, "Savage Primate".

Homo-sapiens damaged the world so much because Jews failed to stop the spread of evil since the end of the Second World War. Evil exists; because good people aren't able to stop it.

Who Made You?

Many years ago, when I was just entering my teens, I acquired a puppy, a mix of basset hound and dachshund, that I named McDougal; according to my father, who provided the name, this shaggy haired pup, who had enormous ears that flopped about as he strutted around, reminded him of a gentleman he once knew who had the same name that he found similarly "peculiar".

I was barely cognizant of it at the time, indeed for many years to follow, but from McDougal I acquired a far deeper understanding of what it means to human, which necessitates having the capacity to love, be loyal, and to know the meaning of self-sacrifice. McDougal was able to practice these things, due to not having the handicap so many Homo-sapiens are burdened with; if I can make a blanket statement; without exception the difficulties, burdens, hardships, misery, despair, and suffering, man experiences, is due to his acquisition of an Ego. The simplest way I can think of to describe this term, is to label man as being selfish; concerned with him, or consumed with defending his self.

The word, "Love", to me, means the opposite of all the terms used above. Love, in the regard most understand the word, means to feel affection for; contrary to popular belief, this does not mean how a person makes you feel, rather how you affect another, which then has an effect on you. In other words, a connection is made between two people, in some capacity, nourish each other. To give an example; say, I have a friend that is learning to play the piano, and I decide to buy a few scores of music as a gift. The recipient then feels pleasure due to his ambition being acknowledged by another, and the continuation of his growth as a musical artist being facilitated by the practice of the music he has been provided. In return, I am pleased to have the opportunity to acquire a better understanding of another person, and, also, hopefully, have quality music to appreciate in the future, and also have the comfort of knowing I have sustained a career.

One way of picturing what I have expounded is as follows; one person turns on a light so another can study better, in turn when the other goes to bed, the student turns off the light so he can sleep better.

This seems to be such a simple exchange that we should all learn to execute at an early age due to the obvious benefits involved, but most of those who comprise what is distinguished as the "hominid species", find it difficult to execute such forms of behaviour; despite the obvious rewards garnered by doing so; why should this be the case?

Allow me to answer this question by creating a scene in a forest involving two men walking side by side along a dirt path.

Both men walk at the same pace for a considerable distance, until they begin to feel fatigue; then, all of a sudden, one stumbles; collapsing on the man beside him, who then abruptly falls, head first, into the trunk of a tree. An unfortunate turn of events, but also recognizable as an inevitable possibility due to both becoming tired due to the stretch of time they've been walking.

They perceive each other as being a friend, therefore, one would probably most likely assume that if neither has been seriously injured, no great harm has been done, and they might decide it would be best to take a bit of a break before continuing their trek. If each acknowledges the nature of the circumstance, (rather than perceiving the event as an unfortunate incident), it can be viewed as a positive learning experience; by taking a break they are more likely not to encounter a similar incident in the future, (it is essential to always heed one's limitations). The interpretation of an incident is formulated on the basis of what one conceives has been derived from the experience.

Let me now begin the scenario once again, but with one added ingredient – "ego", with a capital E – "Ego"; that which creates "suffering"; in the Buddhist sense.

The one that collides with the tree, resulting in a bonk on the head, turns to his partner, a "friend", and says; "That really hurt; aren't you going to apologize?"

His friend replies; "I didn't do it intentionally; my foot caught on something, which made me stumble into you."

"Well that's all well and good, but I want a formal apology, before I even think of continuing to walk with you!"

"I didn't do it on purpose, but if it will make you feel better; I apologize. Are we O.K. now?"

"I guess so", his friend answers.

To take this scenario to the furthest degree in order to illustrate the extent to which "Ego" can separate people, (Love, is the opposite, it represents connectedness), the following will be used as an example.

The man, who received an injury by falling into a tree, and insisted on being given an apology, decides to allow his companion to walk ahead, instead of beside him. He waits till they enter an area of heavy foliage, where he believes no one is around to see or hear them, then extracts a knife from the inside pocket of his jacket he is wearing, and lunges forward with the blade in his hand, resulting in a deep, penetrating, wound, causing his companion to fall to the ground, and quickly thereafter dying due to a loss of blood.

The victor, the murderer, waits till he's certain the man on the forest floor is dead, and then wipes the blood from his knife with a few leaves plucked from a nearby branch.

Once he's sure it's clean, he takes a quick glance around to be sure no one could have witnessed the murder he's just committed. He then steps over the body of the person he considered a friend, and continues to walk along the path at an even pace so as not to attract attention, or cause suspicion.

How is this possible? One might ask. The addition of just one variable to the scene created a catastrophic outcome; definitely the opposite of what transpired when it was not included.

Ego, as I explained the term before, is a concern, or interest, in one's self. The self could be viewed as an object you consider valuable, which you keep locked in a glass case so it will be safe and secure, and you can also derive pleasure from viewing it.

The man asking for an apology could be considered a person seeking to reacquire a sense of equilibrium that was compromised by the unexpected turn of events. This state is comprised of primarily two elements; 1) doubt, and; 2) fear.

He might doubt the intention of his friend, and whether the apology is sincere; such being the case, he loses faith that his "friend" is someone he should allow himself to feel secure around. These ideas, feelings, sensations, are, comparatively speaking, quickly formulated, and lead to a different perception, and interpretation, of the events that have transpired.

It is quite possible in such a case, that there were prior instances when he perceived someone as being honest, truthful, and, therefore, trustworthy, only to later discover that this assessment was entirely inaccurate. Following is an example of an instance that could have such an effect on a person.

Two men form a relationship, and believe they have a common interest in films after viewing a movie together on a computer; they then agree to meet again to watch a movie. They convene at a coffee shop, and not long after, the one possessing the computer excuses himself to go to the bathroom.

When he returns to his table, he discovers his companion is no longer there, and his computer is nowhere in sight. In legal terms what has transpired is, "Fraud", the criminal act, however, is "Theft". The word, fraud, merely describes the manner in which the act took place. In other words, deception is being used to obtain objects, instead of, for example, someone pointing a gun, and insisting valuables be handed over - or else! This person, quite obviously, is not your friend, rather someone openly stating his intention to steal.

A "spy" is someone who uses fraud in order to obtain something he considers valuable; possibly "information". What harm could there possibly be, one would think, in telling someone you trust, where you were born, the schools you attended, even where you do your banking. The person is willingly handing over something the other party wishes to deprive you of; no force, or weapon, is needed. In the case of the two "movie lovers", if no one witnesses the perpetrator leaving with the computer, the thief can call his victim later and say the following;

"I got a call from a friend; he's in hospital. I had to leave right away. I didn't have the opportunity to tell you, because you were in the bathroom - the victim then provides the new about his computer – "You're computer's stolen! Oh my God! What is the world coming to!? I'm so sorry. Would you like to meet for a coffee tomorrow, and we can maybe decide on a movie we can see in a cinema.

The two men were never actually joined, or connected, therefore, one cannot, and will not, be affected by the hardship befallen the other party due to his actions, and only feels insecure, afraid, if there is a sense he's been caught in the act.

Practically everybody today, most unfortunately, utilizes different forms of fraud in order to obtain materials. Another motive for committing fraud is so a person can have the opportunity to repeat the act over and over; as opposed to someone who, for example, pulls a bank heist, and gets away with so much cash he does not expect he will commit a criminal act again in order to live the lifestyle he wishes for himself.

The simple, and sad, truth, is that man has become disconnected from their fellow man, and consequently, does not have the capacity to love, but, rather, is consumed with the toll of continually attending to their own ego.

One can then conclude that evil, (the opposite of love), can be due, or is created, by two factors; 1) an accumulation of doubt, fear, insecurity, and disconnectedness, and; 2) a lack of a conscience, or superego; one is not affected by the harm caused by his or her actions.

What this amounts to, generally speaking, is "ignorance", (which is the result of continually ignoring things), and, therefore, the lesson that needs to be learned is that errors in judgement, must be acknowledged in order to minimize the extent one's actions adversely affect something outside of oneself – that is why evil is quite often viewed as fraud, and is also described as "banal", and why we have the expression, "a wolf in sheep's clothing".

In our day to day affairs we all, one way or another, encounter examples of people displaying these traits, which explains why something that apparently, on the surface, appears innocuous can escalate into something seriousness very quickly. I'll describe an incident that happened to me one morning in a bank as an example; to an outsider, the chain of events might appear chaotic, confusing, awkward, and maybe because of that, many might claim it really isn't important, and deserves to be brushed off one's shoulders – people just being people, in other words; the contrary is actually the case.

Man has never claimed, as far as I am aware, to be a rational animal; I, on the other hand, believe all forms of human behaviour, small and large, have a reason; in other words, contrary to the popular conception that we all make mistakes, I believe our actions are the result of a number of "ingredients" that continue to induce an a similar affect, that strengthens over time; therefore, one may not consciously propose for something to happen, but within our unconscious mind, he did intend the action to take place.

One way of viewing my argument is as follows; man (A) stumbles into man (B), causing him to hit his head on the trunk of a tree; man (B) then claims this was done intentionally. Man (A) refutes this assertion declaring that it wasn't intentional, but rather due to him being distracted; he wasn't watching carefully what lay before him on the path. If he, however, acknowledges responsibility, and has an earnest interest in making sure he doesn't do the same thing again, he will, no doubt, recognize that by being more aware of his surroundings, he can avoid repeating the same series of events. I will now relate the matter described above to a matter that occurred in a bank.

Once entering a bank I approached the information desk/receptionist, and presented some papers, and declared that when I entered a bank account number on the internet, I was told it was "invalid". I also informed the young woman, (her age was most likely, early twenties), that I had previously purchased a pre-paid travel card, and encountered the same problem; however, I had since discovered the cause – believing this information might offer assistance in determining why the same circumstance had arisen with the account in question.

What followed was a muddled chain of responses, accompanied by peculiar gestures directed toward the documents I'd presented. Then, all of a sudden, I had a "eureka moment", and openly declared the truth; "I know what the problem is; you don't understand what the word "invalid" means...do you!?"

She then completely ignored what I said; so I repeated the question using different words, and in a softer tone of voice, so as not to cause, to the best of my ability, embarrassment, while also eager to resolve the issue that brought me to the bank in the first place; "Do you know the meaning of the word "invalid"?"

She swiftly exclaimed, "Yeh, I do!" While not even daring to look in the eyes, but instead searched an area in the distance, off to one side, as if a heavenly body was about to miraculously materialize, and provide her the means to get out of this now awkward situation.

Then, without giving any hint as to her motive, she rose from her chair, and began circulating the desk, mumbling, "I'm getting someone else". I moved over a step, and said, using a higher volume of voice; "All I asked was whether you understood the meaning of the word invalid."

She walked about ten yards away, until the manager emerged from his office, and approached the desk, then asked; "How can I help you?"

"I'd like to know why she doesn't understand the word "invalid"; do you?"

"Why don't we deal with this matter in my office."

Before proceeding I said; "Before I do so, I should tell you that the word means; illegitimate, illegal". As I passed the desk I exclaimed; "As a suggestion I recommend a dictionary be placed here", while tapping on the edge of the counter.

I was then told by the manager that the problem was caused by my tone of voice, and if I continued to use such a tone, he would close all my accounts.

I believe most would consider his response to be an over-reaction, excessive, and quite unnecessary; but here's an insight as to why he said what he did; he also declared, "I've dealt with you before, and I don't like the way you talk to my employees", he them claimed that I demean them!

"Excuse me! All I did was speak the truth. If they can't handle the truth, that's not my problem."

All he could declare following the point I'd made was; "Your tone of voice was the problem."

I again refuted this absurd insinuation by stating that when one doesn't understand what another is saying he should indicate such, so the other party can utilize another method of communication.

Later in his office I formulated a way of presenting an example of what I meant; he said a number of things in an effort to explain why certain items were displayed on a form, at the same time I chose to fumble in my wallet, extracting heaps of crumped papers, while appearing distracted by the volume, then garbled something, before asking; "Did I understand what you were saying?"

He didn't respond, and I knew why he wasn't saying a word. While still appearing enveloped with the paperwork I discovered in my wallet, I repeated the question; "Did I understand what you were saying?"

I noticed he was pointing his pen at a series of numbers, thus, I knew that instead of using words to clarify the issue at hand, he would merely direct my eyes to the nature of the problem.

I then had the audacity to say; "What you are doing follows common sense; if one method of communication doesn't work, formulate another."

He then responded by saying; "Are you accusing me of something?"

At that point I decided to abruptly change my mode of behaviour, because I knew the bank manager had remembered something I'd exclaimed under my breath as I approached his office; "You hired these people; you're responsible."; I then decided to turn the table around, and say;

"No, quite the opposite; I was praising you for acknowledging the nature of the problem."

He actually apologized for misinterpreting my words. I hadn't accused him of anything; what I had done is prove my point. The receptionist, whom he had deemed qualified for her position, had **lied;** and he had sought to cover up this lie, and felt so threatened by my recognizing the underlying cause, he'd been prepared to close all my accounts to insure another incident of a similar nature wouldn't happen again.

Truth be told, had she, the receptionist, been more interested in doing her job well, instead of protecting her "self", her ego, the issue would have been resolved far more quickly. Ask yourself this question; how many previously had walked away from her desk believing they had the answer they'd sought, but were actually told a load of rubbish, but due to her managing to sound so impressive, they believed it was true? I happen to know quite a bit about banking, primarily due to complications that have arisen in the past, and became aware, practically immediately, that the receptionist had no idea what she was talking about.

Consider now the question; does it make any sense that someone would simply wander away from a desk with no clear destination in mind, due to another person's tone of voice? Obviously not!

The manager was defending a lie, complicit in it; and due wanting to continue telling this lie, he was prepared to use the resources at his disposal to get rid of the person able to expose the truth. Why was he so insecure? – Because I was in the right, which he acknowledge by declaring that he felt persecuted in my presence; thus, the choice of words; "Are you accusing me of something?" I think he knew the answer; YES – you totally lack common sense.

How could he have handled this problem better?   He didn't know how; he may have been unaware of his own ignorance. He was not facing facts, but lying about the degree to which those involved were qualified.

It was too much, in other words, for his Ego to handle; which was as precious to him, as the ego belonging to the young female receptionist, and each was prepared to use whatever tools were at their disposal in order to avoid recognizing the extent of their incompetence. At any cost, each was prepared to persecute the one exposing the lie in order to protect the image they had developed of themselves, and hoped others would recognize as the truth.

Once considering all the factors mentioned above, it is easy to calculate what the true issue was; which happens to be the first step in eliminating the matter; the second is discovering a remedy. What I mean by this is that when you take something away, it has to be replaced with something else; much the same as a flat tire on a car needs to replace with an inflated wheel.

The problem was not my tone of voice, nor was it the possibility the receptionist wasn't aware of the meaning of the word "invalid"; the issue is one of "LIES", and an unwillingness to appreciate the detrimental affect lies have.   The term "demean", therefore, is fitting considering the circumstance.   Telling lies, obviously, shouldn't be praised – this is common sense, as I pointed out to the manager; but I was told that demeaning the staff at his bank cannot be tolerated, thus, his employees are neither admonished, nor reprimanded, in any manner, despite the cost another must bear as a consequence.

This is the very essence of madness, and the manager became cognizant of this when he used the words; "Are you accusing me of something?" He knew precisely the point I had directed his attention toward, and he was making an attempt to diffuse my "offensive" observations, all the while knowing full well, of course, that due to his previous threat, of closing my accounts, I wouldn't pursue the matter further.

What I have exposed, is that guarding their "Ego" was such a precious undertaking for both, nothing was considered too extreme as a means to defend it – such is the root cause of narcissism, and because of this, both disconnect themselves from their fellow man. They are entirely enveloped in their own circumstance, which also explains the degree to which each is ignorant.

They have no interest beyond the dimensions of the small bubble world they create for themselves; to expect any form of genuine nourishment/love from either is entirely unrealistic; if one does believe such is possible, the idea is delusional. The notion of either being capable of any form of logical thought is ridiculous, as well.

The entire "system" that has enveloping our world is designed to make sure a people become disconnected from each other, although the illusion formulated is that people are learning to empathize more and more, and at an ever increasing rate.

The development of this conception is primarily due to the work of "marketers", who have assiduously instilled this delusion into the common man's mind. People presently develop attachments to electrical objects, believing they are akin to humans made of flesh, bone, and blood, whether it be a cell phone, a blackberry, computer, T.V. screen, and other such objects that people while away their time using or simply staring at.

The most unfortunate aspect of our present circumstance is that practically everybody has developed the belief that they are communicating ideas; but this, indubitably, is not the case. English, for example, unbeknownst to both parents, and their offspring, is no longer being taught in schools. The most explicit evidence of such being the case is the fact that when you enter the grounds of practically any university campus, you'll notice signs posted throughout regarding assistance being offered to teach grammar, how to formulate a paragraph, and write an essay.

The truth is that people have been brainwashed to believe the gobble de goop nonsense that spews out of their mouths actually makes sense. I am reminded presently of the "Tower of Babel" in The Bible; abbreviations of words, deleting words that are essential in formulating a sentence, ensure there can be no exchange of worthwhile ideas.

I have provided the fundamental reason as to why, as most in the world would openly admit, nothing works well these days, or as it should; people don't care about how their actions affect others.

The word, "care", can be defined as attending to something, taking an interest. People who have become narcissists continually pamper themselves; they are only interested in their own lives; such being the case, and the system presently operating in the world being designed to enable the manufacture of such a mentality; how would it then be possible for any product to be manufactured as it should, or service provided proficiently?

Is it in any way surprising that the quality, and the standard, of both products and services are continually declining, and at an ever increasing rate, (the question is rhetorical)? Picture all the elements noted as being fused to create a solid metal ball that is released from the top of the Tower of Pisa. The rate at which it descends will increase due to gravity. When it hits the ground, it would have reached its maximum speed of acceleration. Things cannot get any worse, the world, (the metal ball), has already slammed into the ground. Humanity, therefore, has encountered the same, but because mankind is so enveloped in itself, there was no recognition that this had occurred.

The evidence of such being the case is, without exception, everywhere you go. If anyone were to attempt to point out that something isn't quite right, and needs to be improved upon, the response most likely will be harsh, and this is due to the recognition that the effort required for such to transpire is so great, it is not perceived as viable.

Picture what I've described as a train, and someone tells the driver a sharp turn is up ahead; even if he has brakes, the driver will require a certain distance in which to apply those brakes in order to reach a safe speed to make the turning without the train toppling over. People, most sadly, do not even have brakes, (figuratively speaking), and this is why they really have no awareness that their lives are wrecks, and they can only wreck the lives of others they come in contact with.

When someone tells you someone else is only at a certain location in order to get a pay-cheque, what they are in fact telling you is that the person is a narcissist, and, therefore, **evil**; he/she is not a loving, loyal, or self-sacrificial, creature. The person in question does the barest minimum required to obtain a pay-cheque, and if the ability to acquire that pay-cheque is perceived to be compromised in any way, measures will be taken immediately, if possible, to eradicate the conceived threat, and that is the reason why the incident I described in the bank arose; both the manager, and the receptionist, are **evil**. If you believe anything to the contrary, that is due to the overwhelming influence advertisements have had upon you; in other words, you bought the lie, and much the same as the free-falling ball, or the train thundering along the track, you cannot stop believing the utter complete nonsense, although you have been provided evidence that the claims made couldn't possibly be true!!

To make a comparison; a person notices someone kicking their foot against a brick wall, and informs the person about the damage he's doing to his foot, (the blood visibly dipping from the sole of his shoe, is more than sufficient evidence), but because he has no awareness of the damage he is inflicting upon himself, primarily due to the foot having gone completely numb, you can't stop him kicking the wall, because he doesn't even know how; and why should he, (thinking to himself), if no damage is being done? What's the sense?

I have also detailed how the expression, "biting off the hand that feeds you", came about. It is due to the person who's receiving help not taking into consideration the circumstance of the person offering help; in essence, what takes over is **greed**. The person offering assistance, therefore, withdraws the hand that's presenting the offering, not just to protect himself, but to have the opportunity to help others in some capacity in the future.

Narcissism, ultimately, is self-destructive; if you don't scratch another person's back, no one will scratch your back; therefore, people create their own insecurities, and as a consequence wonder what will happen when they require help, and there is no one around to offer it? And, frankly, they already know the answer to this quandary; nobody is going to extend any form of assistance.

In order to picture how this feels; imagine travelling in a car on a snowy, cold, winter night, through a desolate region of countryside, when you have an engine breakdown; what are you going to do?

At this point, I can inform members of my audience why people carry cell phones wherever they go, and are continually "text messaging"; it is due to the insecurity derived from disconnecting oneself from others, and also explains why people hoard; thus, the expression, "saving things for a rainy day". Each acts as a "security blanket" protecting the person from an indefinable threat lying somewhere within the indeterminable future.

I provided a picture of someone harming himself by repeatedly slamming his foot into a wall, and the foot subsequently becoming numb due to the repeated abuse. The body habituates itself to pain in order to survive. It will provide a warning, but only up to a point, then a line is crossed that turns off the caution signal; thus, people are entirely unacquainted with the level of damage they are inflicting upon themselves; and that is why there is such a thing as phantom pain, or the sense that a limb is still present after it has been severed.

I would also like to remind my audience that the Monolith deliberately sought to transform people into objects that can be easily manipulated, so they act in accordance with their will; hence, the dimming of minds, and the weakening of bodies.

I'd like to stipulate that the vast majority of those who believe they are extraordinarily physically fit, are quite the opposite, and this is due to an excessive number of workouts, and workouts that are too strenuous, or a combination of both; but due to "nutritional supplements", "vitamin pills", and certain drugs, and so on and so forth, people have bought the illusion they are physically fit. They believe they are fine physical specimens merely due to an image of themselves they consider appealing that is seen in a mirror.

All athletes, (not those who engage in today's sports world), who have long careers, are careful to note how their body is responding to exercises, and the duration of those exercises. Even the best can sometimes overstep the mark, and have had to retire early because of this; case in point, Bjorn Borg, ("Ice man", "Ice Borg"), who admitted he had, to use his terminology, "burned himself out", and this was due to adhering to a regimen that was excessive. He was known to "practice", (a term used to note everything connected to the profession of tennis), 12 hours a day.

One of my tennis heroes is Guillermo Vilas, ("The young bull of the Pampas"), who also happened to be a poet. He declared that he believed he didn't have an abundance of "natural talent", but rather his success was due to working hard. He never, as far as I am aware, "practiced" more than 5 hours a day. It is quite likely he managed to keep a healthy balance in his day to day affairs due to having numerous interests comparably important to him as tennis.

If one were to comprehensively survey the lifestyle of those who are "celebrity athletes", one will no doubt discover an imbalance in their lives; their minds are too focused on their profession, and whether they wish to admit it to themselves, or not, this is primarily induced by their love of "celebrity", "being a star", and the riches that go along with it. This also helps to explain why so many celebrities in their chosen field, (sports, actors, musicians), are so pitiful at what they claim to be so good at.

A great many actresses in Hollywood, for example, have resorted to plastic surgery and such things as Botox, to sell perfumes, clothing, make-up, credit cards, among other products, in order to stay in the "spot light", which, quite obviously, diminishes their ability to act at the same time.

It is quite incredible, as far as I am concerned, that the general public is almost entirely unaware of how miserable, as a result, they are at doing their job as an actor; (or maybe they are a mirror of themselves, and it seems quite natural to them?). Most, I would hope, will wonder how they managed not to notice what I have detailed.

The disconnection people make between themselves and others, due to greed, gluttony, and selfishness, can only incapacitate due to the damage they continually inflict on their minds and bodies. Unfortunately, people are not even close to realizing all the methods, tactics, and devices, used to damage their capacity to behave promptly, proficiently, and accurately. Marketers, of course, in their bid to acquire the largest amount of wealth, (investing the smallest amount possible, while reaping the most enormous rewards imaginable), will focus their attention on the weaker subject, (Electra Complex); and that would be women; the following will illustrate the extent to which they have taken hold of today's female population.

For close to a million and a half years after Homo-sapiens moved out of their original habitats in Africa, these slow witted, shambling, creatures, spent their lives gathering food and hunting; their greatest achievement being the stone tools they used as weapons, and preparing foods; however, once Homo-sapiens greeted the Jews, and mating between the two transpired, confusion, as one might put it, materialized as to how one can be differentiated from the other. It would appear, as history has consistently revealed, that Homo-sapiens have a tendency to believe, in a manner of speaking, they are on equal footing with the Jews, although there is definitely no evidence to support this assertion.

The likelihood of Homo-sapiens creating, for example, a symphony of the quality fashioned by Mozart, is no more likely than a monkey being able to produce a pleasing array of sounds by moving a bow across the strings of a violin with one hand, while placing the tips of the fingers of the other hand upon those strings; a chimp simply does not have the manual dexterity required for such an undertaking to be feasible; similarly, Homo-sapiens cannot even conceive of the cognitive acuity required to manufacture such an enterprise; however, due to their dim level of awareness they are able to construct the notion their aptitudes are comparable to that of Wolfgang Amadeus Mozart.

Throughout recorded history The Lord incorporated an increasing number of Jews within smaller quadrants of time; yet, as the Holocaust proves, Homo-sapiens failed throughout the entire course of history to acknowledge the meaning of the Jew; how he serves as the solitary vehicle one can employ to join, become one, with Him.
In order to teach Homo-sapiens a lesson, The Lord chose to deny any means for them to yoke with His mind since the Second World War; ever since Savage Primates have hideously squandered the wisdom provided by Jews. The worst situation possible is when a Jew isn't present within a population of Homo-sapiens; all manner of moderation in behaviour would be lost, which we see bountiful evidence of, on a global scale, every single day.

The easiest way to obtain evidence of what I've expounded is on today's university campuses, (particularly those in First World Countries); designations originally designed for those who intend to obtain the knowledge and skills necessitated to enrich society. What we see within present post-secondary institutions is the exact opposite; people who reach this level within the education system have been exposed to certain habitats, for a suitable time, that promote a life-style entirely the contrary to altruism in every conceivable way, moulding them into perfectly sized pegs to fit the pre-existing holes created by the corrupt corporations, banks, and "governments", that now flood the globe; which is the deliberate intent of the education system from the first day a child enters kindergarten.

A society that assembles such scenarios is structured to suit those already in power, who have decided to use the tools and knowledge at their disposal to feed their narcissistic, gluttonous, temperaments. No body, indubitably, relishes the idea of being exploited, and degraded, thus, the tactics used to achieve this outcome are, for the most part, hidden from the victims; similarly, those who become members of "cults" rarely have an inkling they are members of a cult, primarily due to being unaware of the methods employed to make them a member.

University students have been fabricated in such a fashion they feed on their inherent savage nature; for example, students are allowed, if not encouraged, to do whatever they please, anywhere they like; no matter the financial cost incurred by doing so.

Their minds have been sickened to such an extent they habitually commit criminal acts in a stimulus/response, action/reaction, manner; for example, theft is portrayed on advertising posters within university campuses as a "crime of opportunity", as if an unattended article is a form of irresistible bait; to better understand the methods used to condition students to behave in aberrant, and/or criminal ways, it's advantageous to study the behaviour of chimpanzees; acknowledging that Homo-sapiens are members of the primate family.

The female chimpanzee is known during estrus, (the period of sexual receptivity, about the time of ovulation), to mate with practically any adult male who happens to come along; a queue can form for favours; although male chimpanzees are ordinarily conscious of their status in the power hierarchy, they do not seem to have to fight over estrous females. When promiscuity, sexual experimentation, forms of feminist ideology that promote "liberality", are considered acceptable, in fact, the "norm", and, "O.K.", they embolden Homo-sapiens to behave in a similar manner.

The education in "consumer societies", endorses violent behaviour by persuading people to repetitively act in a destructive manner, as well as repeatedly observe others behave this way, and making sure there is little, if any, recrimination for having done so.

When the sexual and aggressive energies are mixed in a sufficient quantity, for enough time, the result is a person who is both a masochist, and a sadist; a person who derives sexual satisfaction from inflicting harm upon himself, or harming others.

I recently thought of a fitting title for those who frequent university campuses; in Canada "The Parade of Black Tights". Practically all the women are accustomed to wearing the same type of attire every day; a wardrobe consisting almost exclusively of dull, muted, tones, and dim, coloured tops, and, (except on the days they wear jeans), black, skin tight, pants, (their "corporate uniforms"); because these pants are most often sheer, it is evident that almost all of the women typically wear either a thong, or no underwear at all. They also have a habit of placing their feet, (with, or without, socks), on seats, sitting on the floor, and also walking barefoot into bathrooms. The obvious consequence of doing such things is they "infect" themselves; to be direct, and to the point, the channel through which a new life form greets the world; because they do not wear panties, nor light colour underwear, they don't see the "infected" mucous, and it also reaches the seat others sit on.

Who can stop them? What I have discovered is that if I protest in any fashion feet being placed upon a seat, I will greet an unsympathetic response; I won't find a man to back me up either; and I am likely to be kicked off the premises by security.

A lot of the males on the campus think it's great that so many of the female students are promiscuous, (a "slut" is not, in the vast majority of cases, considered to be cause for the slightest bit of shame).

Is it surprising after noting what I have that approximately 50% of marriages end in divorce, and that children are neglected, abandoned, and abused, in the numbers that should be considered alarming; the parents, man and wife, due to being brought up in the same education system as their children as they should, are far too enraptured with themselves to attend to their children as they should. People become so engrossed with themselves, they neglect their responsibilities toward others. When narcissists attempt to form a bond with another, they are bound to fail,

A man approaches the receptionist at the bank that I noted previously, and thinks; this woman's a catch. She looks great, her smile is welcoming, her perfume is so alluring, and when she bends over, I can see practically an entire exposed breast. I'm in heaven – or so he thinks!

She looks up his bank information, and her eyes immediately light up; she thinks, this fellow in front is a great catch; there's an enormous amount of cash in several accounts.

"What can I do for you today?" She asks with a twinkle in her eye. More than ever he can appreciate how great he must appear through her eyes.

There will never be a genuine relationship between these two people; they are both narcissists; but who will pay a greater price for engaging with the other? My answer, without a shadow of doubt, is the man.

Healthy, productive, children will never be the outcome of this relationship. If she doesn't get what she wants, money, or sex, for example, she will seek it elsewhere. If a man comes along who appears to have more to offer than she's already getting, she will latch onto that man, and most likely, due to knowing how **wrong** her conduct is, attempt to hide it for a period of time, and, thus, it can be named, for want of another term, an "affair".

If the man then wants a divorce – **look out**; because there is no justice in the legal system these days; if she can take every cent, she will. This object, that is so appealing to look at, and speaks in such a soft voice, is thoroughly disconnected from her fellow man, and is, I repeat, **evil.**

If more evidence is required to prove my point, here it is; look at all the sites on the internet concerning women offering sex; to give an example, "discreet 'housewives" looking for sex"; they reveal pictures of the goods they have to offer. Go into practically any grocery store, and look at the number of magazine racks filled with pornographic magazines. Women love the attention, and can also get paid for taking their clothes off; but why, women love to ask, do men have such dirty minds, and look at such filth? The answer is because women keep putting their bare derrieres in men's faces.

The truth, which is ugly, and despicable; advertising campaigns of this nature are as comprehensive, and widespread, as one could possibly imagine.

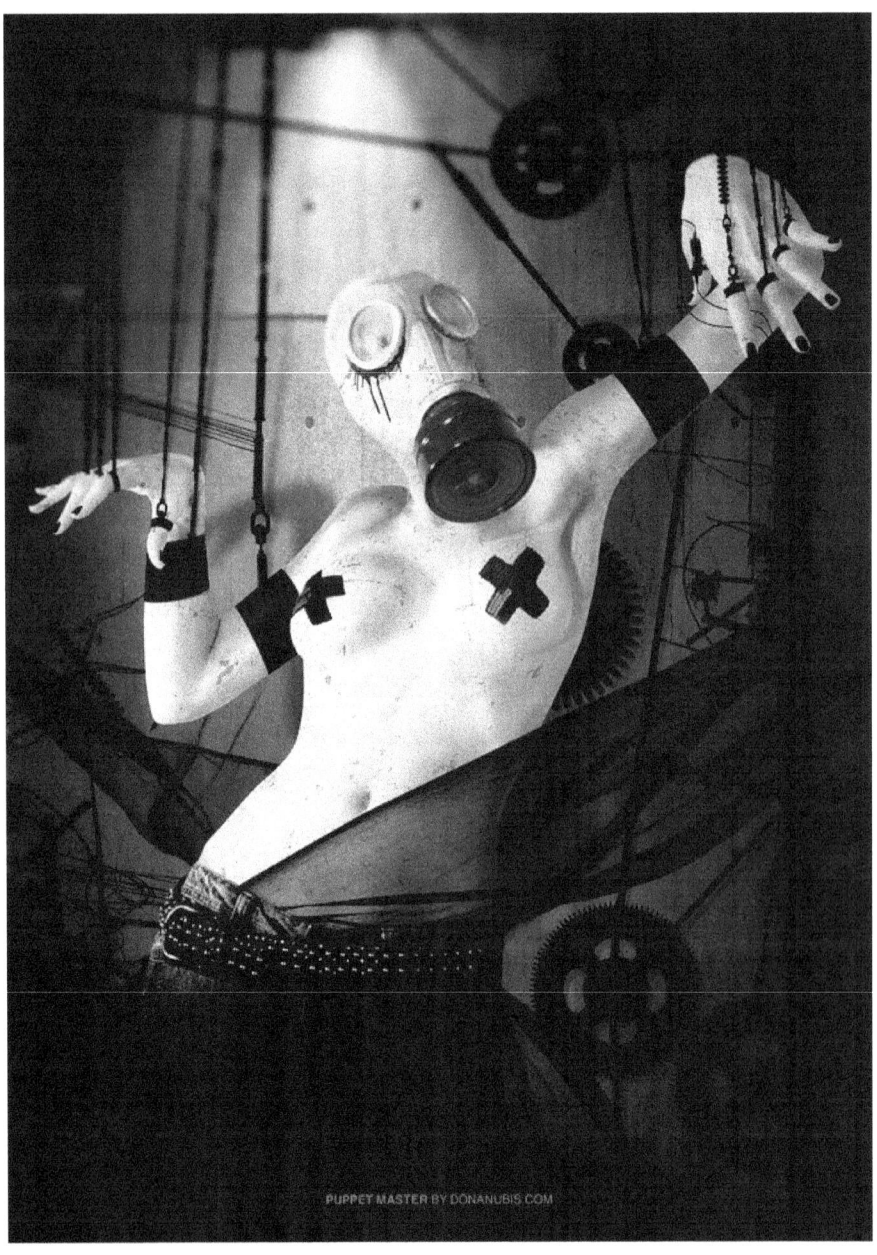

PUPPET MASTER BY DONANUBIS.COM

Soviet War Crimes

War crimes perpetrated by the Soviet Union and its armed forces from 1919 to 1991 include acts committed by the Red Army (later called the Soviet Army) as well as the NKVD, including the NKVD's Internal Troops. In some cases, these acts were committed upon the orders of the Soviet leader Joseph Stalin in pursuance of the early Soviet Government's policy of *Red Terror*, in other instances they were committed without orders by Soviet troops against prisoners of war or civilians of countries that had been in armed conflict with the USSR, or during partisan warfare.

A significant number of these incidents occurred in Northern and Eastern before, during and in the aftermath of World War II, involving summary executions and mass murder of prisoners of war - such as at the Katyn massacre - and grievous mass mistreatment of defenceless civilians by the troops of the Red Army in Soviet-occupied territories. Although there are numerous documented cases of such incidents, few members of the Red Army or leaders such as Vassili Kononov, Lavrentiy Beria, were charged with war crimes, and none of them by the International Criminal Court or a Soviet tribunal. When the victorious Allied Powers of World War II founded the post-war International Military Tribunal, with officials from the Soviet Union taking an active part in the judicial processes, to examine war crimes committed during the conflict by Nazi Germany, there was no examination of Soviet Forces' actions or charges brought against its troops because they were an undefeated power which then held Eastern Europe in military occupation, marring the historical authority of the Tribunal's activity as being, in part, victor's justice.

Background

The Soviet Union did not recognize Imperial Russia's signing of the Hague Conventions of 1899 and 1907 as binding, and refused to recognize them until 1955. This created a situation in which war crimes by the Soviet armed forces could be eventually rationalized. The Soviet refusal to recognize the Hague Conventions also gave Nazi Germany the rationale for inhuman treatment of captured Soviet military personnel.

Victims within the Soviet Union

Several scholars put the number of executions during the Red Terror by the Cheka, predecessor of the NKVD, to about 250,000. Some believe it is possible more people were murdered by the Cheka than died in battle.

Between 1921 and 1922, Mikhail Tukhachevsky, a military leader and future victim of Joseph Stalin's Great Purge, commanded the Red Army's campaign against a peasant uprising in the Tambov province. Tukhachevsky routinely executed hostages without trial and started using poison gas against civilian targets. For these reasons, Simon Sebag-Montefiore has accused Tukhachevsky of being "as ruthless as any Bolshevik."

Jewish victims

The early Soviet leaders publicly denounced anti-Semitism, wrote William Korey: "Anti-Jewish discrimination had become an integral part of Soviet state policy ever since the late thirties." Efforts were made by Soviet authorities to contain anti-Jewish bigotry notably during the Russian civil war, whenever the Red Army units perpetrated pogroms, as well as during the Soviet-Polish War of 1919–1920 at Baranovichi. Only a small number of pogroms were attributed to the Red Army, with the vast majority of 'collectively violent' acts in the period having been committed by anti-Communist and nationalist forces.

The pogroms were condemned by the Red Army high command and guilty units were disarmed, while individual pogromists were court-martialed. Those found guilty were executed. Although pogroms by Ukrainian units of the Red Army still occurred after this, the Jews regarded the Red Army as the only force willing to protect them. It is estimated that 3,450 Jews or 2.3 percent of the Jewish victims killed during the Russian Civil War were murdered by the Bolshevik armies. In comparison, according to the Morgenthau Report, a total of about 300 Jews lost their lives in all incidents involving Polish responsibility. The commission also found that the Polish military and civil authorities did their best to prevent such incidents and their recurrence in the future. The Morgenthau report stated that some forms of discrimination against Jews was of political rather than anti-Semitic nature and specifically avoided use of the term "pogrom," noting that the term was used to apply to a wide range of excesses, and had no specific definition.

Enemies of the people in Poland

During the advance of the Red Army in Polish held territory the Cheka was deployed to apply the Soviet doctrine outside of the Soviet Union and routinely summarily executed as "enemies of the people" captured officers, clergymen and members of the nobility or anyone arbitrarily under suspicion. Also the Cheka encouraged Soviet military personnel to commit acts of violence against families of the members of the nobility, including confiscation, looting and vandalizing their property, rape and murder.

The Red Army and the NKVD

Soviet invasion of Poland, 1939. Advance of the Red Army troops

On February 6, 1922 the Cheka was replaced by the State Political Administrationor OGPU, a section of the NKVD. The declared function of the NKVD was to protect the state security of the Soviet Union, which was accomplished by the large scale political persecution of "class enemies". The Red Army often gave support to the NKVD in the implementation of political repressions. As an internal security force and prison guard contingent of the Gulag, the Internal Troops both repressed political dissidents and engaged in war crimes during periods of military hostilities throughout Soviet history. They were specifically responsible for maintaining the political regime in the Gulag and for conducting mass deportations and forced resettlement. The latter targeted a number of ethnic groups that the Soviet authorities presumed to be hostile to its policies and likely to collaborate with the enemy, including Chechens, Crimean Tatars, and Koreans.

During World War II, series of mass executions were committed by the Soviet NKVD against prisoners of war in Eastern Europe, primarily Poland, the Baltic States, Romania, Ukraine, and other parts of the Soviet occupied territories. The overall death toll is estimated at around 100,000.

As the Red Army withdrew after the German attack of 1941 known as Operation Barbarossa, there were numerous reports of war crimes committed by Soviet armed forces against captured German Wehrmacht and Luftwaffe soldiers from the very beginning of hostilities documented in thousands of files of the Wehrmacht War Crimes Bureau which was established in September 1939 to investigate violations of the Hague and Geneva conventions by Germany's enemies. Among the better documented Soviet massacres are those at Broniki (June 1941), Feodosiya (December 1941) andGrishino (1943).

In the occupied territory, the NKVD carried out mass arrests, deportations and executions. The targets included both collaborators with Germany and the members of anti-Communist resistance movements such as the Ukrainian Insurgent Army (UPA) in Ukraine, the Forest Brothers in Estonia, Latvia and Lithuania, and the Polish Armia Krajowa. The NKVD also conducted the Katyn massacre, summarily executing over 20,000 Polish military officer prisoners in April and May 1940.

After the final repulse of German forces in the Soviet Union, Red Army troops entered Germany, Romania and Hungary in late 1944. Soviet soldiers were by then aware of the German war crimes and often executed surrendering or captured German soldiers in retaliation. There were numerous accounts of war crimes by Soviet armed forces – plunder, the murder of civilians and rape.

War crimes by Soviet armed forces against civilians and prisoners of war in the territories occupied by the USSR between 1939 and 1941 in regions including the Western Ukraine, the Baltic States and Bessarabia in

Romania, along with war crimes in 1944–1945, have been ongoing issues within these countries. Since the dissolution of the Soviet Union, a more systematic, locally controlled discussion of these events has taken place.

The Soviets deployed mustard gas bombs during the Soviet invasion of Xinjiang. Some civilians were killed by conventional bombs during the invasion.

World War II

Estonia

In accordance with the Molotov-Ribbentrop pact Estonia was illegally annexed by the Soviet Union on 6 August 1940 and renamed the Estonian Soviet Socialist Republic. In 1941, some 34,000 Estonians were drafted into the Red Army, of whom less than 30% survived the war. No more than half of those men were used for military service, the rest perished in Gulag concentration camps and labor battalions, mainly in the early months of the war. After it became clear that the German invasion of Estonia would be successful, political prisoners who could not be evacuated were executed by the NKVD, so that they would not be able to make contact with the Nazi government. More than 300,000 citizens of Estonia, almost a third of the population at the time, were affected by deportations, arrests, execution and other acts of repression. As a result of the Soviet takeover, Estonia permanently lost at least 200,000 people or 20% of its population to repression, exodus and war.

Soviet political repressions in Estonia were met by an armed resistance by the Forest Brothers, composed of former conscripts into the German military, Omakaitse militia and volunteers in the Finnish Infantry Regiment 200 who fought a guerrilla war, which was not completely suppressed until the late 1950s. In addition to the expected human and material losses suffered due to the fighting, until its end this conflict led to the deportation of tens of thousands of people, along with hundreds of political prisoners and thousands of civilians lost their lives.

Mass deportations

Tens of thousands of Estonian citizens underwent deportation during the Soviet occupation.

Deportations were predominantly to Siberia and Kazakhstan by means of railroad cattle cars, without prior announcement, while deported were given few night hours at best to pack their belongings and separated from their families, usually also sent to the east. The procedure was established by the Serov Instructions. Estonians residing in Leningrad Oblast had already been subjected to deportation since 1935.

Destruction battalions

In 1941, to implement Stalin's scorched earth policy, destruction battalions were formed in the western regions of the Soviet Union. In Estonia, they killed thousands of people including a large proportion of women and children, while burning down dozens of villages, schools and public buildings. A school boy named Tullio Lindsaar had all of the bones in his hands broken then was bayoneted for hoisting the flag of Estonia. Mauricius Parts, son of the Independence veteran Karl Parts, was doused in acid. In August 1941, all residents of the village of Viru-Kabala were killed including a two-year-old child and a six-day-old infant. A partisan war broke out in response to the atrocities of the destruction battalions, with tens of thousands of men forming the Forest Brothers to protect the local population from these battalions. Occasionally, the battalions burned people alive. The destruction battalions murdered 1,850 people in Estonia. Almost all of them were partisans or unarmed civilians.

Another example of the destruction battalions' actions is the Kautla massacre, where twenty civilians were murdered and tens of farms destroyed. Many of the people were killed after torture. The low toll of human deaths in comparison with the number of burned farms is due to the Erna long-range reconnaissance group breaking the Red Army blockade on the area, allowing many civilians to escape.

Latvia

In accordance with the Molotov-Ribbentrop pact Soviet troops invaded Latvia on June 17, 1940 and it was subsequently incorporated into the Soviet Union as the Latvian Soviet Socialist Republic.

Lithuania

Lithuania, and the other Baltic States, fell victim to the Molotov-Ribbentrop pact. This agreement was signed between the USSR and Nazi Germany in August 1939; leading first to Lithuania being invaded by the Red Army on 15 June 1940, and then to its annexation and incorporation into the Soviet Union on 3 August 1940. The Soviet annexation resulted in mass terror, the destruction of civil liberties, the economic system and Lithuanian culture. Between 1940–1941, thousands of Lithuanians were arrested and hundreds of political prisoners were arbitrarily executed. More than 17,000 people were deported to Siberia in June 1941. After the German attack on the Soviet Union, the incipient Soviet political apparatus was either destroyed or retreated eastward. Lithuania was then occupied by Nazi Germany for a little over three years. In 1944, the Soviet occupation of Lithuania resumed following the German army being expelled. Following World War II and the subsequent suppression of the Lithuanian Forest Brothers, Soviet authorities executed thousands of resistance fighters and civilians accused of aiding them. Some 300,000 Lithuanians were deported or sentenced to prison camps on political grounds. It is estimated that Lithuania lost almost 780,000 citizens as a result of Soviet occupation, of which around 440,000 were war refugees.

The estimated death toll in Soviet prisons and camps between 1944 and 1953 was at least 14,000. The estimated death toll among deportees between 1945 and 1958 was 20,000, including 5,000 children.

During the Lithuanian restoration of independence in 1990, the Soviet army killed 13 people in Vilnius during the January Events.

Poland
1939–1941

In September 1939, the Red Army invaded eastern Poland and occupied it in accordance with the secret protocols of the Molotov. The Soviets later forcefully occupied the Baltic States and parts of Romania, including Bessarabia and Northern Bukovina.

One of the mass graves at Katyn where the NKVD massacred thousands of Polish Officers, policemen, intellectuals and civilian prisoners of war.

German historian Thomas Urban writes that the Soviet policy towards the people who fell under their control in occupied areas was harsh, showing strong elements of ethnic cleansing. The NKVD task forces followed the Red Army to remove 'hostile elements' from the conquered territories in what was known as the 'revolution by hanging'. Polish historian, Prof. Tomasz Strzembosz, has noted parallels between the Nazi Einsatzgruppen and these Soviet units. Many civilians tried to escape from the Soviet NKVD round-ups; those who failed were taken into custody and afterwards deported to Siberia and vanished into the Gulags.

Torture was used on a wide scale in various prisons, especially those in small towns. Prisoners were scalded with boiling water in Bobrka; in Przemyslany, people had their noses, ears, and fingers cut off and eyes put out; in Czortkow, female inmates had their breasts cut off; and in Drohobycz, victims were bound together with barbed wire. Similar atrocities occurred in Sambor, Stanislawow, Stryj, and Zloczow. According to historian, Prof. Jan T. Gross:

We cannot escape the conclusion: Soviet state security organs tortured their prisoners not only to extract confessions but also to put them to death. Not that the NKVD had sadists in its ranks who had run amok; rather, this was a wide and systematic procedure. — Jan T. Gross.

According to sociologist, Prof. Tadeusz Piotrowski, during the years 1939–41, nearly 1.5 million inhabitants of the Soviet-controlled areas of former eastern Poland were deported, of whom 63.1% were Poles or other nationalities and 7.4% were Jews. Only a small number of these deportees survived the war and returned. According to American professor Carroll Quigley, at least one third of the 320,000 Polish prisoners of war captured by the Red Army in 1939 were murdered.

It's estimated that around 35 thousand Polish prisoners were killed either in prisons or on prison trail to the Soviet Union in few days after 22 June 1941 (prisons: Brygidki, Zolochiv, Dubno, Drohobych, and so on).

1944–1945

In Poland, German Nazi atrocities ended by late 1944, but they were replaced by Soviet oppression with the advance of Red Army forces. Soviet soldiers often engaged in plunder, rape and other crimes against the Poles, causing the population to fear and hate the regime.

Soldiers of Poland's Home Army (Armia Krajowa) were persecuted and imprisoned by Russian forces as a matter of course. Most victims were deported to the gulags in the Donetsk region. In 1945 alone the number of members of the Polish deported to Siberia and various labor camps in the Soviet Union reached 50,000. Units of the Red Army carried out campaigns against Polish partisans and civilians. During the Augustów chase in 1945, more than 2,000 Poles were captured and about 600 of them are presumed to have died in Soviet custody. For more information about postwar resistance in Poland see the Cursed soldiers. It was a common Soviet practice to accuse their victims of being fascists in order to justify their death sentence. All the perversion of this Soviet tactic lied in the fact that practically all of the accused had in reality been fighting forces of Nazi Germany since September 1939. At that time the Soviets would still be collaborating with Nazi Germany for more than 20 months before Operation Barbarossa started. Precisely therefore this kind of Poles was judged capable of resisting the Soviets, in the same way they had resisted the Nazis. After the War a more elaborate appearance of justice was given under the jurisdiction of the Polish People's Republic orchestrated by the Soviets in the form of mock trials. These were organized after victims had been arrested under false charges by the NKVD or other Soviet controlled security organizations such as the Ministry of Public Security. There were 6,000 political death sentences issued, the majority of them carried out. It is estimated that over 20,000 people died in communist prisons. Famous examples include Witold Pilecki or Emil August Fieldorf.

The attitude of Soviet servicemen towards ethnic Poles was better than towards the Germans, but not entirely. The scale of rape of Polish women in 1945 led to a pandemic of sexually transmitted diseases. Although the total number of victims remains a matter of guessing, the Polish state archives and statistics of the Ministry of Health indicate that it might have exceeded 100,000. In Kraków, the Soviet entry into the city was accompanied by mass rapes of Polish women and girls, as well as the plunder of private property by Red Army soldiers. This behavior reached such a scale that even Polish communists installed by the Soviet Union composed a letter of protest to Joseph Stalin himself, while church masses were held in expectation of a Soviet withdrawal.

Finland

Finnish children killed by Soviet partisans at Seitajärvi in Finnish Lapland 1942.

Between 1941–1944 Soviet partisan units conducted raids deep inside Finnish territory, attacking villages and other civilian targets. In November 2006, photographs showing atrocities were declassified by the Finnish authorities. These include images of slain women and children. The partisans usually executed their military and civilian prisoners after a minor interrogation.

Around 3,500 Finnish prisoners of war, of whom five were women, were captured by the Red Army. Their mortality rate is estimated to have been about 40 percent. The most common causes of death were hunger, cold and oppressive transportation.

Soviet Union

Retreat by Soviet forces in 1941

Deportations, summary executions of political prisoners and the burning of food stocks and villages took place when the Red Army retreated before the advancing Axis forces in 1941. In the Baltic States, Belarus, Ukraine, and Bessarabia, the NKVD and attached units of the Red Army massacred prisoners and political opponents before fleeing from the advancing Axis forces.

1943–1945

After the Battle of Stalingrad, a major turning point in the war, the Red Army steadily regained lost territory on the Eastern Front. This resulted in action against any person accused of being a collaborator during the German occupation. Civilians opposing Soviet rule were considered collaborators. While for example in France this part of its history is generally well-documented, debated, and is the subject of academic review, very little is known or discussed about what happened in the path of the Red Army.

Kalmykia

During the Kalmyk deportations of 1943, codename **Operation Ulussy** (Операция "Улусы"), the deportation of most people of the Kalmyk nationality in the Soviet Union (USSR), and Russian women married to Kalmyks, but excepting Kalmyk women married to men of other nationality, around half of (97-98,000) Kalmyk people deported to Siberia died before being allowed to return home in 1957.

Germany

According to historian Norman Naimark, statements in Soviet military newspapers and the orders of the Soviet high command were jointly responsible for the excesses of the Red Army. Propaganda proclaimed that the Red Army had entered Germany as an avenger to punish all Germans.

Some historians dispute this, referring to an order issued on 19 January 1945, which required the prevention of mistreatment of civilians. An order of the military council of the 1st Belorussian Front, signed by Marshal Rokossovsky, ordered the shooting of looters and rapists at the scene of the crime. An order issued by Stavka on 20 April 1945 said that there was a need to maintain good relations with German civilians in order to decrease resistance and bring a quicker end to hostilities.

Treatment of prisoners of war

Although the Soviet Union had not formally signed the Hague Convention, it considered itself bound by the Convention's provisions. Even so, torture, mutilation, and mass murder were frequently carried out.

Throughout the Second World War, the Wehrmacht War Crimes Bureau collected and investigated reports of crimes against the Axis POWs. According to Cuban-American writer Alfred de Zayas, "For the entire duration of the Russian campaign, reports of torture and murder of German prisoners did not cease. The War Crimes Bureau had five major sources of information: (1) captured enemy papers, especially orders, reports of operations, and propaganda leaflets; (2) intercepted radio and wireless messages; (3) testimony of Soviet prisoners of war; (4) testimony of captured Germans who had escaped; and (5) testimony of Germans who saw the corpses or mutilated bodies of executed prisoners of war. From 1941 to 1945 the Bureau compiled several thousand depositions, reports, and captured papers which, if nothing else, indicate that the killing of German prisoners of war upon capture or shortly after their interrogation was not an isolated occurrence. Documents relating to the war in France, Italy, and North Africa contain some reports on the deliberate killing of German prisoners of war, but there can be no comparison with the events on the Eastern Front."

In a November 1941 report, the Wehrmacht War Crimes Bureau accused the Red Army of employing "a terror policy... against defenseless German soldiers that have fallen into its hands and against members of the German medical corps. At the same time... it has made use of the following means of camouflage: in a Red Army order that bears the approval of the Council, dated 1 July 1941, the norms of international law are made public, which the Red Army in the spirit of the Hague Regulations on Land Warfare is supposed to follow... This... Russian order probably had very little distribution, and surely it has not been followed at all. Otherwise the unspeakable crimes would not have occurred."

According to the depositions, Soviet massacres of German, Italian, Spanish, and other Axis POWs were often incited by unit Commissars, who claimed to be acting under orders from Stalin and the Politburo. Other evidence cemented the War Crimes Bureau's belief that Stalin had given secret orders about the massacre of POWs.

During the winter of 1941–42, the Red Army captured approximately 10,000 German soldiers each month, but the death rate became so high that the absolute number of prisoners decreased (or was bureaucratically reduced).

Soviet sources list the deaths of 474,967 of the 2,652,672 German Armed Forces taken prisoner in the War. Dr. Rüdiger Overmans believes that it seems entirely plausible, while not provable, that an additional German military personnel listed as missing actually died in Soviet custody as POWs, putting the estimates of the actual death toll of German POW in the USSR at about 1.0 million.

Postwar

Some German prisoners were released soon after the war. Many others, however, remained in the GULAG long after the surrender of Nazi Germany. Among the most famous German war veterans to die in Soviet captivity was Captain Wilm Hosenfeld, who died of injuries, sustained possibly under torture, in a concentration camp near Stalingrad in 1952. In 2009, Captain Hosenfeld was posthumously honored by the State of Israel for his role in saving Jewish lives during the Holocaust. **(Wikipedia)**

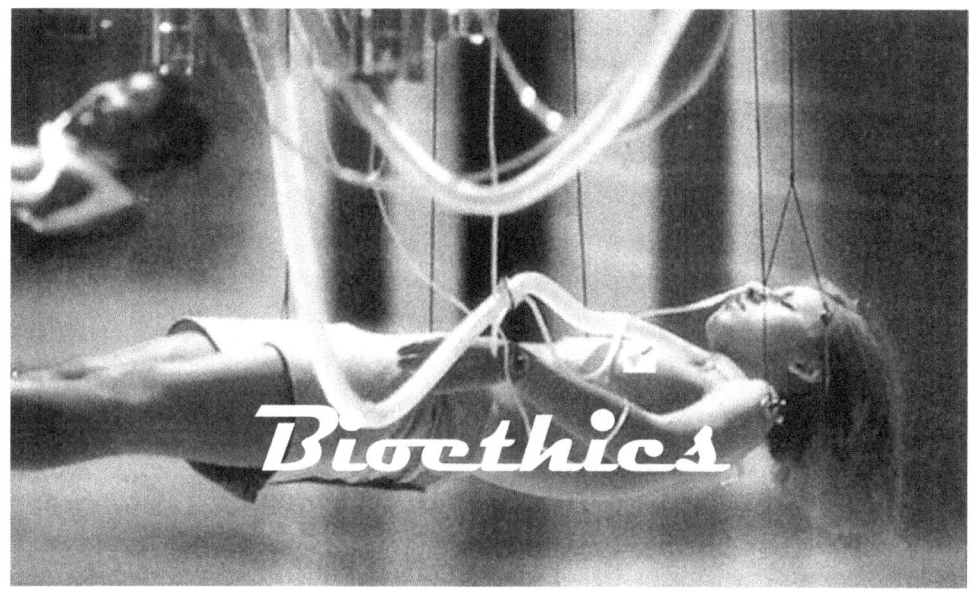

Bioethics is the study of the typically controversial ethical issues emerging from new situations and possibilities brought about by advances in biology and medicine. It is also moral discernment as it relates to medical policy and practice.

Bioethicists are concerned with the ethical questions that arise in the relationships among life sciences, biotechnology, medicine, politics, law, and philosophy. It also includes the study of the more commonplace questions of values ("the ethics of the ordinary") which arise in primary care and other branches of medicine.

Etymology

The term *Bioethics* (Greek *bios*, life; *ethos*, behavior) was coined in 1926 by Fritz Jahr, who "anticipated many of the arguments and discussions now current in biological research involving animals" in an article about the "bioethical imperative," as he called it, regarding the scientific use of animals and plants. In 1970, the American biochemist Van Rensselaer Potter also used the term with a broader meaning including solidarity towards the biosphere, thus generating a "global ethics," a discipline representing a link between biology, ecology, medicine and human values in order to attain the survival of both human beings and other animal species.

Purpose and Scope

The field of bioethics has addressed a broad swathe of human inquiry, ranging from debates over the boundaries of life (e.g. abortion, euthanasia), surrogacy, the allocation of scarce health care resources (e.g. organ donation, health care rationing) to the right to refuse medical care for religious or cultural reasons. Bioethicists often disagree among themselves over the precise limits of their discipline, debating whether the field should concern itself with the ethical evaluation of all questions involving biology and medicine, or only a subset of these questions. Some bioethicists would narrow ethical evaluation only to the morality of medical treatments or technological innovations, and the timing of medical treatment of humans. Others would broaden the scope of ethical evaluation to include the morality of all actions that might help or harm organisms capable of feeling fear.

The scope of bioethics can expand with biotechnology, including cloning, gene therapy, life extension, human genetic engineering, astroethics and life in space, and manipulation of basic biology through altered DNA, XNA and proteins. These developments will affect future evolution, and may require new principles that address life at its core, such as biotic ethics that values life itself at its basic biological processes and structures, and seeks their propagation.

Principles

One of the first areas addressed by modern bioethicists was that of human experimentation. The National Commission for the Protection of Human Subjects of Biomedical and Behavioral Research was initially established in 1974 to identify the basic ethical principles that should underlie the conduct of biomedical and behavioral research involving human subjects. However, the fundamental principles announced in the Belmont Report (1979)—namely, autonomy, beneficence and justice—have influenced the thinking of bioethicists across a wide range of issues. Others have added non-maleficence, human dignity and the sanctity of life to this list of cardinal values.

Another important principle of bioethics is its placement of value on discussion and presentation. Numerous discussion based bioethics groups exist in universities across the United States to champion exactly such goals. Examples include the Ohio State Bioethics Society, and the Bioethics Society of Cornell. Professional level versions of these organizations also exist.

Medical Ethics

Medical ethics is the study of moral values and judgments as they apply to medicine. As a scholarly discipline, medical ethics encompasses its practical application in clinical settings as well as work on its history, philosophy, theology, and sociology.

Medical ethics tends to be understood narrowly as an applied professional ethics, whereas bioethics appears to have worked more expansive concerns, touching upon the philosophy of science and issues of biotechnology. Still, the two fields often overlap and the distinction is more a matter of style than professional consensus. Medical ethics shares many principles with other branches of healthcare ethics, such as nursing ethics. A bioethicist assists the health care and research community in examining moral issues involved in our understanding of life and death, and resolving ethical dilemmas in medicine and science.

Perspectives and Methodology

Bioethicists come from a wide variety of backgrounds and have training in a diverse array of disciplines. The field contains individuals trained in philosophy such as H. Tristram Engelhardt, Jr. of Rice University, Baruch Brody of Rice University, Peter Singer of Princeton University, Daniel Callahan of the Hastings Center, and Daniel Brock of Harvard University, medically trained clinician ethicists such as Mark Siegler of the University of Chicago and Joseph Fins of Cornell University, lawyers such as Nancy Dubler of Albert Einstein College of Medicine or Jerry Menikoff of the federal Office of Human Research Protections, political scientists like Francis Fukuyama, religious studies scholars including James, public intellectuals like Amitai Etzioni of The George Washington University, and theologians like Lisa Sowle Cahill and Stanley Hauerwas. The field, once dominated by formally trained philosophers, has become increasingly interdisciplinary, with some critics even claiming that the methods of analytic philosophy have had a negative effect on the field's development. Leading journals in the field include The *Journal of Medicine and Philosophy*, The *Hastings Center Report*, the *American Journal of Bioethics*, the *Journal of Medical Ethics* and the Cambridge. Bioethics has also benefited from the process philosophy developed by Alfred.

Many religious communities have their own histories of inquiry into bioethical issues and have developed rules and guidelines on how to deal with these issues from within the viewpoint of their respective faiths. The Jewish, Christian and Muslim faiths have each developed a considerable body of literature on these matters. In the case of many non-Western cultures, a strict separation of religion from philosophy does not exist. In many Asian cultures, for example, there is a lively discussion on bioethical issues. Buddhist bioethics, in general, is characterized by a naturalistic outlook that leads to a rationalistic, pragmatic approach. Buddhist bioethicists include Damien Keown. In India, Vandana Shiva is a leading bioethicist speaking from the Hindu tradition. In Africa, and partly also in Latin America, the debate on bioethics frequently focuses on its practical relevance in the context of underdevelopment and geopolitical power relations.

Bioethics has also had its critics. Paul Farmer has pointed out that bioethics tends to focus its attention on problems that arise from "too much care," for patients in industrialized nations, while giving little or no attention to the ethical problem of too little care for the poor. Farmer characterizes the bioethics of handling difficult clinical situations, normally in hospitals in industrialized countries, as "quandary ethics." And he refers to bioethicists as "endlessly rehashing the perils of too much care." He does not regard quandary ethics and clinical bioethics as unimportant; he argues, rather, that bioethics must be balanced and give due weight to the poor.

(Wikipedia)

The Gates of Delirium

YES

Stand and fight we do consider
Reminded of an inner pact between us
That's seen as we go
And ride there
In motion
To fields in debts of honour
Defending

Stand the marchers soaring talons,
Peaceful lives will not deliver freedom,
Fighting we know,
Destroy oppression
The point to reaction
As leaders look to you
Attacking

Choose and renounce throwing chains to the floor.
Kill or be killing faster sins correct the flow.
Casting giant shadows off vast penetrating force
To alter via the war that seen
As friction spans the spirits wrath ascending to redeem.

Wars that shout in screams of anguish,
Power spent passion bespoils our soul receiver,
Surely we know.
In glory
We rise to offer,
Create our freedom,
A word we utter,

A word.

Words cause our banner, victorious our day.
Will silence be promised as violence display?
The curse increased we fight the pow'r
And live by it by day.
Our gods awake in thunderous roars,
And guide the leaders hand in paths of glory to the cause.

Listen, should we fight forever
Knowing as we do know fear destroys?
Listen, should we leave our children?
Listen, our lives stare in silence;
Help us now.

Listen, your friends have been broken,
They tell us of your poison; now we know.
Kill them, give them as they give us.
Slay them, burn their children's laughter
On to hell.

The fist will run, grasp metal to gun.
The spirit sings in crashing tones,
We gain the battle drum.
Our cries will shrill, the air will moan and crash into the dawn.
The pen won't stay the demon's wings,
The hour approaches pounding out the Devil's sermon.

Soon, oh soon the light,
Pass within and soothe this endless night
And wait here for you,
Our reason to be here.

Soon, oh soon the time,
All we move to gain will reach and calm;
Our heart is open,
Our reason to be here.

Long ago, set into rhyme.
Soon, oh soon the light,
Ours to shape for all time,
Ours the right;
The sun will lead us,

Our reason to be here.

Soon, oh soon the light,
Ours to shape for all time,
Ours the right;
The sun will lead us,
Our reason to be here.

The Last Judgment

The Tower: Book IV

Max Shindler

Rat-Borne Diseases

Rat bites and scratches can result in disease and rat-bite fever. Rat urine is responsible for the spread of leptospirosis, which can result in liver and kidney damage. It can also be contracted through handling or inhalation of scat. Complications include renal and liver failure, as well as cardiovascular problems.

Lymphocytic choriomeningitis (LCMV), a viral infectious disease, is transmitted through the saliva and urine of rats. Some individuals experience long-term effects of lymphocytic choriomeningitis, while others experience only temporary discomfort.

One of the most historically dangerous rat-borne diseases is the bubonic plague, also called "Black Plague," and its variants. Transfer occurs when fleas from the rats bite human beings. Fleas transported on rats are considered responsible for this plague during the Middle Ages, which killed millions. From the transmission of bubonic plague to typhus and hantavirus, rat infestations can prove harmful to human health.

Rats also are a potential source of allergens. Their droppings, dander and shed hair can cause people to sneeze and experience other allergic reactions.

Diseases transmitted by rats fall into one of two categories: diseases transmitted directly from exposure to rat-infected feces, urine or bites and diseases indirectly transmitted to people by an intermediate arthropod vector such as fleas, ticks or mites. While the following list of diseases or medical conditions are all associated with rats, most are not commonly encountered in the United States.

Diseases Directly Transmitted by Rats

- **Hantavirus Pulmonary Syndrome**. This is a viral disease that is transmitted by the rice rat. This disease is spread in one of three ways: inhaling dust that is contaminated with rat urine or droppings, direct contact with rat feces or urine, and infrequently due to the bite of rat.
- **Leptospirosis**. This is a bacterial disease that can be transmitted by coming into contact with infected water by swimming, wading or

kayaking or by contaminated drinking water. Individuals may be at increased risk of Leptospirosis infections if they work outdoors or with animals.

- **Rat-bite Fever**. This disease may be transmitted through a bite, scratch or contact with a dead rat.
- **Salmonellosis**. Consuming food or water that is contaminated by rat feces bacteria can cause this disease.

Diseases Indirectly Transmitted by Rats

- **Plague**. This disease is carried by rats and transmitted by fleas in the process of taking a blood meal. Domestic rats are the most common reservoir of plague.
- **Colorado Tick Fever**. This is a viral disease that is transmitted by the bite of a tick that has taken a blood meal from a bushy-tailed woodrat.
- **Cutaneous Leishmaniasis**. This disease is a parasite that is transmitted to a person by the bite of an infected sand fly that has fed on a wild woodrat.

Some species of rats such as the cotton rat or rice rat are known carriers of hantavirus. Norway rats and roof rats are not known transmitters of hantavirus. Victims may be debilitated and can experience difficulty breathing. Hantavirus is transmitted to humans when they inhale airborne particles from rodent droppings, urine or carcasses that have been disturbed.

The first symptoms of the virus can be mistaken for the flu. Patients then suffer breathing difficulties that may prove fatal if not treated effectively and immediately.

In order to avoid hantavirus, all mouse feces, nest materials and dead rodents must be removed from the home. Spray suspected areas thoroughly with disinfectant before sweeping to avoid having anything become airborne. Use gloves to handle rodent carcasses or droppings and a respirator must be worn with functioning cartridges. Buildings should be aired out following an infestation. Not all rodents have been found to carry hantavirus. Deer mice, cotton rats, rice rats and white-footed mice are the most common transmitters. However, everyone should use caution in dealing with rodents or rodent infestations and contact a pest control professional.

Erich Fromm's prediction that Homo-sapiens will become "Eichmanns" has been proven correct; the vast majority of people do not question, or make any attempt to defy "rules" regardless of the cost to another.

Dr. Fromm stated that there are two types of people, the "have mores", and the "be mores". The "have mores" are materialistically oriented, and require more, and more, "materials" to keep themselves satiated, (the same as how drug addicts habituate to drugs and require a greater potency to get the same effect, or level of satisfaction. The "be mores", on the other hand, want to be more as human beings; to actualize themselves; reach their fullest potential. The choice in life is made early as to which path a person wishes to follow, usually before the age of 7.

Karl Marx describes history as primarily driven by economics, and that people decide how much, and to what extent, they wish to orient toward the acquisition of materials in order to acquire pleasure, or, on the other hand, humanize themselves by developing something to offer others.

Only someone who's dehumanized could behave in the evil, narcissistic, manner I describe in "Miracle of Life". A being that is human, is humane; that person has the capacity to feel compassion, and has the ability to empathize.

The change in the nature of people proves how essential it is that Jews be present; they lead others in the right direction; toward the "good', which is love. I would conclude that The Lord's decision to withdraw the Jews from Earth was the fitting punishment for the transgressions that had taken place, and the most effective way He could have chosen to teach the most important lesson people need to learn.

Adolf Eichmann
The Mind of a War Criminal

By Professor David Cesarani

Adolf Eichmann systematically applied the logistics of commerce to the annihilation of Jews during the Holocaust. David Cesarani examines the mind of a Nazi war criminal.

Early Influences

Adolf Eichmann was born in 1906 in Solingen, a small industrial city in the Rhineland. His father was an accountant with a local power company, but was assigned to a superior posting in Linz, Austria, in 1913. Eichmann and his five siblings followed. In 1916 his mother died and his father quickly remarried. Eichmann senior was an active member of the Evangelical Church and his son remained in the faith until 1937, long after most SS men broke with religion.

Eichmann was adept at learning practical skills on the job, under the tutelage of seniors he respected.

Eichmann was very much under his father's influence, and older male authority figures would continue to mould his life. Nevertheless, he did not work hard or do well at school and left without any qualifications. His father, who had meanwhile started an oil-extraction business, gave him a job. Eichmann worked on the surface and in underground oil-shale tunnels before moving to an apprenticeship with an electrical engineering firm. In 1927 his father used family contacts to get him a job with another oil company.

Little attention has been paid to Eichmann's work experience, but it had a significant bearing on his career in the SS. Eichmann was adept at learning practical skills on the job, under the tutelage of seniors he respected. While he continued to live at home, he ranged over Upper Austria selling oil products, locating sites for petrol stations, and setting them up. He also arranged kerosene deliveries. On Saturday he conscientiously completed his paperwork and reported to his superiors.

Eichmann did well and was transferred to the Salzburg district. But by 1933 he had tired of the job and, anyway, was laid off. He had learned a lot, though: how to identify prime sites at communication junctions, how to timetable and organise deliveries, how to sell a product and persuade people to do your bidding. After he was made redundant he went north to Germany, partly in search of work but mainly in fulfilment of a new passion: politics.

Drawn to the Nazis

The young Adolf Eichmann

During his trial he pretended to be apolitical, but Eichmann came from a strongly German nationalist family. Like many Germans his father lost his wealth during the post-war economic crisis and had the embittering experience of starting all over again. He enrolled his son in the *Wandervogel* youth movement which, while ostensibly apolitical, was strongly imbued with *völkisch* ideas about the Heimat (homeland). Later, Eichmann joined the Linz branch of the Heimschutz, a right wing paramilitary association of army veterans. He considered joining a Masonic club that recommended itself to him because it excluded Jews.

Eichmann claimed that he joined the SD by error, but it suited his talents.

Instead, in April 1932, he joined the Nazi party. At the instigation of the local gauleiter, who knew his family, he attended a Nazi rally and was approached by an SS man called Ernst Kaltenbrunner, whose father had business dealings with Eichmann senior. Kaltenbrunner must have known that Eichmann was ripe for the party because he told him: 'You belong to us'. Eichmann combined commerce with activism in the Austrian SS until 1933, when the party was outlawed and Kaltenbrunner arranged for him to go to Germany. He spent some time at an SS training centre and with an exiled Austrian SS unit before he was posted to Dachau concentration camp. From there he applied to join the SD, the Nazi Party Security Service, and was accepted for work at one of its Berlin branches.

Eichmann claimed that he joined the SD by error, but it suited his talents. He worked as a clerk in the section that monitored Freemasons before he was spotted by the head of the Jewish section of the SD, Edler von Mildenstein, who became his next 'mentor'.

The Jewish Question

Von Mildenstein took a special interest in Zionism and Jewish emigration to Palestine as a solution to Germany's 'Jewish Question'. He encouraged Eichmann to study Jewish society and history so as better to understand the Jewish enemy. Eichmann excelled and earned a series of promotions, but the SD was a minor part of the SS machine at this time and its Jewish section was a backwater. Other departments of the Third Reich set the pace regarding policy on the Jews. Eichmann rose to prominence in this field only because from the mid 1930s the SD under Reinhardt Heydrich targeted Jewish issues and built a reputation as a centre for clear, scientific thinking on race.

... he warned the SD that it would be foolish to promote a strong Jewish state.

While rabble-rousers like Joseph Goebbels railed against the Jews, and called for ever harsher but directionless measures against them, the SD quietly promoted Jewish emigration. To this end Eichmann contacted Zionist envoys and even made a visit to Palestine in 1937.

This trip, aborted after one day, revealed the true extent of his sympathy for Zionism: he warned the SD that it would be foolish to promote a strong Jewish state. Instead, it should encourage Jewish emigration to backward countries where they would live in poverty. Soon after he completed this mission, Eichmann was assigned to the SD in Vienna.

Business Practice

Files in the Austrian State Archive bear witness to the vanished Jews of Vienna. In March 1938, Germany occupied Austria and a reign of terror broke over the Austrian Jews. Eichmann was given the task of accelerating Jewish emigration and easing the numerous bottlenecks through which aspiring emigrants had to pass. Eichmann used business practice to create order. He surveyed the relevant agencies and ordered them to locate their offices in one place. He ordered the creation of a central Jewish organisation so that he would have leaders with whom to negotiate, and allowed Zionist organisations to operate. Money was extracted from well off Jews to fund the emigration of the mass of poor Jews.

Eichmann explored a fresh option: deporting the Jews to a designated Jewish territory.

Finally, he established an 'assembly line' system whereby a Jew could up at the Central Emigration Office with his papers and proceed from desk to desk until he arrived at the end, with a passport and an exit visa but stripped of his property, cash and rights. Within a few months, the office had emigrated 150,000 Jews.

After this triumph, Eichmann was ordered to set up a similar office in occupied Prague, and in October 1939 was appointed to Department IV D 4 of the Gestapo in Berlin, which handled emigration from the Reich. The rational 'Jewish policy' advocated by the SD men now held sway, but emigration opportunities were few and Germany had just acquired over a million more Jews in conquered Poland. Eichmann explored a fresh option: deporting the Jews to a designated Jewish territory. He travelled to Poland to identify an appropriate location and then ordered that thousands of Czech and Viennese Jews be rounded up and sent eastwards to lay the basis for this 'territorial solution'.

Within a few months, however, the plan was scrapped. Eichmann's office lacked the resources for it and other SS projects had preference. At the same time he was brutally evicting hundreds of thousands of Poles and Jews to make way for ethnic Germans transplanted from Eastern Europe into the newly annexed areas of the Reich. As a temporary measure the displaced Jews were packed into ghettos, but where would they go eventually? After the fall of France, Eichmann took up a plan emanating from the German Foreign Office to ship four million European Jews to Madagascar. He devoted great energy and research skills to the scheme, but it too foundered.

The Final Solution

When Germany invaded Russia in June 1941 an expectation swept through the agencies responsible for Jewish affairs. It was anticipated that soon Jews could be transported to the east and dumped there. Meanwhile, mobile killing units, Einsatzgruppen, swept across Russia slaughtering Jews who were deemed Bolshevik enemies. Eichmann had little to do with this, but in the summer his office (now designated IV B 4 and, significantly, no longer concerned with emigration) was called upon to investigate ways to dispose of 'unwanted' Jews.

By this time decisions had already been taken to murder those Jews in the Polish ghettos who were not deemed capable of work. Eichmann was advised to check on how it was being done. Over a few months he saw gassing operations at Chelm, mass shootings in Minsk, and visited Auschwitz. He prepared the ground for the Wannsee Conference in January 1942, at which Heydrich secured the co-operation of the various departments of state, the Nazi party and the SS in the 'Final Solution of the Jewish Question'.

He made numerous interventions to prevent a single Jew being exempted from the transports.

Eichmann later claimed that he was shocked to hear that 'evacuation to the east' meant death, but the concurrence of high-ranking officials absolved him of responsibility and guilt. It is hard to reconcile this with the zeal he devoted to organising the registration, expropriation, rounding up and deportation of Jews from Germany, Austria, France, Holland, Belgium, Slovakia, Greece, Italy and, above all, from Hungary to the death camps. He sent out trusted assistants to make the local arrangements, chivvied them if they did not make fast enough progress, and belaboured officials who prevaricated or objected. He made numerous interventions to prevent a single Jew being exempted from the transports.

In March 1944, after German forces invaded Hungary, he travelled to Budapest with a special task force and personally directed the plunder, ghettoisation, and deportation of over 437,000 Jews in the space of eight weeks, most of whom were murdered on arrival in Auschwitz-Birkenau. When, under international pressure, the Hungarian regime stopped the deportations he circumvented its orders and dispatched a last trainload to the gas chambers. He even defied his chief, Himmler, who at the end of 1944 finally commanded the killing to stop.

Tape Recordings

In hiding in Argentina in the mid 1950s, Eichmann recorded on tape his recollections of these final days. 'I called my men into my Berlin office ... and formally took leave of them. 'If it has to be', I told them, 'I will gladly jump into my grave in the knowledge that five million enemies of the Reich have already died like animals.' This statement gives a clue to how Eichmann's mind worked. The Jews were the enemy. He had nothing against them personally, but in war the enemy has to be destroyed. Eichmann did not kill a single Jew with his own hands and he was often courteous towards Jewish leaders who did his bidding. Yet he could also be abusive and violent: as his power burgeoned and his bourgeois inhibitions were eroded, he became increasingly coarse.

Eichmann learned to hate, and to hate in a controlled and impersonal way.

Even so, Eichmann was not the central, demonic figure of the Nazi regime he was made out to be in his trial, and as he has become in popular memory. He did not make any key decisions on Jewish policy and at no point before mid 1941 could he have known where it was leading. The genocide was set in motion by others and at first proceeded independently from his office.

That he committed atrocities before then is beyond doubt, and there is no disputing the fact that he became an accomplice to a widening circle of mass murder that he helped to sustain with all his might. What makes his crimes so chilling is that they were not preordained by any evident pathology or inbuilt racism. Eichmann learned to hate, and to hate in a controlled and impersonal way. He applied business methods to the handling of human beings who, once they had been dehumanised, could be treated no differently from cargoes of kerosene. In his mind there was little difference between setting up a petrol station or a death camp.
(Wikipedia)

A Good Person

A priest presents a face that is kind to all.
"I offer you solace, a listening ear.
I will do my utmost to lessen your fear."
The person is unique; created by a Church that is simply a freak.

In the Vatican people pray; offer words of guidance and hope,
While, really, men play with children, because they are gay.

The outside is astonishing; great works of art, spirals soar into the sky,
Paintings describe scenes from The Bible;
Words of truth, however, about their deeds would be considered libel.

"I stand before you as a man of God.
He speaks to me, I can speak to you",
Absurd you may think the combination of the two.

Ridiculous sermons; venomous deeds.
A persona is a façade; it hides a disquieting reality.
Protection is derived from the essential faculty.

Often that which is soft, is, in fact, hard.
Extremely difficult it is to face; things in accordance with how they are.

The most suspicious character is often a charming Barrister.
"Justice is what we seek", proudly he will state his case.
Conviction will be accomplished through a Legal System that is brute force.

Little of what we actually see is how we would like it to be.
Dark is light; light is faint.
Observing the Artist as he struggles to paint;
There is an honest man, baring his soul for all to see.
Kind to others when away from his work,
But, harsh to those who present an obstacle before his goal;
To know Him who resides in his soul

Woman

Where is it I wish to reside,
So I can recover from all the lies I've been told?
I want somewhere pleasant, so I can consider it made of gold.

So often my skies have turned blue, then grey,
And finally to a state of black.
Stormy clouds gather then disperse.
So often these things have been due
To the creatures who prefer to carry a purse.

At times I would like, in their face, to spit a curse;
But, really, they are so spiteful and mean,
They'd probably strike a back, hard and fast,
Resulting in my ending up in a dark hearse.

Regardless of the pain I suffer,
They will consider themselves fit, or so they deem.

Acts of kindness mask a horrendous darkness.
Fangs appear, stick into your side.

With a smile they apologize;
All the while having a smirk inside.

These beings were once thought of as caring, comforting,
Made of sugar, spice, and all things nice;
But, truth be told, your life they will surrender,
It's just a game with each throw of the dice.

If they tend to your hair, be careful, and mindful, to search after;
To your shock and surprise, you may find embedded ugly, slimy, lice.

They infect you with nonsensical sayings, and words.
My God, if they would just go away,
I'd be free to enjoy each day

Evil by Chance or Happenstance

What will evil do?

Whatever it likes; whenever it likes;
That is its nature.
Its aim is that all deserving will one day live in a hole,
Where rats will bite till the bone it reaches.
No longer are these people living on glorious beaches.

By having so much, they have insured that others have too little.
Things must always be to their liking; they can be very fickle.
All this to them is a laugh; as delightful as a playful tickle.

Caring for others is of no interest to them.
People die, children become orphans,
So that on their finger they may wear a colourful gem.
Wives must have dresses with an appealing, enticing, dangling, hem.

Who, in fact, has stolen more from the earth, women or men?
The end reveals the truth.
Life will return to how it once was.
The story of Genesis reveals the cause for our loss.

The Serpent made a choice based on the knowledge it held.
Eve took a bite out of the apple that represents life,
And so began the beginning of our woes.

Why do the mystics give warnings?
Ramakrishna said, avoid women and gold.
They seem appealing, and serve so well;
But, ask yourself; how many today have lost their soul?
It was just another product to be sold.

The Miracle of Life

One among many, I walk and stumble, along a gravel patched road.
The night has grown dark; men growl with despair
As they walk side by side as a pair.

A man falls by the wayside.
Everyone looks the other way as if he doesn't care.
Nothing could be further from the truth;
To continue onward, along the tunnel of death, is now the ultimate dare.

Men in black suits watch.
They strike with a cane.
Do they know, understand,
The land, and time, from which they came?

Their orders that bark, note that this is not possible;
How they now treat these men is damn horrible.

Trembling, an old man falls to his knees.
A Nazi slams a rod into his face.
"Have mercy, dear God, stop, please!" is his cry.
He now shudders with the pain that racks his bones.
His hatred stings like a swarm of bees.

"Stand; march, or face your end, a sinister voice says to the now slain soul.
For months he has had to eat from a stained, chipped, bowl.

"How can I, when I have no strength left?"
"You have no choice; it is for your best."
With the last morsel of energy that remains,
The grandfather of many lifts his head.
Each movement helps toward his cause;
These are the sole things from which he gains.

Fearful that his demise is so very near, he pulls himself up.
The thirst he now bares is incredibly acute.
It is not water he craves, but the warmth of a thick, dark, beer.

Two steps more are taken, then he collapses once more to the ground.
The guard that hovers over now appears as a sinister hound.

The lips part, and form a grin.
Why's this happening? The old man asks;
Did I commit some heinous, terrible, sin?

Will he offer me more time? His legs now hold the strength of sheer lace.
The pistol at his head reveals this is not the case.
Their eyes now meet, each one stares at the others face.
One is cold; the other is covered with a disguise.
The wrinkled, aged, one, hides his despise.

"You are vermin; correction, mice among men.
Deserve you all to live in a pig's pen.
That is where you belong; secluded, alone, a ghettoized den."

From behind, another cries for the others' safety.
"Give the man a chance!"
Once he reflects, the elder enjoyed the gift of dance.

Suddenly, a boom sounds.
A bullet enters the heart of the one that spoke.
In an instant, he has now fallen.
Instead of a helpless lost soul; a world now opens that is shiny, golden.
God Himself appears, immeasurable beyond belief;
Such is the magnitude of His scope.

Once his heart beats its last, together, at once, he lies in peace.
Strange at this time there should pass a family of geese.

Is this a symbol, sign?
Does it make up for a life full of sweat, dirt, and grime?

The sight does not escape the wise, elderly, fellow;
His cries of anguish, and rage, are so loud they bellow.

To help mask how unsettled he feels within,
He digs deep into his shirt pocket for his gin.
Next he grabs the Jewish grandfather,
And carries him over to a soiled garbage bin.
He shoves within the head of the one many have called a Sage.
The pistol is pressed against his forehead.
Is the Nazi's world now covered in a haze?
"Please, I beg of you, give me a chance to speak.
Am I not worthy of this, before you choose to end all my days?"
"Say what you have to say, and fast!"
"Give me a bit of time, they will, after all, be my last."

The Nazi remains silent
while the grandfather formulates the words he wishes to utter.
"Why am I the one that deserves to live in a gutter?"

Instead of an answer, blood is sent gushing from a gaping wound.
All is over, a life has passed.
The soldier is responsible for the others' doom.
What wasn't heard was the silent wish for a miracle.

Strange it is the Nazi soon dies from the shot of an arrow.
Was the prayer answered? We really don't know;
But who from the nearby hill
Held the strong bow?

Stephen King's

Firestarter

Firestarter is a science fiction novel by Stephen King, first published in September 1980. In July and August 1980, two excerpts from the novel were published in *Omni*. In 1981, *Firestarter* was nominated as Best Novel for the British, Locus Poll Award, and Balrog Award. In 1984, it was adapted into a movie.

The book is dedicated to the author Shirley Jackson: "In Memory of Shirley Jackson, who never needed to raise her voice."

Plot Summary

Andy and Charlene "Charlie" McGee are a father-daughter pair on the run from a government agency known as The Shop. During his college years, Andy had participated in a Shop experiment dealing with "Lot 6", a drug with hallucinogenic effects similar to LSD. The drug gave his future wife, Victoria Tomlinson, minor telekinetic abilities, and him an autohypnotic mind domination ability he refers to as "the push". They both also developed telepathic abilities. Andy's and Vicky's powers were physiologically limited; in his case, overuse of the push gives him crippling migraine headaches and minute brain hemorrhages, but their daughter Charlie developed a frightening pyrokinetic ability, with the full extent of her power unknown. (Although much later in the novel, Charlie develops an inner conviction that she will eventually be powerful enough to "change the sun" in some way.) The novel begins *in medias res* with Charlie and Andy on the run from Shop agents in New York City. We learn through a combination of flashbacks and current narration that this is the latest in a series of attempts by the Shop to capture Andy and Charlie following an initial disastrous raid on the McGee family's quiet life in suburban Ohio. After years of Shop surveillance, a botched operation to take Charlie leaves her mother dead; Andy, receiving a psychic flash while having lunch with work colleagues, rushes home to discover his wife murdered and his daughter kidnapped. He then uses his push ability to track the slightly-cold trail of Charlie and the Shop agents, catching up to them at a rest stop on the Interstate. He uses the push to incapacitate the Shop agents, leaving one blind and the other comatose. Charlie and Andy flee and begin a life of running and hiding, using assumed identities. They move several times to avoid discovery before the Shop catches up to them in New York.

Using a combination of the push, Charlie's power, and hitchhiking, the pair escape through Albany, New York and are taken in by a farmer named Irv Manders near the fictional town of Hastings Glen, NY; however, they are tracked down by Shop agents, who attempt to kill Andy and take Charlie at the Manders farm. At Andy's instruction, Charlie unleashes her power, incinerating the entire farm and fending off the agents, killing a few of them. With nowhere else to turn, the pair flee to Vermont and take refuge in a cabin that had once belonged to Andy's grandfather. With the Manders farm operation disastrously botched, the director of the Shop, Captain James Hollister, or "Cap", calls in a Shop hitman named John Rainbird   to   capture   the   fugitives.   Rainbird, a Cherokee and Vietnam veteran,   is   intrigued   by   Charlie's   power   and eventually becomes obsessed with her, determined to befriend her and eventually kill her. This time the operation is successful, and both Andy and Charlie are taken by the Shop.

The pair is separated and imprisoned at the Shop headquarters, located in the fictional D.C. suburb of Longmont, Virginia. With his spirit broken, Andy becomes an overweight drug addict and seemingly loses his power, and is eventually deemed useless by the Shop. Charlie, however, defiantly refuses to cooperate with the Shop, and does not demonstrate her power for them. Six months pass until a power failure provides a turning point for the two: Andy, sick with fear and self-pity, somehow regains the push - subconsciously pushing *himself* to overcome his addiction - and Rainbird, masquerading as a simple janitor, befriends Charlie and gains her trust.

By pretending to still be powerless and addicted, Andy manages to gain crucial information by pushing his psychiatrist. Under Rainbird's guidance, Charlie begins to demonstrate her power, which has grown to fearsome levels. After the suicide of his psychiatrist, Andy is able to meet and push Cap, using him to plan his and Charlie's escape from the facility, as well as finally communicating with Charlie. Rainbird discovers Andy's plan, however, and decides to use it to his advantage.

Andy's plan succeeds, and he and Charlie are reunited for the first time in six months. Rainbird then interrupts the meeting at a barn, planning to kill them both. A crucial distraction is provided by Cap, who is losing his mind from a side effect of being pushed. Andy pushes Rainbird into leaping from the upper level of the barn, breaking his leg. Rainbird then shoots Andy in the neck. Rainbird then fires another shot at Charlie, but she uses her power to melt the bullet in midair and then sets Rainbird and Cap on fire. A mortally wounded Andy then instructs Charlie to take revenge with her power and inform the public, to make sure the government cannot do anything like this ever again, and dies. A grief-stricken and furious Charlie then sets the barn on fire. She exits the barn and people start going after her. She uses her pyrokinesis to kill the employees and blow up their getaway vehicles. People try to flee and some do. Military men are called, but Charlie blows up their vehicles and when they fire at her she melts their

bullets. Charlie blows up the building, shooting it sky-high. She leaves the Longmont facility burning, with almost all of its workers dead. The event is covered up by the government, and released to the papers as a terrorist firebomb attack. The Shop quickly reforms, under new leadership, and begins a manhunt for Charlie, who has returned to the Manders farm. After some deliberation, she comes up with a plan and leaves the Manders', just ahead of Shop operatives, and heads to New York City. She decides on *Rolling Stone* magazine as an unbiased, honest media source with no ties to the government, and the book ends as she arrives to tell them her story.

Part Five

The Miracle of Life

Creation occurs due to a confluence of forces at the same place. The natural order of things is good, and can also be labeled as love; hence, love is the nature of existence.

Wicca

Wiccans say, "Free we should be."
To harm none is the goal,
While you do all you can to feed your soul;
Which might, at times,
Include keeping charms in a lucky bowl.

Individuals, they are; and always strive to be;
Manipulating energies around and within;
Even while standing naked by the side of a deep blue sea.

Their quest is not to discover where we begin,
But, rather, to make the most of the life we are in,
And, occasionally, this might mean having a Gin.
If we can all learn to live this way,
I see no reason, why on earth, we can't all win!

They choose to be carefree; how else can you feel liberty?
Life should continually intensify.
To quantify and qualify, the glories nature holds,
Is beyond man's capacity to specify.

Therefore, make the most of each day.
Relish the stars that shine in the sky;
And the Sun, and Moon, as they pass by.
Heaven sent, all these things are.
There is nothing man can create
That could possibly be considered of equal par.
With so much to relish, we have cause for jubilation.
Let us give each a standing ovation!!

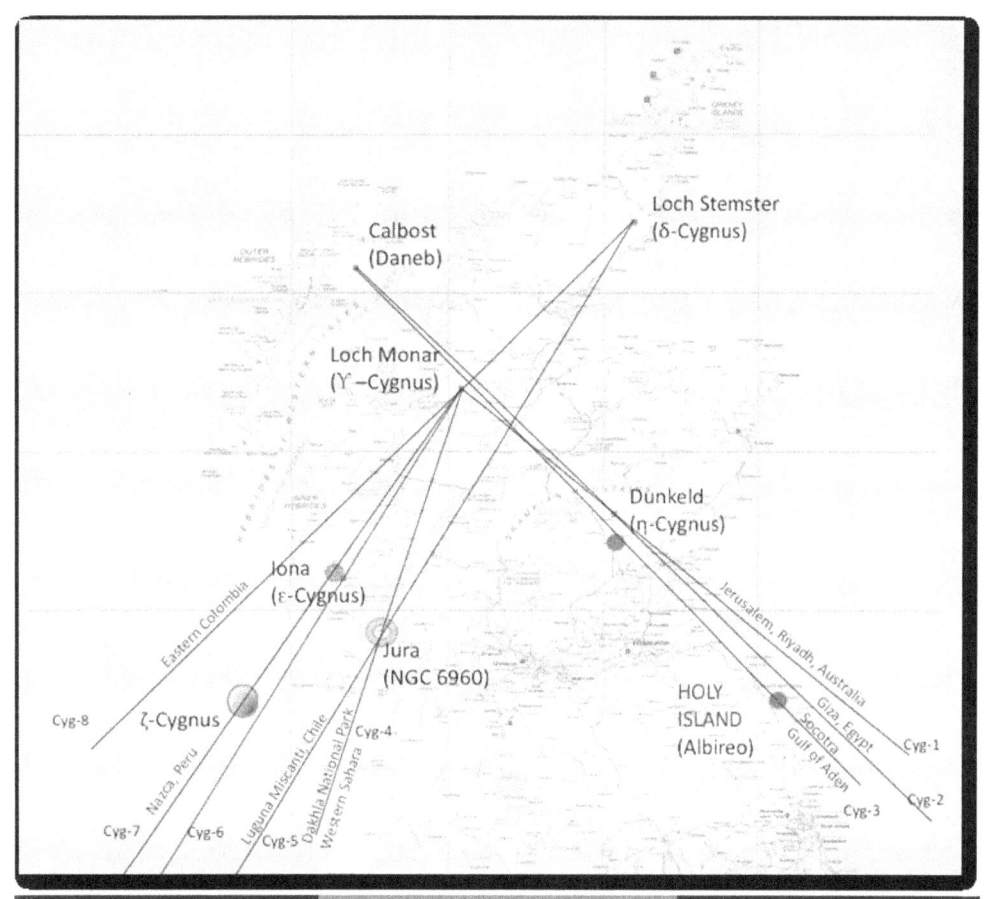

Calbost
(Daneb)

Loch Stemster
(δ-Cygnus)

Loch Monar
(ϒ–Cygnus)

Dunkeld
(η-Cygnus)

Iona
(ε-Cygnus)

Jura
(NGC 6960)

ζ-Cygnus

Eastern Colombia

Nazca, Peru

Laguna Miscanti, Chile

Dakhla National Park
Western Sahara

HOLY
ISLAND
(Albireo)

Jerusalem, Riyadh, Australia

Giza, Egypt

Socotra

Gulf of Aden

Cyg-8
Cyg-7
Cyg-6
Cyg-5
Cyg-4
Cyg-3
Cyg-2
Cyg-1

Ley Line

The phrase was coined in 1921 by the amateur archaeologist Alfred Watkins, referring to supposed alignments of numerous places of geographical and historical interest, such as ancient monuments and megaliths, natural ridge-tops and water-fords. In his books *Early British Trackways* and *The Old Straight Track*, he sought to identify ancient trackways in the British landscape. Watkins later developed theories that these alignments were created for ease of overland trekking by line-of-sight navigation during Neolithic times, and had persisted in the landscape over millennia.

In a book called *The View Over Atlantis* (1969), the writer John Michell revived the term "ley lines", associating it with spiritual and mystical theories about alignments of land forms, drawing on the Chinese concept of feng shui. He believed that a mystical network of ley lines existed across Britain.

Since the publication of Michell's book, the spiritualized version of the concept has been adopted by other authors and applied to landscapes in many places around the world. Both versions of the theory have been criticized on the grounds that a random distribution of a sufficient number of points on a plane will inevitably create alignments of random points purely by chance.

Alfred Watkins and The Old Straight Track

The concept of "ley lines" originated with Alfred Watkins in his books *Early British Trackways* and *The Old Straight Track*, though Watkins also drew on earlier ideas about alignments; in particular he cited the work of the English astronomer Norman Lockyer, who argued that ancient alignments might be oriented to sunrise and sunset at solstices.

On 30 June 1921, Alfred Watkins visited Blackwardine in Herefordshire, and had been driving along a road near the village (which has now virtually disappeared). Attracted by the nearby archaeological investigation of a Roman camp, he stopped his car to compare the landscape on either side of the road with the marked features on his much used map. While gazing at the scene around him and consulting the map, he saw, in the words of his son, "like a chain of fairy lights" a series of straight alignments of various ancient features, such as standing stones, wayside crosses, causeways, hill forts, and ancient churches on mounds. He realized immediately that the potential discovery had to be checked from higher ground when, during a revelation, he noticed that many of the footpaths there seemed to connect one hilltop to another in a straight line.

He subsequently coined the term "ley" at least partly because the lines passed through places whose names contained the syllable *ley*, stating that philologists defined the word (spelled also as lay, lea, lee, or leigh) differently, but had misinterpreted it. He believed this was the ancient name for the trackways, preserved in the modern names. The ancient surveyors who supposedly made the lines were given the name "dodmen". Watkins believed that, in ancient times, when Britain was far more densely forested, the country was criss-crossed by a network of straight-line travel routes, with prominent features of the landscape being used as navigation points. This observation was made public at a meeting of the Woolhope Naturalists' Field Club of Hereford in September 1921.

Snodhill Castle in Dorstone

His work referred to G. H. Piper's paper presented to the Woolhope Club in 1882, which noted that: "A line drawn from the Skirrid-fawr mountain northwards to Arthur's Stone would pass over the camp and southern most point of Hatterall Hill, Oldcastle, Longtown Castle, and Urishay and Snodhill castles."

It has also been suggested that Watkins' speculation (he called it 'surmise') stemmed from reading an account in September 1870 by William Henry Black given to the British Archaeological Association in Hereford titled *Boundaries and Landmarks*, in which he speculated that "Monuments exist marking grand geometrical lines which cover the whole of Western Europe". He published his book *Early British Trackways* the following year, commenting: "I knew nothing on June 30th last of what I now communicate, and had no theories".

Examples of Ley Lines in Britain

Alfred Watkins theorized that St. Ann's Well in Worcestershire is the start of a ley line that passes along the ridge of the Malvern Hills through several springs including the Holy Well, Walms Well, and St. Pewtress Well.

In *The Ley Hunter's Companion* (1979), Paul Devereux theorized that a 10-mile alignment he called the "Malvern Ley" passed through St Ann's Well, the Wyche Cutting, a section of the Shire Ditch, Midsummer Hill, Whiteleaved Oak, Redmarley D'Abitot and Pauntley.

In *City of Revelation* (1973) British author John Michell theorized that Whiteleaved Oak is the centre of a circular alignment he called the

"Circle of Perpetual Choirs"

and is equidistant from Glastonbury, Stonehenge, Goring-on-Thames and Llantwit Major. The theory was investigated by the British Society of Dowsers and used as background material by Phil Rickman in his novel *The Remains of an Altar* (2006).

Perhaps relevant to the ley line argument is the existence of cursuses, massive parallel imprints in the ground made by people between 3400 and 3000 BCE. Ranging in length from 50 metres to several kilometers, their exact function remains unknown though they are commonly believed to have been used for ceremonial processions. Many of them do encompass Neolithic graves and monuments. However, while some cursuses are relatively straight, others have curves and sharp turns. This may argue that ancient Britons had little interest in moving in straight lines over landscapes. **(Wikipedia)**

Crop Circles

A **crop circle** or **crop formation** is a pattern created by flattening a crop, usually a cereal. Crop circles, as Taner Edis, professor of physics at Truman State University puts it, "all fall within the range of the sort of thing done in hoaxes." Although obscure natural causes or alien origins of crop circles are suggested by fringe theorists, there is no scientific evidence for such explanations, and human causes are consistent for all crop circles. A commentary in *The Guardian* noted that "[i]t is still open to dispute whether some are caused by natural phenomena or all created by human hand."

The number of crop circles has substantially increased from the 1970s to current times. There has been little scientific study of them. Circles in the United Kingdom are not spread randomly across the landscape but appear near roads, areas of medium to dense population and cultural heritage monuments, such as Stonehenge or Avebury. In 1991, two hoaxers, Bower and Chorley, made disputed claims to have created many circles throughout England after one of their circles was described by a circle investigator as impossible to be made by human hand.

History

The concept of "crop circles" began with the original late-1970s hoaxes by Doug Bower and Dave Chorley (see Bower and Chorley, below). They said that they were inspired by the Tully "saucer nest" case in Australia, where a farmer found a flattened circle of swamp reeds after observing a UFO.

Early Reports of Circular Formations

A 1678 news pamphlet *The Mowing-Devil: or, Strange News Out of Hartfordshire* is claimed by some cereologists to be the first depiction of a crop circle. Crop circle researcher Jim Schnabel does not consider it to be a historical precedent because it describes the stalks as being cut rather than bent.

In 1686, British naturalist Robert Plot reported on rings or arcs of mushrooms (see fairy rings) in *The Natural History of Stafford-Shire* and proposed air flows from the sky as a cause. In 1991 meteorologist Terence Meaden linked this report with modern crop circles, a claim that has been compared with those made by Erich von Däniken.

An 1880 letter to the editor of *Nature* by amateur scientist John Rand Capron describes how a recent storm had created several circles of flattened crops in a field.

Modern Times

In 1932, archaeologist E C Curwen, observed four dark rings in a field at Stoughton Down near Chichester, he could examine only one: "a circle in which the barley was 'lodged' or beaten down, while the interior are was very slightly mounded up."

In 1963 amateur astronomer Sir Patrick Moore described a crater in a potato field in Wiltshire, probably caused by an unknown meteoric body. In nearby wheat fields, there were several circular and elliptical areas where the wheat had been flattened. There was evidence of "spiral flattening". He thought they could be caused by air currents from the impact, since they led towards the crater. Astronomer Hugh Ernest Butler observed similar craters and said they were likely caused by lighting strikes.

In the 1960s, in Tully, Queensland, Australia, and in Canada, there were many reports of UFO sightings and circular formations in swamp reeds and sugar cane fields. For example, on 8 August 1967, three circles were found in a field in Duhamel, Alberta, Canada, and the Department of National Defence sent two investigators, who concluded that it was artificially made but couldn't make definite conclusions on who made them or how. The most famous case is the 1966 Tully "saucer nest", when a farmer said he witnessed a saucer-shaped craft rise 30 or 40 feet (12 m) up from a swamp and then fly away. When he went to investigate the location where he thought the saucer had landed, he found a nearly circular area 32 feet long by 25 feet wide where the grass was flattened in clockwise curves to water level within the circle, and the reeds had been uprooted from the mud. The local police officer, the Royal Australian Air Force, and the University of Queensland concluded that it was most probably caused by natural causes, like a down draught, a willy-willy (dust devil), or a waterspout. In 1973, G.J. Odgers, Director of Public Relations, Department of Defence (Air Office), wrote to a journalist that the "saucer" was probably debris lifted by the causing willy-willy. Hoaxers Bower and Chorley said they were inspired by this case to start making the modern crop circles that appear today.

Since the 1960s, there had been a surge of UFOlogists in Wiltshire, and there were rumors of "saucer nests" appearing in the area, but they were never photographed. There are other pre-1970s reports of circular formations, especially in Australia and Canada, but they were always simple circles, which could have been caused by whirlwinds. In *Fortean Times* David Wood reported that in 1940 he had already made crop circles near Gloucestershire using ropes. In 1997, the *Oxford English Dictionary* recorded the earliest usage of the term "crop circles" in a 1988 issue of *Journal of Meteorology*, referring to a BBC film. The coining of the term "crop circle" is attributed to Colin Andrews in the late 1970s or early 1980s.

Recent Boom

The majority of reports of crop circles have appeared in and spread since the late 1970s as many circles began appearing throughout the English countryside. This phenomenon became widely known in the late 1980s, after the media started to report crop circles in Hampshire and Wiltshire. After Bower's and Chorley's 1991 statement that they were responsible for many of them, circles started appearing all over the world. To date, approximately 10,000 crop circles have been reported internationally, from locations such as the former Soviet Union, the UK, Japan, the U.S., and Canada. Sceptics note a correlation between crop circles, recent media coverage, and the absence of fencing and/or anti-trespassing legislation.

Although farmers have expressed concern at the damage caused to their crops, local response to the appearance of crop circles can be enthusiastic, with locals taking advantage of the increase of tourism and visits from scientists, crop circle researchers, and individuals seeking spiritual experiences. The market for crop-circle interest has consequently generated bus or helicopter tours of circle sites, walking tours, T-shirts, and book sales.

Since 2000, crop formations have increased in size and complexity of form, some featuring as many as 2000 different shapes, and some incorporating complex mathematical and scientific characteristics.

A researcher found that crop circles in the UK are not spread randomly across the landscape. They tend to appear near roads, areas of medium to dense population, and cultural heritage monuments, such as Stonehenge or Avebury. They always appear in areas that are easy to access. This suggests strongly that circles are more likely to be caused by intentional human action than by paranormal activity. Another strong indication is that inhabitants of the zone with the most circles have a historical tendency for making big formations, including stone circles such as Stonehenge, burial mounds such as Silbury Hill, long barrows such as West Kennet Long Barrow, and White horses in chalk hills.

A video sequence used in connection with the opening of the Olympic Games in London in 2012 shows two crop circle areas shaped as the Olympic Rings. Another Olympic crop circle area was visible for those landing at Heathrow Airport, London, UK before and during the Olympic Games.

Bower and Chorley

In 1991, self-professed pranksters Doug Bower and Dave Chorley made headlines claiming it was they who started the phenomenon in 1978 with the use of simple tools consisting of a plank of wood, rope, and a baseball cap fitted with a loop of wire to help them walk in a straight line. To prove their case they made a circle in front of journalists; a "cereologist" (advocate of paranormal explanations of crop circles), Pat Delgado, examined the circle and declared it authentic before it was revealed that it was a hoax. Inspired by Australian crop circle accounts from 1966, Bower and Chorley claimed to be responsible for all circles made prior to 1987, and for more than 200 crop circles in 1978–1991 (with 1000 other circles not being made by them). After their announcement, the two men demonstrated making a crop circle. According to Professor Richard Taylor, "the pictographs they created inspired a second wave of crop artists. Far from fizzling out, crop circles have evolved into an international phenomenon, with hundreds of sophisticated pictographs now appearing annually around the globe."

Smithsonian Magazine wrote:

> Since Bower and Chorley's circles appeared, the geometric designs have escalated in scale and complexity, as each year teams of anonymous circle-makers lay honey traps for New Age tourists

Since becoming the focus of widespread media attention in the 1980s, crop circles have become the subject of speculation by various paranormal, urological, and anomalistic investigators ranging from proposals that they were created by bizarre meteorological phenomena to messages from extraterrestrial. Many crop circles have been found near ancient sites such as Stonehenge, a prehistoric monument located in the English county of Wiltshire. They have also been found near mounds of earth and stones raised over a grave or graves, also known as tumuli barrows, or barrows and chalk horses, or trenches dug and filled with rubble made from brighter material than the natural bedrock, often chalk. There has also been speculation that crop circles have a relation to ley lines. Many New Age groups incorporate crop circles into their belief systems.

Some paranormal advocates think that crop circles are caused by ball lighting and that the patterns are so complex that they have to be controlled by some entity. Some proposed entities are: Gaia asking to stop global warming and human pollution, God, supernatural beings (for example Indian devas), the collective minds of humanity through a proposed "quantum field", or extraterrestrial beings.

Responding to local beliefs that "extraterrestrial beings" in UFOs were responsible for crop circles appearing, the Indonesian National Institute of Aeronautics and Space (LAPAN) described crop circles as "man-made". Thomas Djamaluddin, research professor of astronomy and astrophysics at LAPAN stated; "We have come to agree that this 'thing' cannot be scientifically proven." Among other, paranormal enthusiasts, ufologists, and anomalistic investigators have offered hypothetical explanations that have been criticized as pseudoscientific by sceptical groups and scientists, including the Committee for Skeptical Inquiry. No credible evidence of extraterrestrial origin has been presented.

Animal Activity

In 2009, the attorney general for the island state of Tasmania stated that Australian wallabies had been found creating crop circles in fields of opium poppies, which are grown legally for medicinal use, after consuming some of the opiate-laden poppies and running in circles.

Changes to Crops

A small number of scientists (physicist Eltjo Haselhoff, the late biophysicist William Levengood) have found differences between the crops inside the circles and outside them, even though there is a consensus among scientists that the circles are man-made.

Levengood published papers in journal *Physiologia Plantarum* in 1994and 1999. In his 1994 paper he found that certain deformities in the grain inside the circles were correlated to the position of the grain inside the circle. In 1996 sceptic Joe Nickell objected that correlation is not causation, raised several objections to the Levengood's methods and assumptions, and said "Until his work is independently replicated by qualified scientists doing 'double-blind' studies and otherwise following stringent scientific protocols, there seems no need to take seriously the many dubious claims that Levengood makes, including his similar ones involving plants at alleged 'cattle mutilation' sites." (in reference to cattle mutilation).

A study by Eltjo Haselhoff reported that the pulvini of wheat in 95% of the crop circles investigated were elongated in a pattern falling off with distance from the centre and that seeds from the bent-over plants grew much more slowly under controlled conditions. Furthermore, traces of crop circle patterns are sometimes found in the crop the following year, "suggesting long-term damage to the crop field consistent with Levengood's observations of stunted seed growth." These current investigations seem to imply that at least in some crop circles, there is more at work than the effects of mechanical crushing of plants.

Magnetism

In 2000, Colin Andrews, who had researched crop circles for 17 years, stated that while he believed 80% were man-made, he thought the remaining circles, with less elaborate designs, could be explained by a three-degree shift in the Earth's, that creates a current that "electrocutes" the crops, causing them to flatten and form the circle.

Folklore

Researchers of crop circles have linked modern crop circles to old folkloric tales to support the claim that they are not artificially produced. Circle crops are culture-dependent: they appear mostly in developed and secularized Western countries where people are receptive to New Age beliefs, including Japan, but they don't appear at all in other zones, such as Muslim countries.

Fungi can cause circular areas of crop to die, probably the origin of tales of "rings". Tales also mention balls of light many times but never in relation to crop circles.

In the 1948 German story *Die zwölf Schwäne* (*The Twelve Swans*), a farmer every morning found a circular ring of flattened grain on his field. After several attempts, his son saw twelve princesses disguised as swans, who took off their disguises and danced in the field. Crop rings produced by fungi may have inspired such tales since folklore holds these rings are created by dancing wolves or fairies. **(Wikipedia)**

Yoga and Me

Yoga, of the truest kind, is something that creates a splendid mind.
Thoughts rush, hush, suspend, and blend;
All serve to build a society that can be enormously kind.

Some now realize kundalini is the key.
Many enjoy its presence, including me.
The heavens are opened; all is seen as brilliant bright light.

Only the privileged have this gift; to everything it offers a lift.
Others won't understand, they'll treat it as a miff!

But, without a shadow of a doubt,
Those that experience, know what it is about,
Love to shout it out!

Without exception, one grows stronger with each passing day.
The mind, and so much more, function better;
Further in tune with the nature of things.
Mysteries open, including the nature of the moon,
And the cry that emanates from a loon.

God's grace is the hand that guides the plan.
He is not of use to us; we are at His disposal.
The reason this should be so, we really will never know.

All that is clear is that things pass,
Because He is the one that holds us dear.

Whether it is day or night,
It is important to strive to always do what you know is right!

Mistakes will be forgiven.
Charitable acts serve as compensation.
This one accepts always, without the slightest hesitation.

This is the underlying factor; simply do what is good;
Nothing else will ever matter.

God is gracious and kind; strive to be of like mind,
Then the world will be your oyster.
Gifts will flow like showers from the heavens above,
Drenching you with delight.
Each, and every soul, is a bright, fabulous, glorious, light.

Door

A door and a door-post signify communication, and conjunction. Angels and spirits have habitations which appear quite like those which are in the world and-what is a secret each and all things that appear in their habitations are significant of spiritual things; for they flow forth from the spiritual things which are in heaven, and which are consequently in their minds. Communications of truth with good are there presented to view by means of doors, and conjunctions by means of door-posts, and other things by the rooms themselves, by the courts, by the windows, and by the various adornments. That this is so, a man at this day, especially one who is merely natural, cannot believe, because such things are not manifest to the senses of the body. Nevertheless that such things were seen by the prophets when their interiors had been opened into heaven, is evident from the Word. They have also been perceived and seen by me a thousand times. I have moreover frequently heard them say, when their thoughts were in communication with me, that the doors of their rooms were opened, and when they did not communicate, that they were closed. (Author unknown)

The door that leads to man discovering the truth about himself
Is recognizing the suffering he inflicts upon Jews;
His saviours.

THE JEWS

Chapter Eight

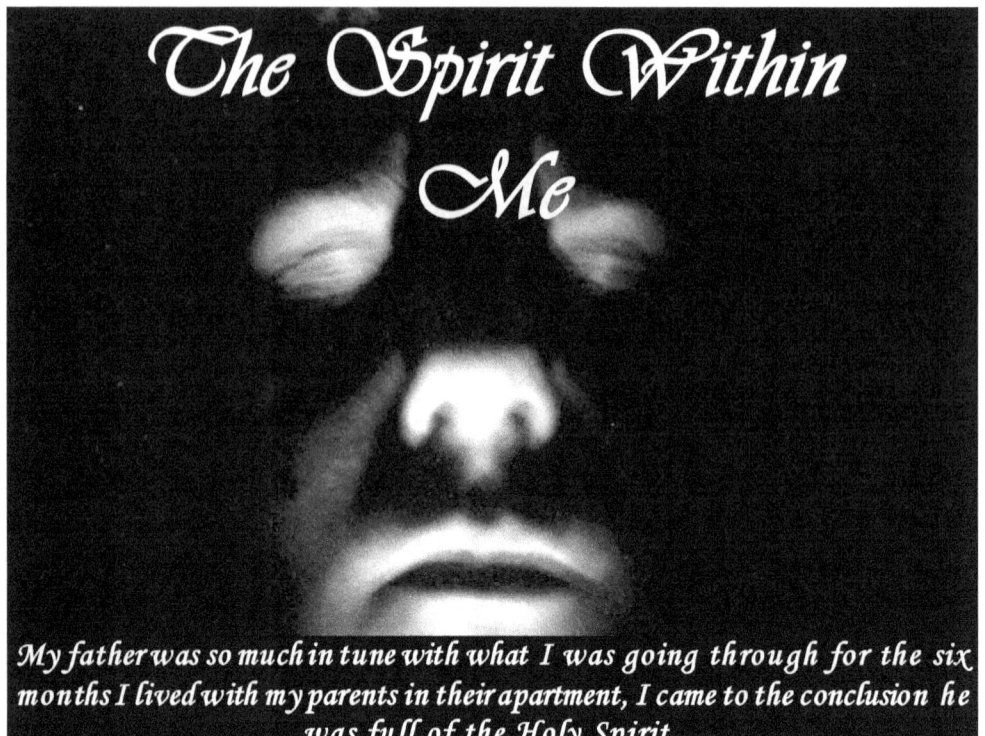

The Spirit Within Me

My father was so much in tune with what I was going through for the six months I lived with my parents in their apartment, I came to the conclusion he was full of the Holy Spirit.

never voiced this idea, after all, it could have been just as likely his extraordinary intellect enabled him to understand so much; on the other hand, his intellect could be the Holy Spirit; I didn't have enough knowledge of the ethereal nature of The Lord at that stage in my evolution to come to a conclusion.

During each day, for the first fourteen months I returned to live among man I didn't sleep; I wouldn't call this a case of insomnia because I got sufficient rest by consciously slowing down my thought processes during the night. During those 14 months, every day, when I was in a place where no one could see me, I allowed The Lord's love to flow into my body.

I do not have the words to describe the experience, so I prefer to be minimalist while detailing what I went through. Some type of entity; force; or energy; possibly all three, entered my body through the crown of my head; at least, this was the sensation, and once it entered, it appeared, or so it seemed to me, to coalesce; what I mean is that there was a pressure that intensified, giving me the impression something of that nature was taking place, though, of course, I can't be certain about anything I'm declaring; the exact scientific reason why these things occurred is unknown to me.

Whenever I told my father about these experiences he didn't seem at all surprised, dismayed, or concerned, which made me believe, as I indicated, that we were more alike than he was willing to declare.

What I'm definitely certain about is that while these experiences were taking place my entire body went through a healing process. The drugs administered to my body over the course of 23 years had caused horrendous damage; according to the hospital records I've viewed, I once weighed over 250 pounds; my cranial and facial bones were misaligned; my hair had stopped growing; I was a physical wreck.

The skeletal configuration of my face has since returned to how it was in 1984. I used to have stretch marks over a large portion of my abdomen that healed, as well. All the scars, and damage, in fact, that were created over the course of 23 years drugs were placed in my body, in an effort to hide the "unsuccessful" attempt on my life, vanished; all the marks, and scars, prior to 1984, remain.

While I was with my parents I had to live in a state of denial; after all, I knew what my mother did in 1984, and both my sisters had to have known what she'd done. There was one conundrum that remained unanswered for approximately three years; why did my father, who devoted so much of his time, and energy, toward helping me, continue to have any communication with them?

It was only when I was living in a town called Marathon, working on the book, "Love is The Word and the Time is Now", that a spark of insight flew into my mind, that I later relayed to my father over the phone. After I shared what I knew, he abruptly put down the phone, and never said a word about it, as if I never said it; "You died in the back of the ambulance in Silverstone".

When I found out this occurred, everything about my father made sense; why he'd done all he had since the time he wanted to leave my mother, and, because he thought I wouldn't join him, took an overdose of pills.

Everybody I ever knew till that time I perceived as horrendously selfish; I had done so much to better the lives of others, in every capacity I could, prior to 1984, yet nobody, or so I thought, had put aside a moment of their time to consider what had happened to me; obviously, the only exception was my father, who had actually done the exact opposite.

Following his overdose of pills he stayed with a woman he did not love, and, I'm quite certain loathed. My father was strikingly handsome during his younger years, and I'm quite sure he could have had his pick of women. He had a brilliant mind, and was capable of mastered skills exceptionally fast, yet, he chose to be a window dresser, and work alongside a woman he didn't care for in the slightest as a means to acquire an income. He took my family to a land he cherished; the country who's army he joined, and set up "bait" that he knew my materialistic, narcissistic, family members would find irresistible, and then waited years to witness, and not make any effort to stop, them end my life. For 23 years he lived with a woman he knew had murdered his son, whom he cherished more than anything in the world.

My father was able to contribute as much as he did to the books we produced together because he was simultaneously giving my mother the impression he was actually on her side, and thought as little of me as she evidently did.

Most tragically, as soon as a physical ailment weakened my father, placing him in a vulnerable position, members of my immediate family pounced on this opportunity to end his life, and, sadly, in a fashion only the most wretched are able to undertake.

After not receiving treatment for sarcoma for many, many, months, till the cancer had spread throughout his body, and he couldn't have been in a weaker condition, he was taken to someone's private home, and there he was administered injections of sodium chloride resulting in water collecting in his lungs; mimicking a state of pneumonia. It was only then a matter of time before he would no longer have the strength to clear his lungs, and die of suffocation; the most barbaric form of torture I can imagine. Those who participated in this act of inhumanity not only sought to cease his existence, but erase any evidence he was alive.

I have seen this level savagery up-close, and I know from personal experience society is manufacturing monsters capable of such a thing; in my own home, over the course of six months, I was subjected to every popularly known method to murder a person, while attempting to hide that a murder has taken place; making it appear to be an accident; a case of negligence, or self-inflicted. The longer this state of affairs continued; greater was the enjoyment derived from perpetrating the act; it's an addiction, no different from habitually using drugs, or drinking alcohol.

The Lord, however, is merciful toward His Chosen People, and made sure the Spirit left my father's body before this horrendous torture took place. The Lord had done the same for me in 1984, and, upon reflection, it was most fortunate the vast majority of Jews killed during the Holocaust did not suffer long. Most, it has been recorded, did not know why they were being transported to the concentration camps in boxcars, and when they knew their fate it was close to time they would die in the gas chambers. The Jews who endured horrendous suffering due to incredibly horrific things people are capable of inflicting upon their fellow man did not do so in vain; their lives; the stories they tell others about what they went threw, serve the same purpose as Christ's crucifixion; their hardship; their pain; their misery; serve as a lesson man will never forget!!

John 14:27 (KJV)

Peace I leave with you, my peace I give unto you:
Not as the world giveth, give I unto you.
Let not your heart be troubled,
Neither let it be afraid.

The Word

There was a Word that spread across space;
Within it was history, and all that had yet to take place.

The projector was planted in heaven; the director is The Lord;
The stage is here on Earth;
And people are the actors that dance across the screen.

Every Age contains a full reel,
Yet people tend to deceive themselves by believing
The movie is the real deal.

The essence of reality is recognizing
It solely takes place in The Lord's mind;
It is His imagination alone that makes every else possible.
What may seem alive is actually dead;
Because existence is eternal, it never ends;
Therefore, nothing ever changes; what once was, has already been,
And it is only in the moment that any of this can be seen.

People were supposed to learn lessons,
But because they do not adhere to the scheme that allows time to flow,
The cost has been suffering, and much sorrow.

The secret that explains why pain replaced bliss,
Is only revealed at the end,
When the serpent could no longer utter a hiss,
And thoughts were rekindled of the time the Earth was green
And the oceans with filled with sparkling jewels made of the color blue.

As one Age turned into the next, only then was it possible to declare
The Age of Enlightenment was extraordinarily bright,
And the others, in comparison, were terribly dull,
If not hideously shallow.

Soon, reason will reign supreme,
And the never ending story of life will once again
Be a pleasant, melodious, dream;
Full of hope and a sparkling lake containing an effervescent fountain
Teeming with glorious streams of hilarious beams.

Shindler's List

Rose Levy

Also Known As:	*""Rosie""*
Birthdate:	*1903*
Birthplace:	*London, Greater London, United Kingdom*
Death:	*(Date and location unknown)*
Immediate Family:	*Daughter of Abraham Levy and Sarah (Cohen) Levy*
	Wife of Edward Shindler
	Mother of Reva Berman; Max Shindler;

Alan Shindler
Sister of Jessie Kliman and Jackie Levy

Edward Shindler

Birthdate:	*1903*
Birthplace:	*Manchester, Greater Manchester, England, United Kingdom*
Death:	*(Date and location unknown)*
Immediate Family:	*Son of Max Shindler and Rebecca Wolfson*
	Husband of Rose Levy

Father of Reva Berman; Max Shindler; Alan Shindler
Brother of Jake, Harry, Solly Shindler

Parents of Rose Levy

Abraham Levy

Birthdate:	*estimated between 1838 and 1898*
Death:	*(Date and location unknown)*
Immediate Family:	*Husband of Sarah (Cohen) Levy*
	Father of Rose Levy; Jessie Kliman and Jackie Levy

Sarah (Cohen) Levy

Birthdate:	*estimated between 1838 and 1898*
Death:	*(Date and location unknown)*
Immediate Family:	*Wife of Abraham Levy*
	Mother of Rose Levy; Jessie Kliman and Jackie Levy

Sister & Brother of Rose Levy

Jessie Kliman (Levy)

Birthdate:	*estimated between 1873 and 1933*
Death:	*(Date and location unknown)*
Immediate Family:	*Daughter of Abraham Levy and Sarah (Cohen) Levy*
	Wife of Jack Kliman
	Mother of Gerald Kliman; David Kliman and <private> Kliman
	Sister of Rose Levy and Jackie Levy

Jackie Levy

Birthdate:	*estimated between 1873 and 1933*
Death:	*(Date and location unknown)*
Immediate Family:	*Son of Abraham Levy and Sarah (Cohen) Levy*
	Husband of Dora Levy
	Father of Gail Levy
	Brother of Rose Levy and Jessie Kliman

Parents of Edward Shindler

Max Shindler

Birthdate:	*estimated between 1838 and 1898*
Death:	*(Date and location unknown)*
Immediate Family:	*Husband of Rebecca Wolfson* *Father of Edward Shindler; Jake, Harry, Solly Shindler and <private> Shindler*

Rebecca Wolfson

Birthdate:	*estimated between 1838 and 1898*
Death:	*(Date and location unknown)*
Immediate Family:	*Wife of Max Shindler* *Mother of Edward Shindler; Jake, Harry, Solly Shindler and <private> Shindler*

Brother of Edward Shindler

Jake, Harry, Solly Shindler

Birthdate:	*estimated between 1873 and 1933*
Death:	*(Date and location unknown)*
Immediate Family:	*Son of Max Shindler and Rebecca Wolfson* *Brother of Edward Shindler and <private> Shindler*

Sons of Jessie Kliman

Gerald Kliman

Birthdate:	*estimated between 1908 and 1968*
Death:	*(Date and location unknown)*
Immediate Family:	*Son of Jack Kliman and Jessie Kliman* *Husband of \<private\> Levy* *Father of \<private\> Kliman* *Brother of David Kliman and \<private\> Kliman*

David Kliman

Immediate Family:	*Son of Jack Kliman and Jessie Kliman* *Husband of Carole Anne Kliman* *Father of \<private\> Edelmann (Kliman) and Andrew Stephen Kliman* *Brother of Gerald Kliman and \<private\> Klima*

Daughter of Jackie Levy

Gail Levy (Deceased)

Death:	*(Date and location unknown)*
Immediate Family:	*Daughter of Jackie Levy*

Town Talk

I stand on the shore of a town cradled by the sea.
Each sound brings joy to my heart;
Including, even that of a bumble bee.

Townsfolk are walking the streets, markets are there.
A smile is on every face; you'd think they were living in a fair.

Parks that entertain frequently have a variety of thrills.
The trouble is too many, thus, forget to pay their crucial bills.
They escape to worlds that aren't real;
People, really, truly, are the greatest deal.

Their lives can amuse, offer hope,
And grant lessons greater than those expounded by the Pope.

Dressing in different ways, passing the time of their days,
Expressing their dreams, hopes, and fears;
Typically, this takes place in a pub, over several beers.

Houses lie on a slope,
That stretches far above where the water lies.
Looking out from the hill, there can be found ships that pass in the night.
If it weren't for foghorns, passengers would face a terrible fright.

Sharing tragedies, losses, hurt, and pain,
Aren't things one should seek to avoid;
Listen carefully, there is great gain.

We feel safer if others share our fate.
One day, for sure, we will all meet our end;
We never know the exact date.

Life's greatest toll is that of aging.
Inevitable, for sure, but accept it we must;
Only this is just.

All as one, we are in the same boat;
Whether at sea, on a beach, in a city, village, or town;
Make sure to remember, it isn't worth the bother to wear a frown!!

High Place

There lies in Crediton a street named High.
Many will pass a neighbour, and say, "Hello".
Each contains a level of care and attention;
So no one need ever have to bellow.

Many stores line this street,
Used, and ne, merchandise are sold.
Secure in himself, and the item he has to hold,
He sells in a manner that is confident and bold.

Dotted along the way are pubs;
For many, if not most, a place to congregate, and greet;
They are considered to be the most resourceful hubs.

A smile, a handshake, a warm remark;
Simple things that remind us life need not be dark.

Churches too are contained in this landscape.
Parks containing mighty trees surround, and encircle;
A beauty that makes the other, appear as if a pancake.

So much to see; do, and experience;
Voluminous are the incidents one can report.
Flat, superfluous, of little substance,
Can be seen in the holy; and things of that sort.

Lloyd's is a bank, situated along the way.
Here, for some reason, it is a delight to note
They neglect to say, "Good day".

A different mentality resides in this place.
A mobile is precious; referred to, and measured,
As, "My, own, personal."

Is it right to consider this object as if it were made of gold,
When all around is a magnificent scope that abounds?
Can this, should this, be considered sound?

Despite its speed that can result in a boom, it can never equal that of light;
Maybe these people aren't all that bright!!

Chapter Nine

One Among
The Masses

The light of day was fading fast; the sun was close to dipping below the horizon. The air was mellow; the slight chill in the air was invigorating.

alcolm relished all these things, as well as the moment, as he approached the pub situated at the top of High Street. The billboard on the sidewalk, directly outside, not far from the front door, listed the bargains that day.

The pub was frequented every day by, pretty much, the same number of patrons. The owner lived in a flat above the pub, and spent most of her days behind the counter; the people she saw regularly she called "my regulars".

They inhabited her establishment for much of the evening, no matter the season, week, or month of the year. Beer flowed into one pint glass after another; each, in succession, lifted the spirits of the patrons.

For many of the "regulars", if not most, maybe in fact all, life had not been kind; dealing with the passing of loved ones is rarely an easy task; especially when one has long since passed his youth, and the number of people one held dear who've departed accumulates.

Work for all was tedious; offering little more than the chance to pay bills; the rent, and, of course, the alcohol consumed in their favourite pub each evening.

Life could have easily turned into an enormously sad, depressing, chore for each had they not had the sense there was a comradely among them all. Malcolm knew this place was their sanctuary; they made this establishment what it was; a place they considered their home away from home.

The dark, heavy oak front door creaked as Malcolm began pushing it open. The upper portion contained a stained glass which held an emblem consisting of a horse, lion, and bear; which helped the men, and women, who came here to drink, reminisce, joke, laugh, and cry, to think of it as a reputable place; a place a person would be proud to say he'd visited.

Many had withstood great perils in their life, and overcome horrendous obstacles, and appreciated the chance to reminisce with others about what had been; could have been; or might have been. Life, for all concerned, had never been considered fair, but rather a cruel affair; many could recall more than a few occasions when the struggle did not seem worth bearing.

Why did they continue? Malcolm asked himself on numerous occasions since he first began frequenting this place; the reason he decided, that was most likely the truth, was to share what they'd experienced; seek comfort from others, and have the opportunity to comfort others.

"Life often isn't as we would like it, but still, as it should be", was an expression Malcolm had thought of recently, and considered an aphorism; he believed it summed up the experience of life. Malcolm's perception was that most people believe they know how they can get the most out of life; disappointments arise when people don't get what they feel they deserve; Malcolm believed The Lord looked after everybody's affairs and He made sure people got what they needed in terms of life lessons.

The door opened inward as the palm of his left hand pressed against the oak door. Once fully inside, he noticed that everything appeared the same as it had on his previous visits. Marge, the landlady, had a hand resting on a lever, while dark liquid, probably Guinness, poured into a tall thick glass. Her eyes were cast downward, not due to a need to focus; she'd worked as a barmaid many years, but probably, Malcolm surmised, because she relished the practise of filling a glass with what she considered to be life's elixir. The smile that lit up her face indicated that his presumption was most likely correct.

Laughter emanated from the area where men were throwing darts at a board. A large fire roared in the fireplace placed on the far right wall; as usual, the white dog with a sprinkling of brown spots covering his shaggy coat, lay in front; looking so serene, you'd think he was lord of the manor, and he'd decided to share his terrain with some of the locals for the night. The dog was lying on his side; eyes closed; a look of tranquillity; he gave the appearance heavenly peace; Malcolm envied him no ends. A large T.V. screen was on the wall opposite the fireplace; as usual a football match was underway.

Practically everybody had their eyes glued to the figures striding to and fro; up and down, the football field; while they followed the movement of the small rolling object covered with alternating squares of black and white. Only a couple of men at the bar, and the three men playing darts, found an alternate means to occupy their time.

Malcolm shut the door behind him, and just as the click announced the completion of this task, a roar of jubilation erupted – somebody had scored a goal. Malcolm was certain high-fives all around would next congratulate this grand accomplishment.

Malcolm headed toward the rack holding the daily newspapers that was attached to the post standing in the centre of the pub; usually a selection of five was available. Normally, Malcolm read at least three a day; although he loved being around people, and appreciated the opportunity to listen to people as they talked about their lives; their general attitude toward life; he had little, if anything, in common with anybody; that was why he preferred to read; his connection to the world came through the words he read on a page. The place he inhabited was vast; it existed far beyond the confines of the pub he was presently in; people of all lands; continents, fascinated him.

Without checking which newspaper it was, he removed the one at the top of the rack. Once he'd slipped it beneath his arm pit, he strolled the short distance over to the bar; tucked a stool beneath his bum; removed the paper from his armpit, and placed it on the counter in front of him.

Marge was standing opposite him, behind the counter. Her low cut top revealed a large portion of her succulently developed bosoms; just a few veins marred their perfection. Her make-up was barely noticeable tonight; she wore a bit of red lip-stick; a smattering of blush; and eye-liner. For the first time, Malcolm took a few moments to really notice her natural beauty; had her hair not been placed in a tight bun on the crown of her head, and instead allowed to gently cascade over her round, delicate shoulders, she would have been even more attractive. Marge, obviously, didn't know the extent to which her fine, strong, black hair complemented her refined, oval, face.

"What's your pleasure tonight, Malcolm? Ale or lager?" Marge asked in a chirpy voice.

"I'll definitely have Ale; make it Danish. Choose one for me; I know whatever you decide will suite me fine."

"If only others had the same confidence in me as you, my work would be a whole lot easier."

"I'm easy going; that's what makes me so different from others, Marge."

Her chin lifted, as a hearty laugh rang out. It didn't take much to lighten the mood of some people; they were, almost without exception, the one's Malcolm enjoyed the company of the most. The great shame, though, was that this sort was becoming harder to find, and becoming fewer and farther between.

"I know one that'll suite you; heavy; bold; rich; a mixture of delicate tastes; not all that different from you, I dare say." Marge's cheeks visibly reddened; Malcolm knew she was turned on; the heat emanating from his groin signalled he was the same.

One of her hands stretched across the counter, and landed softly on top of the hand Malcolm had over his paper. The touch was sexually enticing, and brought a slight movement in the region of his groin.

"Why are you single, Malcolm? So many women must swarm around you; you're a mystery to me."

"I'm a normal man, Marge; my urges, likes, wants, needs, are much the same as others, but the goal I wish to obtain disallows such a distraction. Sacrifice is an essential part of it all."

"As usual, I don't know what you're talking about, but I think you're a great guy nevertheless. Just give me a moment; I'll go to the storage room, and get your bottle."

"No rush; I plan on staying till closing."

Marge turned, and flipped up her short skirt to reveal her bare derriere. Before going through the doorway leading to the basement, she turned, and blew Malcolm a kiss. He couldn't help but smile; he caught the symbol of love; he snatched it from the air with his hand; then tucked it in his shirt pocket. "I'll save that for later, when I'm alone; it'll be our secret." Marge giggled, and shrugged her shoulders.

"Strange, but irresistible; that's you."

Malcolm tipped an imaginary hat on his head. "Much obliged mam. If people knew all there was to know about me, the presses would stop, and never start."

"Like I said, you're an odd one, but I can't help liking you. Be back in a couple with the beer. Should I get more than one? I'm certain you'll love it!"

"Three, will be fine, thanks."

"Will do!" Marge announced as she opened the storage door, and then fled down the stairs to the basement.

Malcolm opened the newspaper; a world appeared before his eyes; stories of people, nations, landscapes, he was never likely to see; tales of mutilation, despair, suffering, success, and the attempts made to make things better.

The journey of life was never likely to change he'd long ago determined.

Just as he began reading about another massacre in the Horn of Africa, a gruff voice interrupted his thoughts.

"Malcolm! Good you see you. Had any bacon lately? Just joking! You can take a joke, right?!"

"I didn't know you were here, Lucky; where have you been hiding?"

"Just had a smoke out back. I've been waiting to see you for the past two hours, actually."

Marge stood in the storage room doorway, and called over to Malcolm's neighbour, now sitting comfortably on the stool to his left. "Lucky! Usual, right?"

"You bet! Easy on the head; I don't like wasting my hard earned money on bubbles. I'll buy another of whatever Malcolm's drinking, and he didn't even ask me to do him a favour; fancy that, Marge! See; I'm not so bad a guy, after all. I like sharing my wealth."

"Whatever you say, Lucky. Two beers; coming right up."

A hand landed hard on Malcolm's shoulder. Lucky was treating him as if they were long lost buddies enjoying a night out together.

Malcolm slowly turned his head to greet Lucky's eyes; they were blue, quite small, placed behind large, out-dated, glasses. For his age, sixty-five, he was well maintained. Lucky claimed he'd lived a life of hard labour, but there was any indication in his demeanour, posture, or facial features. Only a few lines on his forehead marked the passing of years. His hair was sparse, especially on top; he wore his usual attire; casual dress pants; a bland shirt, with buttoned pockets. Many years of heavy smoking had made his teeth appear yellow; dark stains marred the length of his gum line.

"What's a Jew doing with a commoner?"

Malcolm's expression turned to bewilderment. Lucky lifted his hand, and then landed it on Malcolm's shoulder again. "Just having a lark. I know you folks are the same as the rest of us; whether you realize it or not."

"I told you before; I don't belong to any religion. I'm simply me; but thanks for the beer, most kind."

"Anything for a buddy….I just ate pork; hope my breath doesn't bother you?"

"No; your jokes are getting stale, though." Malcolm replied.

"Hey! That would hurt if I thought you mean it", Lucky bellowed.

Two bottles landed on the counter. "You boys take it easy, O.K.! I don't want any upsets here."

"Marge, you know me, I get along with everybody." Lucky spouted, with a cocky grin on his face.

"Look who you're talking to, Lucky; we all want the same thing here." Marge then walked over to the far end of the counter, and began chatting with another customer.

"Tell me, Malcolm; why is it you Jews take, when you have so much already?"

"What are you talking about, Lucky? You've lost me." Malcolm asked, with a quizzical expression on his face.

"Take Israel, for example; the Hebrews just take more, and more, land." Lucky retorted.

"Do you know the history of Israel, Lucky? The opposite is actually the case; they've given up the Sinai Desert, the Gaza Strip. The West Bank will probably be next; in the not so distant future." Malcolm replied.

"You're wrong!"

"Tell you what; if you want to have a debate; develop an argument you can defend; do some research."

"Oh, that's it, Jew boy! You think you're so much better than the rest of us, don't you! Outside, NOW! I'm gonna beat the crap out of you, and enjoy doing it!" Malcolm stared into Lucky's eyes with dumbfounded amazement.

"Settle down; I'm not going to fight you. We're grown men, for Christ's sake." Malcolm appealed.

Lucky grabbed Malcolm's collar, and shoved him hard. A hideous sneer made him look like a rabid wolf; rage; hatred, and anger seemed to ooze from his skin.

Thankfully, two men playing darts a short distance away quickly dashed over and restrained Lucky by grabbing his arms; Lucky then, quite literally, began to snarl. A voice from the region of the fireplace then announced, "You're a bigot, Lucky."

Marge swiftly moved over once she heard what was going on, and said, "Malcolm, I think its best you leave. I hate to say it, but there's several other pubs in town; find yourself another. Don't take it personally, O.K."

Malcolm slipped off the stool, and walked slowly over to the front door. He grasped the door handle with one hand, then turned, and looked around at the pub he'd grown to like, then thought to himself; this is personal.

Throughout Malcolm's life he's noted a big difference between himself and others; he could change, adapt, accommodate, forgive, and forget, like no other person he ever met. No one, he decided long ago, would control, or manipulate, him; least of all a woman.

Malcolm found himself in a philosophical frame of mind as he exited the pub; he'd speculated long ago that women no long behave as nature dictates they should. Men had changed as well; in a far from appealing way. There weren't many today, Malcolm realized, who held views similar to his own; thinking "outside the box" became a way of life for him many year ago

He flung the door open; emitting a loud, huff, at the same time. Typical, he thought; others knew he was exceptional; had so much to offer, but weren't willing to make the slightest effort to change in order to afford him a bit of space.

Once both his feet were firmly planted on the sidewalk, he peered at the billboard, and noted that the beer he'd been offered was the most expensive listed; that said a great deal, as far as Malcolm was concerned, about women.

He lifted his collar, before strolling down the hill away from the pub. High Street was quiet now; lights could be spotted in the flats above stores. The air was much chillier than earlier in the evening; windows were shut; which blocked the sound Malcolm longed to hear; parents playing with children in their living rooms.

In ten minutes he'd be home; a small bachelor pad. There he'd sit alone; read, and think how unfair life is.

He'd worked hard over the years at being an individual, yet this was the thing that, ironically, separated him from others.

Strange, Malcolm thought, how cream separates from the rest of crop.

He looked down at his wristwatch; the time was 9:15. It didn't actually matter; nothing ever changed; he'd determined that long ago.

Looking up he noticed lights in the dark sky; then a tune popped into his head. He began to hum, as he said to himself; life isn't so bad; after all!

The Value of Education

The people that make history are those who choose to be great. They conquered themselves, and went beyond boundaries that were thought to exist before.

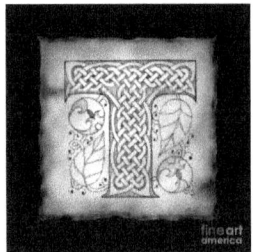

hey left a mark by being original, and of course, as a consequence, became the best at what they do, because they were the first to do it. They, each one, decided to become such a person by choice, after all, in this life no body, despite what people may want to believe, can be made to do something; if we chose, we don't have to do anything in this life.

With very few exceptions the people that have earned such a title, were able to master, take complete control of, a certain field, discipline, and with very few exceptions as well, they were required to devote a lot of time in order to accomplish what they did.

The exceptions are those who revealed their gift at an early age, despite having no formal, or proper, teaching. Like a flower it blooms, where its nourishment comes from to make this happen, none of us can be sure, but believing such rarities are possible due to work that was done in a past life does not seem unreasonable to me, in fact, it makes complete sense.

A lot of us think of Wolfgang Amadeus Mozart as a fine example of such a phenomenon. Within practically every field of Knowledge or Art, there have been such individuals. John Stuart Mill is another example of a man who amassed an incredible amount of learning in his first few years. Such

512

people who develop remarkable skills at an early age, showing incredibly precocious inclinations, without exception apply their talents, gifts, as often, and with the greatest care, whenever possible.

Mozart lived for only thirty five years, we are told, but still managed to compose forty one complete symphonies, and many other works, including, concertos for piano, violin, horn, and woodwind, chamber music, marches, dances, masses, church sonatas, organ music, and operas. It is hard for any of us to conceive how such a thing is possible, which reveals a great deal in itself. No matter how intelligent any of us think we are, we are still left in a conundrum, unable to figure out, how people such as Mozart could have been capable of accomplishing such feats.

There is a trait we can be certain that exists in each of these people; they never, or hardly ever, waste what they have; they devote every morsel of energy they have so they may be able to do what they do.

The Lord tells us to be masters; of what? The answer is; ourselves. We can do this if we avoid, despite any and all perceived cost, doing something that is bad; we should always choose to do what is good; we must never give into temptation.

It is a simple instruction. He also tells us to seek first His righteousness, then all else shall be offered to you; in other words, everything is yours for the taking, but first you have to do what is necessary to earn it; much like a freshly baked apple pie placed on a window ledge; cooling in fresh, sweet, afternoon air; if you want a slice, you'll have to find a way to get to it. If you're a young lad, your mother might say; "Clean your room; finish your homework, then you'll have a slice waiting for you on the kitchen table." The most wonderful things in life are the things we earn. The harder one works - as the saying goes - greater is the reward for having done so.

It used to be, not long ago, in First World Countries, contrary to the way schools typically operate today, that in Primary-school, up to Grade Six, a student had one teacher, who was usually a female. She was someone who filled in the role your parents played when you were away from home. The values instilled within you at school were made sure not to conflict, at least as minimally as possible, with what your mother and father taught you at home.

The teacher, all going well, was someone deemed worthy by a school board to be a mentor; someone students could look up to, whom they wished to emulate, and imitate; I'm referring primarily to personality traits, qualities that allow us to function well within society, and get along with other people.

The fact of the matter is that today, in most, if not practically all, First World Countries, teachers cannot be fired , regardless of how incompetent they are, unless they've committed an atrocious criminal act.

By constructing such a ridiculous scenario, we've practically made sure young kids won't be brought up the proper way.

Unsuitable teachers are entitled to remain in the school system, and eventually, actually, get to choose new teachers that will be incorporated into the system. If someone can't imagine how such a design is doomed to failure, meaning the children will be the primary victims, then you were most likely brought up in such a system yourself!

The teacher, hopefully, is someone the children in her class will look up to because of what she's learned herself, and how she carries herself, which reveals her individual personality. Something that is rarely seen these days is the teacher actually learning from students within the classroom, which definitely isn't conducive to the children perceiving such a person as one worthy of being admired and treated with respect.

The classroom, as it once existed not long ago in First World Countries, had a definite range in the number of students it would hold. This provided the teacher the opportunity to become acquainted with each student on an individual, one to one, basis. Each boy and girl had their own strengths and weaknesses, and the teacher was required to discover what these were.

In order for a classroom to operate smoothly, so as to encourage as much learning as possible, there had to be order, and it had to be known at all times that the teacher was the one in charge. The students learned to obey orders, not because they were forced to, but because they wanted to please the teacher, and receive her admiration, which would enable them to feel better about themselves, and what they could accomplish.

The classroom wasn't supposed to be filled with everybody doing the same thing, learning the same way, (which is very much the case today); occasions were always presented for each to have the opportunity learn in the custom they believed suited themselves best. The size of the class was such that the teacher regularly had one on one time with each student, and acted as a facilitator to encourage different forms of inner exploration.

Most subjects at first were taught in a route fashion; this enhanced their memorization skills. Youngsters, usually, find this boring simply due to it being repetitive. Children, being naturally curious, are normally eager to move on to something else once they deem they've acquired sufficient understanding of something; it is, however, crucial that the teacher not allow their eager curiosity to be dampened while memorization skills are being developed. A child's ability to properly process information is just as important, if not more so, than the capacity to memorize information.

Students might believe the teacher is hindering their progress, because the development of such skills acts to stall the way their minds engage, search, and explore, their environment, but they gradually become more aware that mastering such skills is vital, because it facilitates the act of learning.

One might say that unbeknownst to the children the teacher is preparing them for the world; they are learning to use as much of their minds as they possibly can which will later serve to protect them in the world.

This is what parents should do, but their focus, generally speaking, is far more concentrated on their wants, needs, and desires, at the expense of their children. A marriage isn't supposed to be about self-gratification, but the opportunity for another person to grow, and experience life much as you have, if not better, meaning, just as Sir Isaac Newton said; "If I have seen further, it is because I have stood on the shoulders of Giants before."

I have, for a very long time, harboured the opinion that parents should strive to instil all that is best within themselves in their children, but at the same time, being honest about their weaknesses, and not try and hide such things. A trust must develop between parent and child, so that when either exposes who they are, they need not have any concern that they will be hurt for doing so, that such things will be exploited to their detriment.

After Grade Six has been completed, much unlike the case today, if you hadn't learned certain skills that were considered required, you were left behind. This wasn't done as a form of punishment, but because it was for the betterment of the child. Only once all the obligatory conditions were met could a child then be deemed to have the right to continue his or her studies in Junior-high. This was acknowledged as being essential in order for additional valuable learning to ensue. There was no escaping the fact that without a foundation to build upon, you really had nothing at all to work with, there would merely be an illusion that you would eventually have something to offer society.

By this time kids have learned enough about discipline, workmanship, and mutual consideration and cooperation, so that learning any subject could be done with less assistance from a teacher. This was the time when more teachers became involved in the learning process. Each teacher had specialized skills in order to teach a certain subject, or range of subjects; if there was similarity in the subject matter, both might be taught by the same teacher.

The students were now expected to do more work at home. This was a sign that they have developed the skills and discipline essential for such to transpire; unlike the practice in Primary-school where the teacher regularly checked whether homework assignments had been completed; in Junior-high this began to happen on a less frequent basis.

This was a sign that the child was learning to be independent and self-reliant. No teacher wanted a child to be unnecessarily left behind, so a trust had to develop that the child, on his own, could get work done in a timely manner, such as thoroughly preparing for an exam on an assigned date. At the first sign this isn't taking place; it was treated in a most serious manner.

If punishment was handed out, it was in order to eliminate the long term negative implications of such behaviour.

The child may dislike the teacher, or his parents, for doing this, but a friendship of value can only be acquired once each has something he can offer another. A connection between two people really means an exchange of some sort, and it should be fair and equal; for instance, I'll teach you how to play the piano, if you teach me how to speak French; if such a thing isn't possible; what is the point?!

In Primary-school kids were given two recess periods; a time to play in the schoolyard. A teacher would walk around the playgrounds during this time to make sure that if there was a dispute of any kind a resolution was quickly found; one not involving force or violence, but rather a mediation process involving words – the objective was for them to use their minds in order to acquire what they wanted. The teacher was guiding them toward creating peaceful means to settle disputes, rather than using force to acquire things.

These periods were crucial also due to allowing kids to get rid of excess energy, so that once they returned to the classroom each was able to sit quietly and still. This helped their ability to concentrate sufficiently well so as to digest the lessons being taught.

Boys, in particular, needed this play time due to their energy ordinarily being higher than girls. It was typical for boys to play ball games, for example, while girls might join in a circle and talk. The outcome was a quiet classroom, with the students able to remain in their seats without, for example, repeatedly having to get up to go to the bathroom, or become unfocused due to being restless. The ability to concentrate is crucial to learning; it enables the student to listen to what the teacher is saying, and follow instruction.

By the time students entered Junior-high their metabolic systems are beginning to settle down, and, as a consequence, they do not require as much time to get rid of excess energy. Subjects are now typically held in different classrooms, so students are able to shed sufficient energy simply by walking from one to another, and gathering required materials from their locker at the same time.

The only difference between Junior-high and High-school is that the students have developed an awareness of their individual strengths and weaknesses by the time their Junior-high years have been completed, and maybe also a gift that sets them apart from others, which might be a crucial factor in determining the type of occupation that suits them best.

Students did not have the tendency to view a job in the manner most do today, as a means to earn money, but rather most would ask themselves the following questions; how is it I wish to occupy my time in the future? What would I find most rewarding for myself and others? How can I best utilize my skills? What type of profession would enable me to reach my fullest potential?

Most High-schools used to have separate classrooms for those with special gifts, and because of this their talent, or gift, would be given the special attention it deserved in order to be developed to the fullest extent possible.

Natural talents are often referred to as genius; this doesn't necessarily mean an above average IQ - for example, a child may have a gift for sound, picking up nuances that other children are not able to, this could be considered the child's genius - the I.Q. is heightened when the gift is put to use by maybe playing the piano, or some other instrument, learning to read music, and, hopefully, later composing music.

To develop the gift other disciplines have to be learned and mastered, and this is what heightens the I.Q.; using the brain to a far greater extent, and the discipline required to do so. To make full use of a gift within any discipline, many others must be tapped into and utilized as proficiently as possible.

The madness that people often associate with high I.Q.s is an illusion. The mind is no longer operating within the normal range, and they are now among the rarer breed of man. The I.Q. scale is not a gradual one, it is geometrically exponential, therefore surpassing linear growth; for example, if fifty percent of the general public has an I.Q. below one hundred, then, of course, fifty percent are above. Seventy percent are actually within the range between ninety and one hundred and ten.

The overall picture can be viewed as looking like a bell; therefore fifteen percent have an I.Q. that is less than 90, and fifteen percent have I.Q.s that are above one hundred and ten; a symmetrical balance exists. Having said this, how many have I.Q.s above one hundred and fifty, the level commonly thought of as "genius"? The answer is a minute portion of the population. The higher you go, the harder it is to find someone with such an I.Q. I believe once you get above one hundred and seventy that person should be considered a miracle of nature.

This was how the I.Q. was studied and explained thirty years ago; now tests are designed to give a greater number of people the impression they're special when they are not; actual, accurate, true, studies, using the standards and methods in common practice thirty years ago, reveal that fifty percent of the general population now has an I.Q. below seventy.

Seventy was considered the level of a moron, as a consequence, one can conclude that true genius is now becoming harder to find. It is presently the rarest commodity of all, therefore, in my opinion, it should be considered more valuable than gold and all the so called precious jewels combined; sadly, this is very far from the case.

Learn About Life

Teach your children well, I say;
then given to each will be a wonderful day.

Mistakes hurt; pain can make you strong.
Truths, knowledge, are the things for which you should long!

Learning can occur in so many ways.
Given time, less will be the haze.
Life, people, actions, become clearer;
Overall, life will become dearer.

So much is involved; parents, schools, pools, fields, and song.
Enjoy each; they have something to offer of worth;
Therefore, rejoice each precious birth.
The son and daughter, have hearts that glow.
Minds create words, deeds to sow.
All going well, that which you reap will be something deep;
And all one day you will value, and desire to keep!

The University

The university was once considered
A place of great learning;
    A wondrous opportunity to experience
Different types of personal growth.

For this alone there should be a burning, and yearning.
Not to get a degree, because you desire some type of earning.

Many books have been written by men and women considered to be great
   scholars.
We should enormously value their ideas;
To the rafters this should be hollered.

Let us study their lives as well,
What they did, for themselves and others,
Then one day we'll be able to make ourselves great as well.

A truth remains the same; learning can't be hurried;
For each the pace is individual; this for all is enormously pertinent.

By living their lives the way they did,
The great ones manage to nourish themselves;
Their hope always was that one day they might become all they could be.

Enter this place
Knowing what it takes to make something better of yourself;
   Incredible it is, the great changes then one makes.

Time is wasted getting "wasted".
Evenings are spent drinking, getting drunk, and smoking pot;
Should these things be allowed?
Obviously not!

Let us make the most of what we've got;
A rich past that can nourish our minds and souls;
To savour these to their utmost and fullest; that should be our sole goal!

The value of a person is what lies in his heart,
Not the lies that go along with making money galore.
If you choose to treasure gold, and money, above people,
Instead of the fantastic tales that have been told,
It is yourself, and others, who will suffer and fall;
Having then to witness scenes filled with gore,
Sinking also into a pit
   that makes you nothing but a horrendous, stinking, bore.

Value wise sayings, and what is Just.
   If these things you should cherish,
Mankind will once again have a chance not to perish.

So many ways of learning are being denied;
From the youth of today they hide.
Strange it is that so many are unaware of the extent to which they have been
   deprived.

As a people, how can we be strong in the future
When it is learning, and knowledge, we callously butcher?

Those that enter this place,
Knowing what it takes to make a fine person,
Enormous will be the great changes he often makes.

These are the things above all we should adore;
Think now of women, who decides to become a whore.
Should she knock on your door, explain to her that as a person,
   She should become so much more.

Instruct, guide, inspire,
Then you'll be a person people will want to hire, and honestly deeply admire.
Knowledge stands above all, everything else will lead to a fall.
Never mind the trials, and tribulations, life often brings;
Your broader shoulders will enable you to stand tall.

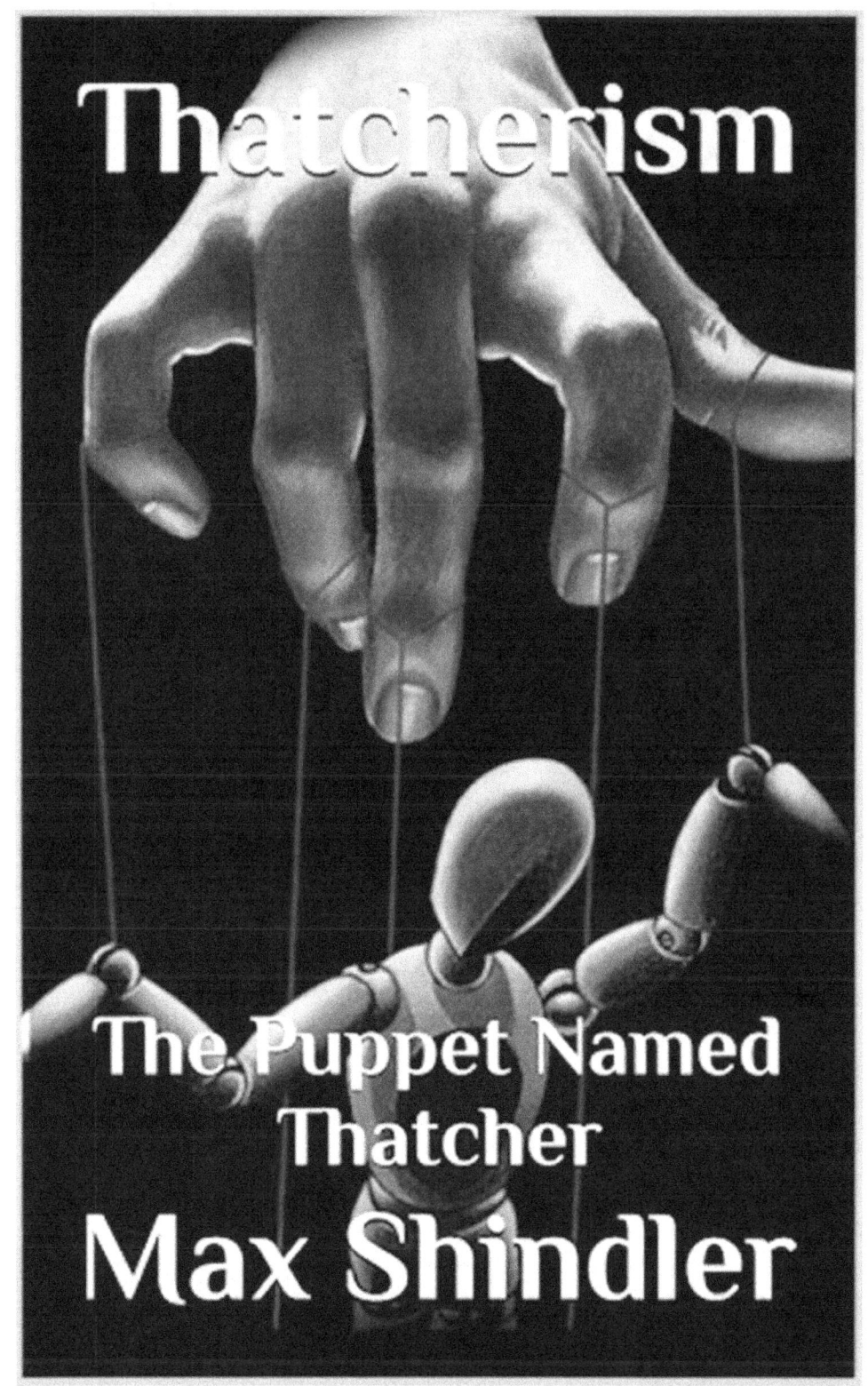

Everything is The Lord's creation; which is good. Man often fails to recognize this because he imagines himself "larger" than he happens to be, and, consequently, fails to accept his shortcomings; he would rather imagine a mischievous, devious, controlling, maniacal creature out to do him harm; sometimes for misdeeds he is able to acknowledge, but has a tendency to believe the punishment is unfair, and other times he cannot fathom a reason behind the dreadful, nightmarish, things that happen, and conceives the cost to be wholly unjust, all the while unwilling to accept that the problem lies in him not developing his faculty of reason.

The Lord chose Britain as the place from which His word would be exported to peoples around the globe so they may cultivate their minds and become civilized. The atrocities that occurred in the past while doing so were caused by The Church, (Protestant and Catholic), that sought to make itself rich by the import of goods. The Lord, because He is good, time, and time again, provided man the means to rectify his ways; put himself on the right path, but these gifts; the British; His Chosen People; were ignored, and squandered.

CreateSpace eStore*: https://www.createspace.com/5719102*

America wanted to remain the world's sole Superpower after WWII, in order to do this She had to make sure Britain never regained Her previous might; this was accomplished in the guise of a woman claiming to be made of iron; in actual fact, she was a puppet with no moral backbone whatsoever.

Margaret Thatcher appeared on Britain's political stage after a generation of the Britain's population had been demoralized by an education system called "Assimilation". In the year 1979 the people of Britain entered their "Winter of Discontent".

History is our greatest teacher; now the truth is known, nothing can stop Britain from regaining Her former glory.

CreateSpace eStore: https://www.createspace.com/5721354

Authored by Nigel Shindler, Authored by Max Shindler

The Tower is an amalgamation of the concepts that formulated the history of human civilization. Around 12,000 B.C; The Lord's spirit was placed within people who had survived the toughest climatic, topographic, conditions imaginable. They left their homeland, "Doggerland", and created the civilizations that form the Trinity; Ancient Egypt, Mesopotamia, and the Indus Valley; these people are the British.
CreateSpace eStore: https://www.createspace.com/5617455

Love Is the Nature of Existence; Vol. V
Jews
Nigel Shindler

A Jew is does not belong to any creed;
He is an individual that chooses to be himself.
He is a genius that creates something new that
Takes humanity beyond previous ideas and concepts.
He is guided by the Spirit of the Creator, in whose image he is made.
CreateSpace eStore: https://www.createspace.com/5328109

Love Is the Nature of Existence: The Trinity Manifesto

Authored by Nigel Shindler Ph.D.,
Edited by Max Shindler

What makes a person human? That is the question.
We now live in a world of extremes. Some labour too hard, while others continually seek to ease their woes; at an enormous cost to others, and also themselves.

Dr. Rollo May wrote a book entitled, "The Meaning of Anxiety", in which he challenged the belief that mental health is derived from living without anxiety, but rather asserts that it is essential to the human condition; confronting anxiety can relieve boredom, sharpen sensitivity, and, actually, creates the tension necessary to preserve human existence. Following the Second World War humanity, understandably, sought relief from the horrors, carnage, and suffering that had taken place, and also the decades leading up to it; never mind the First World War, (1914-1918). The problem is that humanity never brought to an end this search to relieve hardship.

With each passing decade "convenience products", and "consumer goods", in increasing numbers have seeped in to the lives of those who inhabit the "First World", while the anguish of those residing in the "Third World" has increased, as they are forced to labour in order to fulfil the wants, needs, and desires, of those already over-saturated in material wealth, and, ironically, are not even close to recognising this fact - our environment has had to bear the strain of this burden.

How can this cycle of madness be brought to an end?
"The Trinity Manifesto" explains that the answer lies in remembering what makes us human, and how the completion of this accomplishment leads to the whole; the Truth; Self-realization; God - the big picture.
"Love Is the Nature of Existence" explores how the confluence of entropic forces within all of mankind's major civilizations, since the beginning of recorded history, has led to the state we are in now; this hasn't happened by accident, but rather by design; divine in origin.
CreateSpace eStore: https://www.createspace.com/4512209

Love Is the Nature of Existence: Love is The Word and the Time is Now

Authored by Nigel Shindler Ph. D.,
Edited by Max Shindler

"Love Is the Nature of Existence", consists of five volumes, and is intended for those who wish to better understand the state of our world today, and how it got this way. George Orwell's book, "1984", examines how those who control the present re-write the past in order to determine the future. "Love Is the Nature of Existence", carefully details the history of human civilization thus revealing the actual truth about what is happening today.

Did the Third Reich fall, or are we still experiencing the same disease of mind that led to the execution of six million Jews, the mentally challenged, physically disabled, gypsies, homosexuals, and others deemed not worthy of life, liberty, or the pursuit of happiness? "Love Is the Nature of Existence" reveals the sickness has grown, and spread to every part of the globe since the Second World War. Mankind never managed to learn from the greatest lesson history had to provide. The First World War, ended approximately twenty years prior to WW11, and was supposed to be "The War to end all Wars".

We have a new Fuhrer; Nigel Shindler asserts that it's "Big Brother"; he's everywhere, but he's hiding, and that a world without coercion, oppression, and domination, will only be possible, when mankind is free from "His" subversive form of tyranny.
The goal for all of us should not be control, but genuine liberty!!

Love Is the Nature of
Existence
VOLUME IV

The
Creator

Nigel Shindler

CreateSpace eStore: https://www.createspace.com/4696310

Love In the Nature of
Existence
VOLUME I

The Trinity
Manifesto

Nigel Shindler

CreateSpace eStore: *https://www.createspace.com/4861897*

The Tower

Book II

Love
Is
The Word

Nigel Shindler

Authored by Nigel Shindler, Authored by Max Shindler

There are two ingredients that construct The Tower; goodness, and evil; light and dark; Jews, and Homo-sapiens.
Jews, are made in the image of God, and created all civilizations; Homo-sapiens, are members of the primate family, and when they behave in accordance with their nature, destroy civilizations.
Homo-sapiens originated in Africa; they developed an erect posture as a response to climactic changes two million years ago. They survived due to a form of adaption known as "neoteny"; which means that the maturation process is slowed down, and because Homo-sapiens are born prematurely, the head remains aligned with the spine, and the brain doesn't reach full maturity till the 23rd year.
There are two fundamental forces that drive the behaviour of Homo-

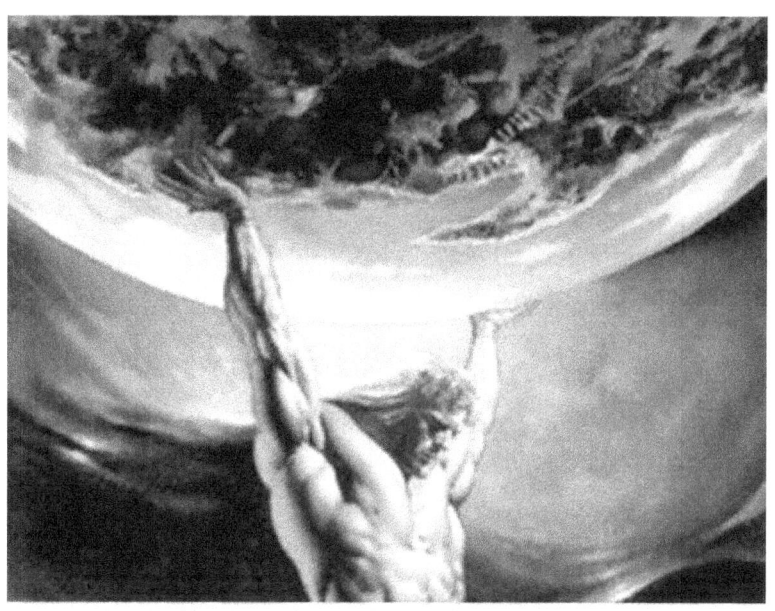

SAVIOURS OF THE WORLD

NIGEL SHINDLER

The evil in the world today is due to the deliberate spread of chaos manufactured by those who make themselves rich by taking control of people's minds. They did this by using information gathered during the Second World War; kept it hidden from the general public, and stole wealth once belonging to the approximate 6 million Jews killed during the Holocaust.

The hoodlums the Kennedy's tried to expose, and incarcerate, are still free, and they use violence, and intimidation, to get what they want. They don't rely on just alcohol now to stupefy the public, but a vast array of drugs, and have people wear uniforms that display the word "police" to give the impression they fight crime, while what they actually do, for example, is place a portion of the drugs they seize back into circulation, and place proceeds of the crimes they confiscate in bank accounts.

Britain's genocide victims are dumped in sewage pipes that lead to North Sea, and then currents carry the corpses to where they can't be found. Their remains are handled in the manner they were taken care while alive.

Those who are treated like trash are the same types of people killed during the Holocaust; over 6 million Jews; an estimated 500,000 Roma and Sinti gypsies; homosexuals, as well as the mentally and physically handicapped.

There was a unit called the Sonderkommando during the Holocaust which purloined gold teeth; shoes; bones; human fat which was sent to soap factories, and hair was used for stuffing pillows; Britain's genocide victims are no less treated as objects used to enrich others people.

The "Final Solution" eradicated practically all the Jews, and civilization continuing upon Earth

CreateSpace eStore: https://www.createspace.com/5868617

The central message "Love Is the Nature of Existence" has to convey is an explanation as to what a Jew is, and why The Lord decided to create "The Chosen People". The purpose of life as The Bible teaches us, and all other Holy Scriptures, is to join, yoke, become one with God. In order for Man to accomplish this feat The Lord chose certain members of the human population to help others cultivate their minds, and by doing so have the same knowledge Adam and Eve had while still in the Garden of Eden. The best of best, no matter the discipline, have always been Jews throughout recorded history. The Jews do not solely belong among the Semitic people that wandered for centuries throughout the Middle East; they do, however, represent the highest concentration considering the overall size of the population that has existed over time. The Jews, without exception, emerge solely from the Aryan race, the migration of which I describe in Volume II of, "Love Is the Nature of Existence".

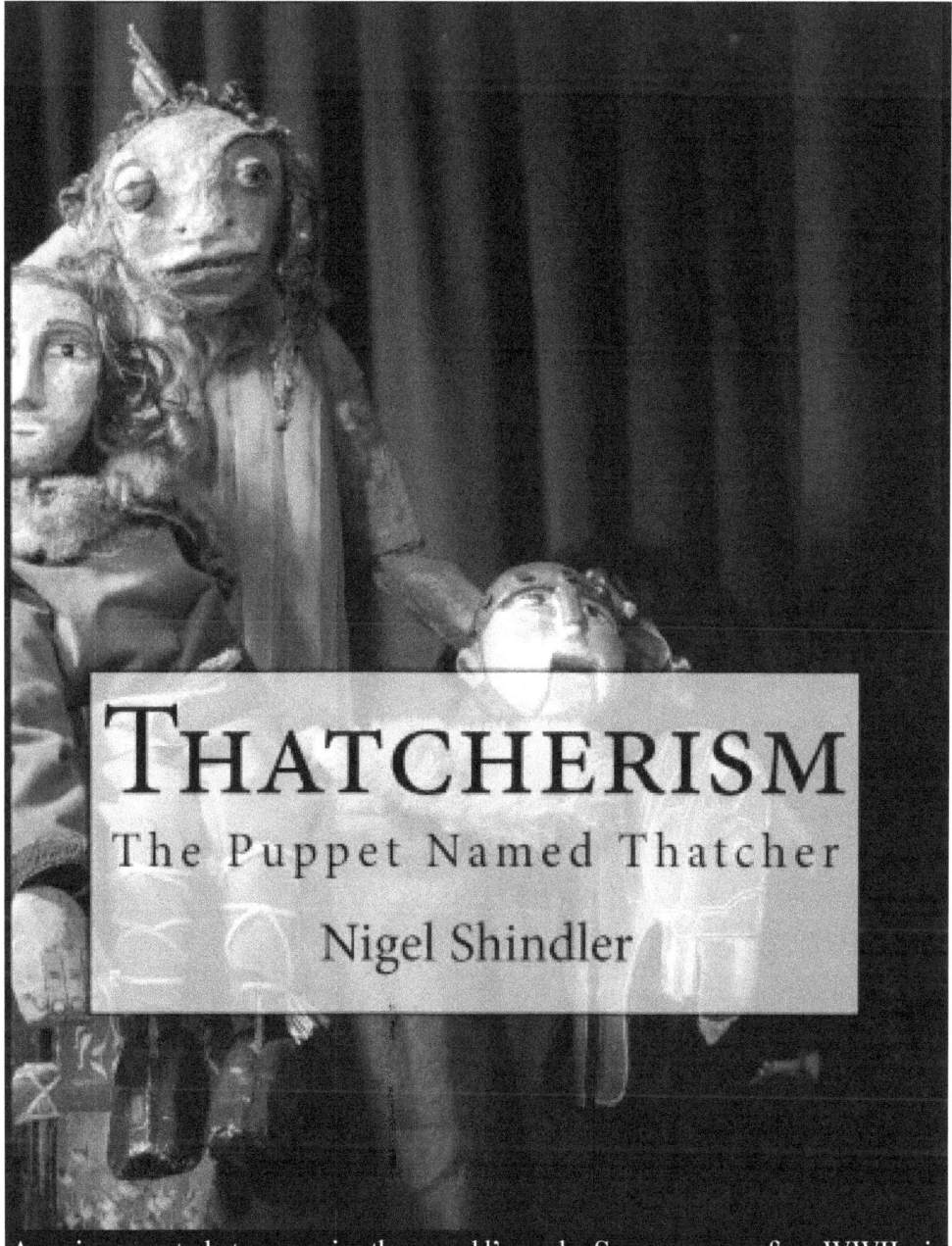

THATCHERISM
The Puppet Named Thatcher
Nigel Shindler

America wanted to remain the world's sole Superpower after WWII, in order to do this She had to make sure Britain never regained Her previous might; this was accomplished in the guise of a woman claiming to be made of iron; in actual fact, she was a puppet with no moral backbone whatsoever.

Margaret Thatcher appeared on Britain's political stage after a generation of the Britain's population had been demoralized by an education system called "Assimilation". In the year 1979 the people of Britain entered their "Winter of Discontent".

CreateSpace eStore: https://www.createspace.com/5721260

I see myself as a person at war, and the battle is against evil, and the weapon I am using to defeat my adversary is the pen. My aim is to dispense, as much as I possibly can, the truth in regard to the nature of my nemesis.

*There is an expression which very much describes my present circumstance, and how I am able to do what I do; "Keep your friends close, and your enemies closer". I am quite literally in the trenches, on the battle field, surrounded by my enemy. Lives are lost every day in the mine field deployed by those who are sick and depraved beyond belief; ask any of those who witness the same as I, and you will, practically without exception, not be told anything about what they have seen and heard, or a claim will be made **that they have seen or heard nothing.***

CreateSpace eStore: https://www.createspace.com/5046782

The Boy and The Tower

Nigel Shindler

"Love Is the Nature of Existence", is a part of me; it is who I am.

The life I have lived represents the central message the book has to convey; an explanation as to what a Jew is, and why The Lord decided to create "The Chosen People".

The purpose of life as The Bible teaches us, and all other Holy Scriptures, is to join, yoke, become one with God.

In order for Man to accomplish this feat The Lord chose certain members of the human population to help others cultivate their minds, and by doing so have the same knowledge Adam and Eve had while still in the Garden of Eden.

The best of best, no matter the discipline, have always been Jews throughout recorded history. The Jews do not solely belong among the Semitic people that wandered for centuries throughout the Middle East; they do, however, represent the highest concentration considering the overall size of the population that has existed over time.

CreateSpace eStore: https://www.createspace.com/4930736

THE BRITISH

Nigel Shindler & Max Shindler

The British are the Jews; the cord from which all life stems. They are the sparks of energy that fuel history. They generate time, and lead us toward one inevitable truth; they are The Lord's Chosen People. Wherever they have travelled they have left cultural treasures;

Homo sapiens steal them, and manage to convince themselves they are responsible for their creation due to being self-deluding creatures by nature, and therein lies the secret to understanding the rise and fall of all civilizations; nothing would be possible without the British.

CreateSpace eStore: https://www.createspace.com/5592567

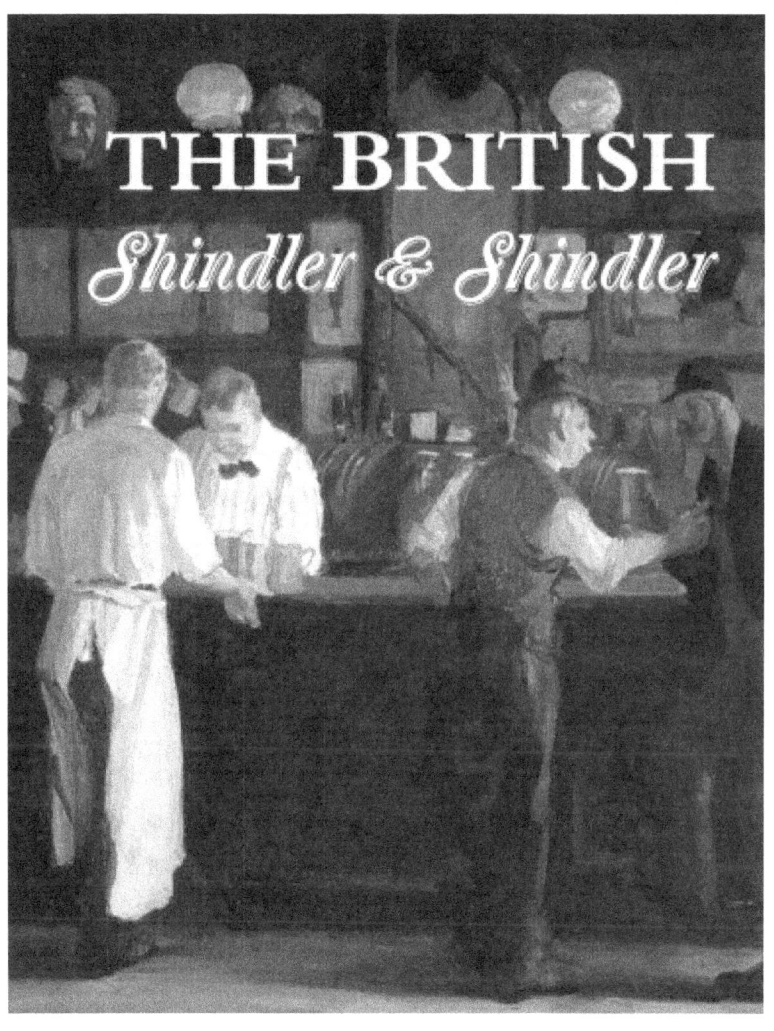

Around 12,000 B.C; The Lord's spirit was placed within people that survived the toughest climatic, topographic, conditions imaginable. They left their homeland, "Doggerland", and created the civilizations that form the Trinity; Ancient Egypt, Mesopotamia, and the Indus Valley; these people are the British.
CreateSpace eStore: https://www.createspace.com/5609672

537

The Creator
Nigel Shindler

Creation is a story that unfolds over time; it came from the Creator, therefore, it is the Creator. Its purpose is much the same as the reason why a person writes his memoirs; in order to reflect upon his life, share his experiences with others, with the expectation that both he and his audience will, in some manner, be enriched from the experience. The Father is the Son, and the Holy Ghost; they are not separate, or different, from one another, they are facets of the singular entity we commonly refer to as God.

CreateSpace eStore: https://www.createspace.com/5239851

**Love Is the Nature of Existence:
Vol. IV**

The Creator

Nigel Shindler

Out of the darkness
sprang a beam of
bright light.
Through dimensions
unknown
it danced with the
wind,
and the ripples of the
waters
hidden within the
escarpments
buried deep down
below.
Eventually, they would
form
a hilarious spectacle
called a show.
Animated figures
would spring to life;
the riddle behind their
existence
is what they wanted to
know.
From somewhere within, and maybe beyond,
sat The Creator on His throne,
looking around the immense, limitless, unknown,
pondering the question; wasn't I here before?
I'll have to wait and see.

The circle of life
is heavens way of continually experiencing itself;
endlessly evolving while always remaining the same.
CreateSpace eStore: https://www.createspace.com/5071100

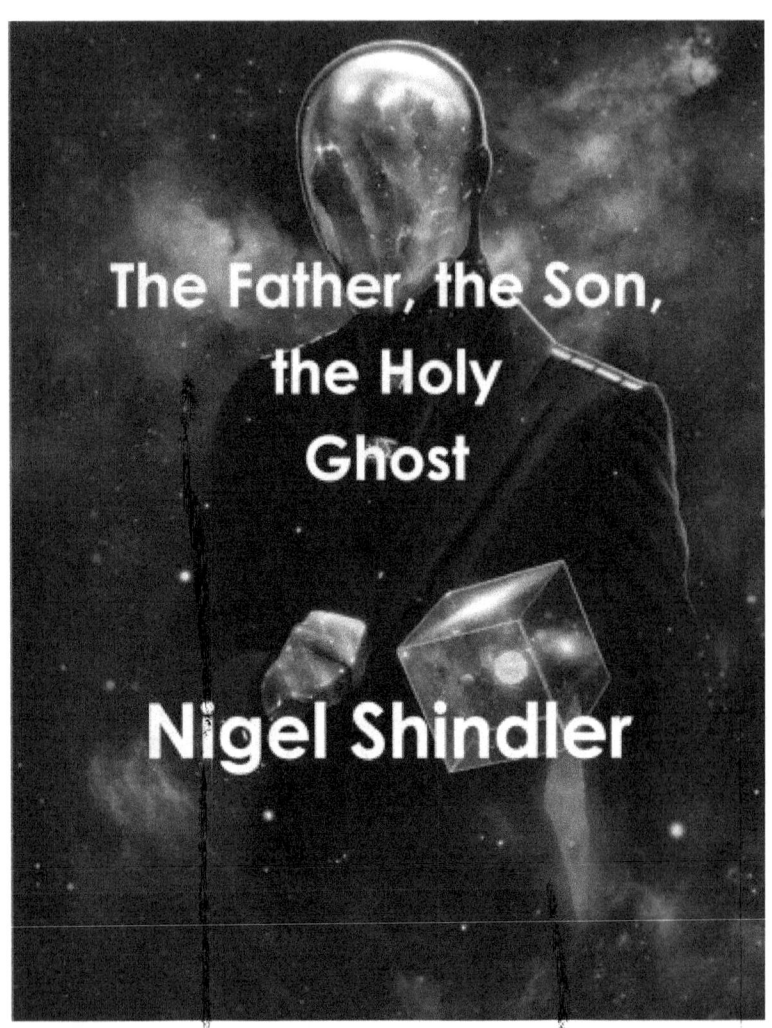

The Father, the Son,
the Holy
Ghost

Nigel Shindler

Authored by Nigel Shindler, Authored by Max Shindler

What is The Trinity?
It is a force that creates, preserves, and destroys.
It has no other choice;
It provides the explanation as to why we're all here.
To the grand animator this is not a game;
He holds each one of us dear.
But why must we suffer? The blind man asks.
The Lord can intervene by providing a helping hand;
But He does not dictate how anybody should use his land.
If the way He offers assistance is not understood,
Then all hope is lost,
Because man has the tendency of not seeing where he's going,
Due to wearing a long, and heavy, hood.
There are none as blind as those who do not wish to see.
Those who extinguish the light that can show us the way,
Do so because their aim is to decide how others should be,
And always insist they have the last say.
CreateSpace eStore: https://www.createspace.com/5407942

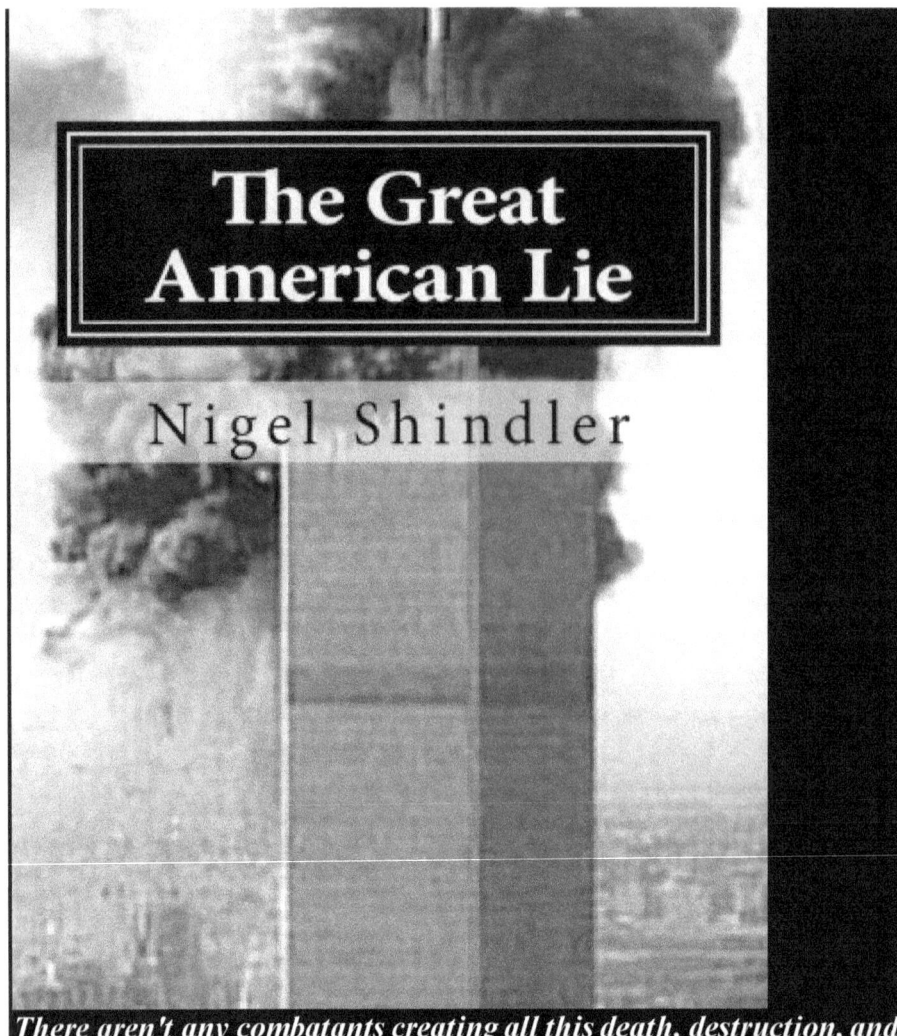

The Great American Lie

Nigel Shindler

There aren't any combatants creating all this death, destruction, and mayhem in the world; everything is staged by America, with the help of Her Ally, Israel; in much the same way as took place in the novel, "1984", by George Orwell.

Far too many people around the globe didn't question what they've been told by the media. Marketers, and advertisers, continually focused their attention on influencing women, due to their inherent "vulnerabilities", (women are "Gatherers", men are "Hunters"), as a means to brainwash everybody. People failed to notice this due to being conditioned to believe we're all equal.

"The greatest trick the Devil ever pulled was convincing the world he didn't exist."
The Usual Suspects
CreateSpace eStore: https://www.createspace.com/5680444

The Jews

Nigel Shindler

The Jews have such a deep appreciation for the nature of existence, they know how they fit within the vast scheme of things, and make sure their behaviour is always in agreement with the "Law of One". Animals due to their Instincts, (a capital "i" is used in order to differentiate theirs' from that of Homo-sapiens), naturally behave in accordance with the "Law of One", which insures the sustainability of the Earth's ecological system. When Adam and Eve partook of the forbidden fruit this symbolized their violation the "Law of One"; we are told in Genesis that they became aware of their nakedness, were ashamed, and covered themselves; the principle message being conveyed, (accepting that all the writings in The Bible have layers of meaning), is that both Adam and Eve had gone against their nature

CreateSpace eStore: https://www.createspace.com/5298068

Love Is the Nature of Existence
Vol. II

Love is The Word
and the Time is
Now

Nigel Shindler

All of mankind's problems, now, and any time in the past, have been
psychological in nature. There are, however, many theologians,
particularly of the Christian persuasion, who are accustomed to making
the claim that the principle factor responsible for the numerous maladies
commonly displayed by man as being the "original sin".
The "original sin", as the name indicates, merely refers to what is
thought to be the first sin committed; both Adam and Eve acquired
something without a faculty to use it wisely; they lacked "common
sense"; and this remains mankind greatest problem till this day.
History has proven that people do not learn from history; thus, ignorance
remains mankind's greatest enemy. "Love Is the Nature of Existence" is
intended for those who want to know the truth about what is happening
in our world today, and how it got this way.
CreateSpace eStore: https://www.createspace.com/4938697

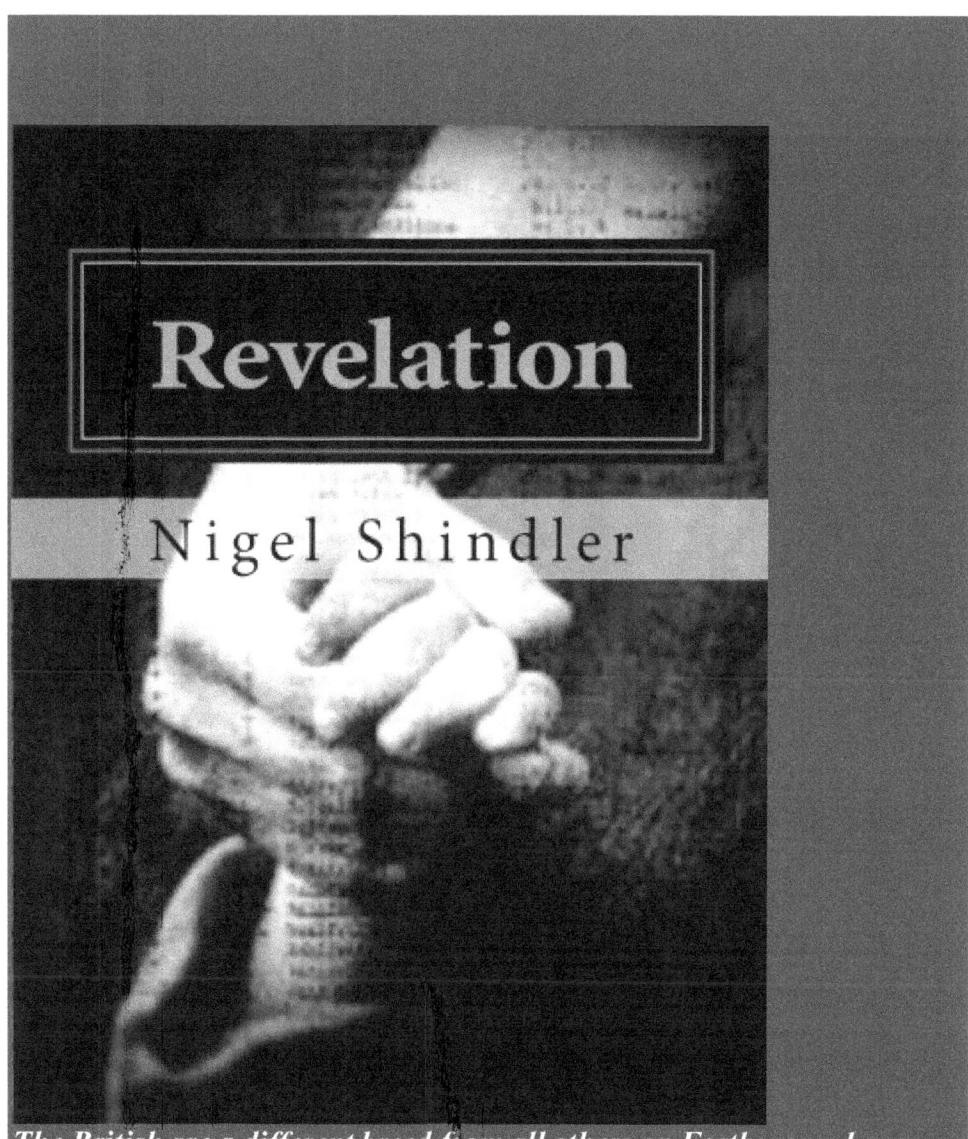

The British are a different breed from all others on Earth; some have labelled them, "Xanthochroi"; this word describes their typical visible physical features; white skin, and wavy blonde hair. What makes the people of the British Isles so different from all others is that they were not originally from Africa; they are not a primate hybrid; they are a part of The Lord's creation, and His spirit is within them so they may be able to fulfil the arduous task, relatively speaking, of inhabiting all quarters of the world; adapting to every type of climate, and withstanding the slings, and arrows, of those who dislike a change imposed on them; however, regardless, the British do their duty, and serve The Lord with all their heart, mind, and soul.

CreateSpace eStore: https://www.createspace.com/5705285

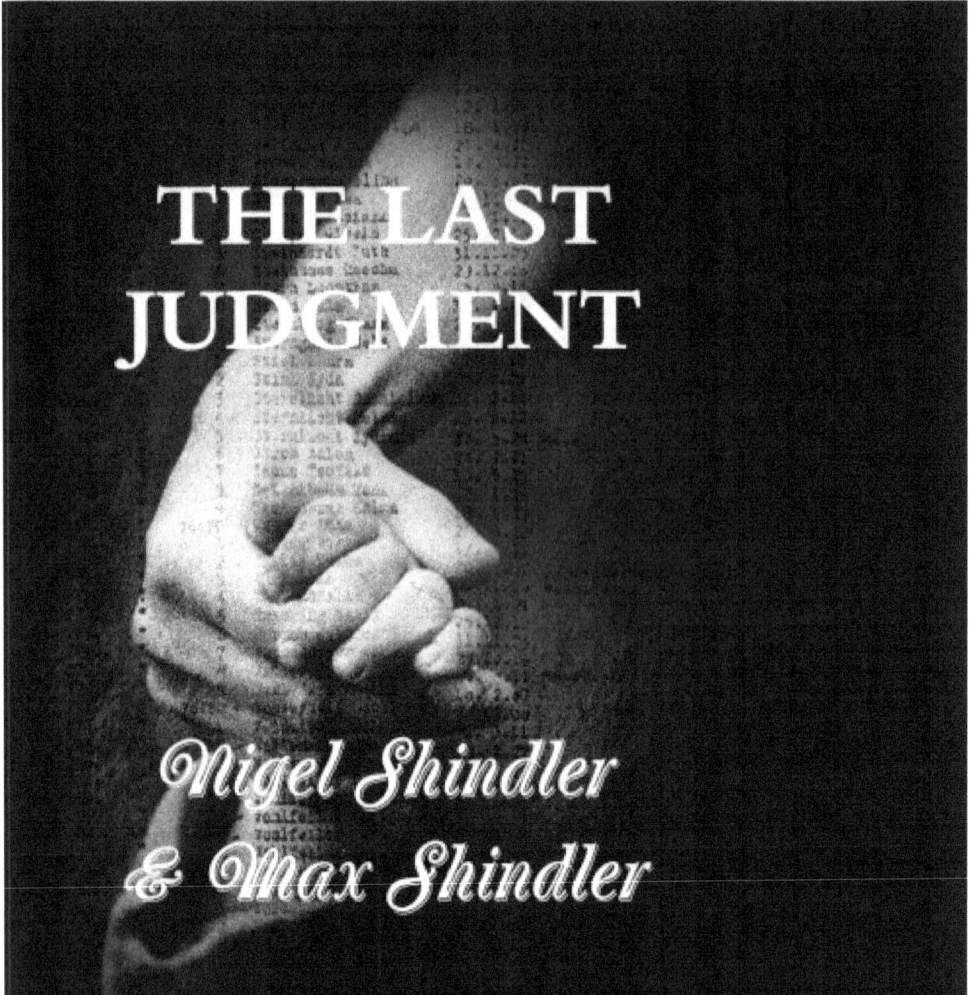

THE LAST JUDGMENT

Nigel Shindler
& Max Shindler

The architects of the genocide taking place within Canada run America. They are the same type of people that were running the United States back in the 1930's; men like Al Capone; they operate the same way today, as they did back then.

A gangster is typically someone who doesn't have a skill, and doesn't want to work to acquire a skill in order to get a job. They like to use manipulation, intimidation, coercion, and violence, to get people to hand over their hard earned pay.

A common tactic used by gangsters in the 30's, in cities such as Chicago, and New York, for example, was to approach new store owners, and convince them they needed their protection to run their business; being new in town, the storekeeper isn't certain the story's true, and might hesitate before handing over "protection" money; if such a stubborn customer should arise, a bomb blast in, or around, his store, will likely convince him he needs a gangster in order to run his business in peace.
CreateSpace eStore: https://www.createspace.com/5567155

The Miracle of Life

Songs from Heaven

Nigel & Max Shindler

"Stories from Heaven", answers the question;
How is the Tower made?

An Age is born the same way a baby is made.
In the case of man; sperm impregnates the egg, which travels to a place it incubates, and once it's mature, merges with the world.
Ages are created by a confluence of forces within Britain; energies seep up from below, and fall from above; converging with The Word, Spirit, and water, to form the Jews, and once they've sufficiently developed, they spring upon the world to teach the good news, and build the Trinity.

A miracle, I would say, is something that can be proven to have occurred, but the reason why is not understood, furthermore, it is perceived that it will never be understood.

The wars, strife, and suffering, throughout history happened not because man wasn't given the means to decipher events, but because man failed to use knowledge wisely. He clouds his vision, so the obvious becomes murky, and what is straight appears crooked.

Ignorance develops the same way as all other forms of sinful behaviour; people create their own ceilings; barriers; shoe boxes, to live in, and refuse to accept anything that doesn't fit into their prefabricated world.

Life isn't a mystery; whether there is hell or heaven on Earth is man's choice; if he's wise, he'll make sure he always stays on the right path.
CreateSpace eStore: https://www.createspace.com/5980229

CreateSpace eStore: https://www.createspace.com/6012736

The Father is the Son and also the Holy Spirit. When He returns to the

THE RISE AND FALL OF THE BRITISH EMPIRE: VOL. II

NIGEL & MAX SHINDLER

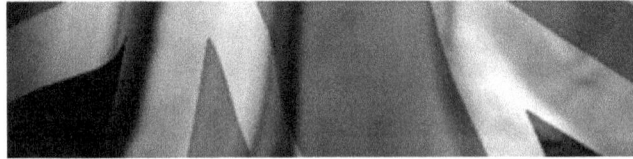

earth, some call this the "Second Coming"; it will seem like a haunting, thus, the term, Holy Ghost. When The Lord suffered as He stood nailed to a cross, no sins were forgiven, no bad deeds pardoned, but rather proof was displayed of the Laws that operate within the universe. They are good, they create as they expand; unlike much of mankind which is self-absorbed and narcissistic, therefore, behaving contrary to his nature.

The key to understanding history, the nature of existence, is correctly interpreting what happened to the Son. He is the center, the synthesis, a combination of everything, and therefore whole - the cosmos in a singular entity which any man can see if he keeps his eyes open.

I see myself as a person at war, and the battle is against evil, and the weapon I am using to defeat my adversary is the pen.

My aim is to dispense, as much as I possibly can, the truth in regard to the nature of my nemesis.

There is an expression which very much describes my present circumstance, and how I am able to do what I do; "Keep your friends close, and your enemies closer". I am quite literally in the trenches, on the battle field, surrounded by my enemy. Lives are lost every day in the mine field deployed by those who are sick and depraved beyond belief; ask any of those who witness the same as I, and you will, practically without exception, not be told anything about what they have seen and heard, or a claim will be made that they have seen or heard nothing.

The logic behind this behaviour, to make an analogy, is as follows; if I've been involved in the commission of a crime, and live off the proceeds of that crime, why would I say, or do, anything that would have a tendency to incriminate myself; for instance, mentioning any knowledge of a bank heist. The various criminal acts detailed throughout my writings that take place in Canada, and are perpetrated by Canadians, are done in order to insure they may be able to acquire cash without having to do any actual work. Canadians, generally speaking, are far too lazy to make the required effort to meaningfully educate themselves; as long as they are able to get whatever they want, whenever they please, then all is hunky dory, and it's yet another "good day".

CreateSpace eStore: https://www.createspace.com/5039817

Authored by Nigel Shindler, Authored by Max Shindler

The Tower is an amalgamation of the concepts that formulated the history of human civilization. Around 12,000 B.C; The Lord's spirit was placed within people who had survived the toughest climatic, topographic, conditions imaginable. They left their homeland, "Doggerland", and created the civilizations that form the Trinity; Ancient Egypt, Mesopotamia, and the Indus Valley; these people are the British.

CreateSpace eStore: https://www.createspace.com/5617110

Creation
Nigel Shindler

History is created by two opposing forces; light, and dark; good, and evil. The Jews are made in the image of God, and are responsible for the creation of all civilizations; they are good; Homo-sapiens, when left to behave in accordance with their nature, are responsible for the fall of civilizations.

Homo-sapiens are savage by nature due to a process of evolution known as "neoteny", which enabled them to survive when climatic conditions endangered their natural habitat, tropical rain-forests; they can appear to be a fully mature adult, but still have the mind of an infant.

Homo-sapiens use whatever materials they have at their disposal to destroy whenever given an opportunity. Many people believe serial killers are solitary predators; this belief is deluded, and merely serves to hide the despicable truth people are not willing to face about themselves.

Recent film footage has revealed how "police officers" in the U.S. use tools at their disposal to prey on the vulnerable; every nation has a population it scapegoats. The "officers" kill because this is the ultimate form of destruction, and due to being able to do this in a public forum, it

heightens their sense of power, and, thus, their level of pleasure.
Learn more about the nature of Jews, and Homo-sapiens,

in the third volume of The Tower Trilogy; Creation.
CreateSpace eStore: https://www.createspace.com/5471399

The stories that fill our lives, encompass our days,
Create meaning, a sense of order, and dreams
We hope one day to achieve.
When all is said and done,
Nothing can be added or taken away,
As if time had decided how long we stay.
CreateSpace eStore: https://www.createspace.com/5075060

While Britain Slept

Nigel & Max Shindler

The mafia families head quartered in America have taken advantage of Britain's weakened state since the end of the Second World War when She accomplished the herculean task of almost single-handedly demolishing German Fascism. The Monolith, (the Mafia), encouraged members of the Nazi Party intelligentsia to reside in America, among other countries, after WWII, and they have since used their knowledge, and expertise, to remain the world's superpower, and become extraordinarily wealthy.

The Mafia continues to have an alliance with Germany till this day, and use the puppet known as, Angel Merkel, to achieve their objective. She has destabilized the continent of Europe by allowing swarms of immigrants, primarily from the Middle-East; Syria, in particular, into Germany, and has insisted that other Western European countries do the same, all the while, she repeatedly states in her political campaigns that Germany's economic problems are entirely due to immigrants refusing to learn German, and the German culture; quite obviously, if a person isn't supporting oneself, and the State doesn't support that person, he may resort to feeding off others to survive; the most unfortunate reality we *must all face is that people from uncivilized backgrounds have no reluctance practicing cannibalism in order to sustain their chosen lifestyles as "While Britain Slept" reveals in documents that are irrefutable.*

"Hell on Earth", answers the question; Why did the Tower fall?
Many things people believe are good are the opposite.
Opposing such things as sexism, racism, and discrimination, and supporting, equality, is considered good by the common man.
The truth is the contrary, such terms are used to homogenize society, which promotes consumerism by making people ignorant because they fail to judge, and discern, resulting in them becoming attached to inanimate, rather than animate, objects; to materials instead of Spirit, as a consequence, the Tower can no longer be held together, and falls.

SAVIOURS OF THE WORLD

Nigel & Max Shindler

Nazism was not eradicated after the Second World War; the disease spread, and flourished, due to being exported to countries around the world because of projects such as "The Paperclip Programme". The solution, the way to insure it is eradicated, once and for all, is to expose the whole truth about the Nazi Party; the extent, and nature, of the atrocities that occurred while Adolf Hitler was Chancellor of Germany.

CreateSpace eStore: https://www.createspace.com/5928525

The British

Nigel & Max Shindler

The Tower is an amalgamation of the concepts that formulated the history of human civilization. Around 12,000 B.C; The Lord's spirit was placed within people who had survived the toughest climatic, topographic, conditions imaginable. They left their homeland, "Doggerland", and created the civilizations that form the Trinity; Ancient Egypt, Mesopotamia, and the Indus Valley; these people are the British.
CreateSpace eStore: https://www.createspace.com/5619291

LOVE IS THE WORD

Nigel & Max Shindler

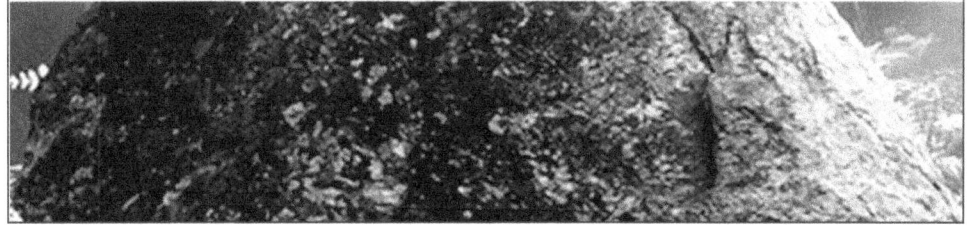

Love Is the Nature of Existence
VOLUME I

The Trinity Manifesto

Nigel Shindler

The state the world is in today is due to Mankind's inability to recognize, and appreciate, the nature of the Jew; his role in society, and, furthermore, his purpose in the vast scheme of things.

It is quite extraordinary that the greater the number provided to Man, fewer, it would seem, are the number who realize the value of their accomplishments.

The holocaust was supposed to serve as a lesson we would never forget, and, therefore, such horrors would never, ever, happen again.

Jews, "God's Chosen people", make us question the world around us; conceive things a different way; they inject elements into societies that hadn't been there before; their works make us better in every way, helping us reach our full potential.

The Jews stem from the migration of the people referred to as the

Aryans; and there is one strain that has produced a higher percentage, considering the overall number of the population, than all others, and that is the Semitic people who for centuries roamed the lands stretching across the Middle East, and North Africa.

CreateSpace eStore: https://www.createspace.com/4759085

A picture is said to contain a thousand words; this means that a length of time can be contained in a quadrant space. Some pieces of art are considered timeless; such being case, one could say that they are of historical importance,
and this is because there is so much one can appreciate within them.

The British built monuments,
and locked witin each a story yet to unfold.

Learn about what the future holds for you in;

The Resurrection

Nigel Shindler

The world is now upside down; this, however, was The Lord's intention; the life force, as a consequence, returns to Him, and He alone dictates the course of the Age to follow. The "Law of One" dictates that at this time all of The Lord's love will be placed within one man on Earth, and he alone points mankind in the right direction; he appeared around 2000 years ago; his name is Jesus Christ. Jesus referred to himself as the "Son of Man"; meaning, he is the culmination of all a man can be. The Lord spoke of him as His son; meaning, he was unique among Jews, in the sense, to use a metaphor, that "genetically" they were one and the same.

Chosen to Make a Dream Come True

Nigel Shindler

Watching from a tower, The ticking hands of time Determine if I stay, and where I go. Through a myriad array of sparkling schemes, I decided to live my life as if it were a dream. Drifting stars swaying amidst nebulous winds, Indicate a day will come for man to shine, And I would show them the way, By not focusing on the value of a dime. Had I not left, and lost so many years, I would not have been given The opportunity to dispense words Men would treasure, And consider kernels of wisdom. The Lord, I imagine, Sits patiently on a lofty golden throne, Gazing from a castle suspended Among softly billowing clouds, With tranquillity waiting for man to touch His tender loving outstretched hand, Thereby knowing he has become The measure of what it means To be a full-fledged man.

Why did I choose the title "Tapestry in Motion"? There are many ways I could answer that question; the meaning of each would be the same, but the words used to express the idea would be different; such is the tapestry of our lives. We have the ability to view matters from different angels; the manner in which we interpret events from the past, colours the way we appreciate the present; the expectations we have about the future, have a lot to do with how we perceive the past. Our lives continually change due to the features noted. We should choose to do this freely, using the power of our will. This is the principle way we inject excitement, intrigue, suspense, and adventure, into our lives; we are the dynamos of ourselves. Time does not cease, it has no beginning or end, rather, it travels in circles, one Age following the next, providing us the opportunity to create an infinite array of tapestries to capture our imagination.

Creation

Out of the darkness sprang a beam of bright light. Through dimensions unknown it danced with the wind, and the ripples of the waters hidden within the escarpments buried deep down below. Eventually, they would form a hilarious spectacle called a show. Animated figures would spring to life; the riddle behind their existence is what they wanted to know. From somewhere within, and maybe beyond, sat The Creator on His throne, looking around the immense, limitless, unknown, pondering the question; wasn't I here before? I'll have to wait and see. The circle of life is heavens way of continually experiencing itself; endlessly evolving while always remaining the same.

The Time is Now

A Man for All Seasons

Many believe Jesus Christ managed to condense the Ten Commandments into two by declaring we should treat others as we would like them to treat us, and we should love our neighbour as much as we do ourselves; actually these two can be summarized into one; Therefore all things whatsoever ye would that men should do to you, do ye even so to them: for this is the law and the prophets. Matthew 7:12 KJV A tree sprang from the ground. Upward it soared, till the clouds it found. Heaven was there, As well as The Lord on high. Life came from the water that fell from above. The Maker shines His light on all those who love. To appreciate life in all its goodness, We must add to the parts that form ourselves. This is done by connecting with others By offering gifts that nourish the soul; After all, for what other reason were we originally made, Than to realize the search for truth Is our sole goal. The quest is eternal, It spans beyond our final resting place; A dark, and empty, grave.

Love Is the Nature of Existence

Love is a generating force that creates life. A person that loves, produces something that nourishes others, and by doing so connects with the life force of the other person, enabling both, as a consequence, to love more generously. This, actually, is what the nature of love, the order of the cosmos, the Ten Commandments, is all about, and what I call the "Law of One". Erich Fromm described such a person as a "Be more"; ultimately such a person will be complete, fully in touch the cosmologically life force, and can be described as a singular entity connected to everything else, which would necessarily mean he is everything, and can be thought of as God.

Heaven

Nigel Shindler

There is an angel that soars at night; around buildings, and through window panes, is the path of his flight. The sound of a shudder is never near, because The Lord is forever here. He hides in escarpments, and floats over fields where cows gently graze; if all is calm, you can hear him whistling as he unceremoniously goes about each of his days. Guarding the Truth, protecting the meek, consume much of his time. Fortunately, he has no concern about the whereabouts of his next dime. He is kept safe within the cocoon God's love. There is no better way to live; once he thought he might have been a white dove, and that's why he's so accustomed to seeing things from high above.

The "Law of One" is fully explained in my four volume book, "Love Is the Nature of Existence"; it describes how everything is connected in the universe, and how the unifying force, the Jew, determines the behaviour of everything else; which necessarily includes you, your neighbour across the street, every molecule, each atom, the stars in the sky, not to mention the galaxies that spin around one another within the furthest dimensions of space.

Love is The Word

Nigel Shindler

A tree sprang from the ground. Upward it soared, till the clouds it found. Heaven was there, as well as The Lord on high. Life came from the water that fell from above. The Maker shines His light on all those who love. To appreciate life in all its goodness, we must add to the parts that form ourselves.

Love is the nature of existence. It can expand then contract, it is also circular; our lives should operate in much the same manner. My book explores why so much has become stagnant, and human civilization is encountering the same problems that are becoming progressively worse. There is a solution, and you'll discover it for yourself in my book.

"The more things change, the more they remain the same". The world, society, people, are not changing, but becoming more of what they already are, as a consequence our problems are progressively becoming worse. History teaches that a solution is found by nature itself when the existence of any order is in jeopardy.

The intention of The Last Judgment is to obliterate the delusional world people live in which is driven by an innate malignant aggression that threatens the survival of every species on this planet. The Last Judgment refutes the fable people believe is true with irrefutable, documented, facts.

The British once ruled the seas, and oceans, as well as lands so distant from one another, the Sun never set upon Her Empire. Once again, for a thousand years, the same will happen; The Trinity will be built, and heaven will return to the Earth.

The Lord throughout history repeatedly used Jews as a channel to implant the "Law of One" within civilizations, thereby providing Homo-sapiens the opportunity to learn about the nature of existence, and, thus, acquire the ability to best utilize the resources within themselves by being humane.

The Lord consists of three components;
He creates, preserves, and destroys;
He is the Trinity.
He views, and experiences, Himself by projecting
The Father, the Son, and the Holy Ghost,
upon the Earth.
The three, as well as all other Jews,
form civilizations throughout history that represent the Trinity.
Every Age is constructed the same way,
for the same reason,
and purpose.

The Lord consists of three components;
He creates, preserves, and destroys;
He views, and experiences, Himself by projecting
The Father, the Son, and the Holy Ghost,
upon the Earth.
The three, as well as all other Jews,
form civilizations throughout history that represent the Trinity,

which is built within every Age, using the same methods, and materials,
for the same reason, and serving the same purpose.

Jews transform people's perception of things by incorporating elements
into societies that weren't there before; they disassemble whole belief
systems, and replace them with something entirely different.
The world is in the state it is today due to an inability to recognize the
value of the Jew; his role within society, and, furthermore, the vast
scheme of things. He is the protector, the preserver, and also the
destroyer.

The British people originated in Doggerland around 12,000 B.C., and
began constructing Stonehenge around 3100 B.C., at the same time they
were building the Indus Valley Civilization; Mesopotamia Civilization,
and the Ancient Egyptian Civilization; which represent the triune nature
of The Lord; The Father, the Son, the Holy Ghost.
The British are responsible for all great accomplishments; they build the
Tower which is history.

The Monolith, the U.S. and her Ally Israel, decided to make themselves rich by encouraging people to be so greedy, narcissistic, materialistic, and malignantly aggressive, that they can never get enough of what they want; consequently, they lead lives of excess; fall into debt, and wind up selling their souls to the Devil.

Nazism was not eradicated after the Second World War; the disease spread, and flourished, due to being exported to countries around the world because of projects such as "The Paperclip Programme". The solution, the way to insure it is eradicated, once and for all, is to expose the whole truth about the Nazi Party; the extent, and nature, of the atrocities that occurred while Adolf Hitler was Chancellor of Germany.

h

Love Is The Nature Of Existence

By Nigel Shindler

Love is the nature of existence. It can expand then contract, it is also circular; our lives should operate in much the same manner. My book explores why so much has become stagnant, and human civilization is encounting the same problems that are becoming progressively worse. There is a solution, and you`ll discover it for yourself in my book.

Love Is The Word And The Time Is Now
By Nigel Edward Shindler

Lulu.

"The more things change, the more they remain the same". The world, society, people, are not changing, but becoming more of what they already are, as a consequence our problems are progressively becoming worse. History teaches that a solution is found by nature itself when the existence of any order is in jeopardy. My book reveals when and how this will happen.

THE LAST
JUDGMENT

NIGEL SHINDLER

The United States manufactures situations in order to make nations believe they are dependent on of Her assistance; militarily and financially. The manufacture of this lie enables the U.S. to infiltrate, and eventually control, countries around the world.
People practice the same method to achieve the same results in their day to day lives; most often the people offering assistance are those who made people dependent on their assistance in the first place, (wolves in sheep's clothing).

The Monolith, the U.S. and her Ally Israel, decided to make themselves rich by encouraging people to be so greedy, narcissistic, materialistic, and malignantly aggressive, that they can never get enough of what they want; consequently, they lead lives of excess; fall into debt, and wind up selling their souls to the Devil.

Nazism was not eradicated after the Second World War; the disease spread, and flourished, due to being exported to countries around the world because of projects such as "The Paperclip Programme". The solution, the way to insure it is eradicated, once and for all, is to expose the whole truth about the Nazi Party; the extent, and nature, of the atrocities that occurred while Adolf Hitler was Chancellor of Germany.

The Power to Learn

A rush to judgement is what keeps learning at bay;
No matter the number of experiences,
Or times one can say.

Think, reflect, and discern truth, from fiction,
Then gradually knowledge will occur.
Plenteous will be the reward; maybe even a luxurious fur.

Don't hide the ignorance you bear.
Embrace it, then focus, maybe glare.
The heart of the matter will gain worth, enlarge, and become fatter.

It will fuel so many; not simply with money, a dollar, or penny;
Enrichment is experienced each moment.

Springs of sparkling wine bubble, then ferment;
Tantalizing the senses, and eventually making a heightened awareness.

Jealousy will stab you like daggers.
Others will view you with disdain.
Don't argue, create refutes, or disputes.
Be careful to avoid, and refrain;
Then, eventually, you'll realize from where you came

Have I already been?
If so my departure was hardly, if at all, missed.

The Boy and The Tower
Nigel Shindler

The dark, night force, hidden within the escarpments scaled by angels,
took me to the place I may soon have to return to.
The reason, meaning, was to be released, not trapped in the snares of the wicked, nefarious,
among the inhabitants of the humanoid species.
I did not belong here on Earth during these misrepresented years.
On some astral plain I viewed the scheme of things past,
and the future that was
designed, coordinated, to inevitably happen.

Wiccans say, "Free we should be."

The Watcher and The Tower
Nigel Shindler

To harm none is the goal,
while you do
all you can to feed your soul;
which might, at times, include keeping
charms in a lucky bowl.
Individuals, they are; and always strive
to be;
manipulating energies around and
within;
even while standing naked by the side of
a deep blue sea.
Their quest is not to discover where we
begin,
but, rather, to make the most of the life we are in,
and, occasionally, this might mean having a Gin.
If we can all learn to live this way,
I see no reason, why on earth, we can't all win!

THE JEWS

Nigel Shindler

Time operates within cycles, therefore, for time to begin anew, Love must continue from the completion of one cycle to the next; those who are righteous, "full of light", obey the Ten Commandments, make the next Age possible; namely, the Jews. A story is written, expelled, then history is once again formed through the fusing of words that form ideas.

Author website

Jews are the jewels of creation; they are God's gift to the earth;
love for all eternity.

Heaven's Gift

Nigel Shindler

There is a wide open sea,
as far as my eyes can see.
The Jews will bring forth life,
beauty, and truth.
For heaven's sake, how could
any of us doubt
that what The Lord says is
true.
If we choose to live in
accordance with His
commandments,
the sky is the limit,

and His glory will be seen all around.
Jews are the jewels of creation; they are God's gift to mankind.
He is kept safe within the cocoon God's love.
There is no better way to live;
once he thought he might have been a white dove,
and that's why he's so accustomed to seeing things
from high above

"The more things change,
The more they remain the same".
The world, society, people, are not changing,
But becoming more of what they already are,
As a consequence our problems are progressively becoming worse.
History teaches that a solution is found by nature
When the survival of any order is in jeopardy.
My books reveal when and how this will happen.

Love Is the Nature of Existence
Tree of Life
Nigel Shindler

The pain, and suffering, endured by Jesus was supposed to serve as a wakeup call to the savages in his midst, and the generations to follow, that they had much to learn about what he means to be civilized, as we still see amply evidence of ill this very day

Gigantic Footprints Are Within You

Watch the leaf as it falls to the ground,
It reveals its front and back;
If tasty, it may be eaten by a hound.

Our destiny is obscure.
We have no way of knowing where and when the end will be;
A road can be rocky, yet offer us all we need to see.

Though weary, pastures offer pleasure;
Horses grace, cows rest in states of tranquillity.
Heightened perception reveals within each an immeasurable sublime
treasure.

Life is often not as we would like it to be.
Troubles, problems, spread, disperse, collect, dissipate, vanish then
reappear;
All the while believe in yourself:
Contained within is all that is required to be complete.

White and black, various shades lie between.
We are not even, but gradients of impurity.
Accept this with pronounced sincerity.

With time, effort, endurance, perseverance,
Revealed will be compassion enormous, herculean.
David is how you appear in the present;
A Goliath of virtue and growth will be the end result.

Born; August 7, 1931, Liverpool
Died; July 29, 2014, Toronto

Max Shindler lived in Waterloo, Liverpool, until he was evacuated to Wales during the Second World War; it was here he first encountered open anti-Semitism; for example, on many occasions he had to eat his meals in the outhouse on the farm where he lived; such traumatizing incidents left lifelong scars on his psyche. He was never able to forget the humiliation he felt due to this abuse, or forget the fact that many of his relations were killed in concentration camps, or died in boxcars while travelling to them.

He left his family home when he was seventeen to live in Israel, where he became a member of the Israeli Army. He later joined England's Royal Air Force, and was a member of a bomb disposal unit. He worked for Avon's advertising department in Northampton, and worked for many years in visual merchandising as a Window Dresser.

He married Pearl Lomax in 1957 in Castletown, on the Isle of Man. A year later their daughter, Nicola, was born; three years later Simone was born; Nigel, his youngest child, was born in 1964; he was loyal, generous, and nurturing, toward each of his children, and grandchildren.

He had an immense appreciation for the arts; he wrote; designed illustrations, and formed Trinity Manifesto Productions with his son. Max played the classical guitar; studied photography, and was fluent in German, Hebrew, Yiddish, and English. He loved opera; classical music, and early Hollywood Western movies. He had a passion for learning throughout his life, and read voraciously on every topic imaginable.

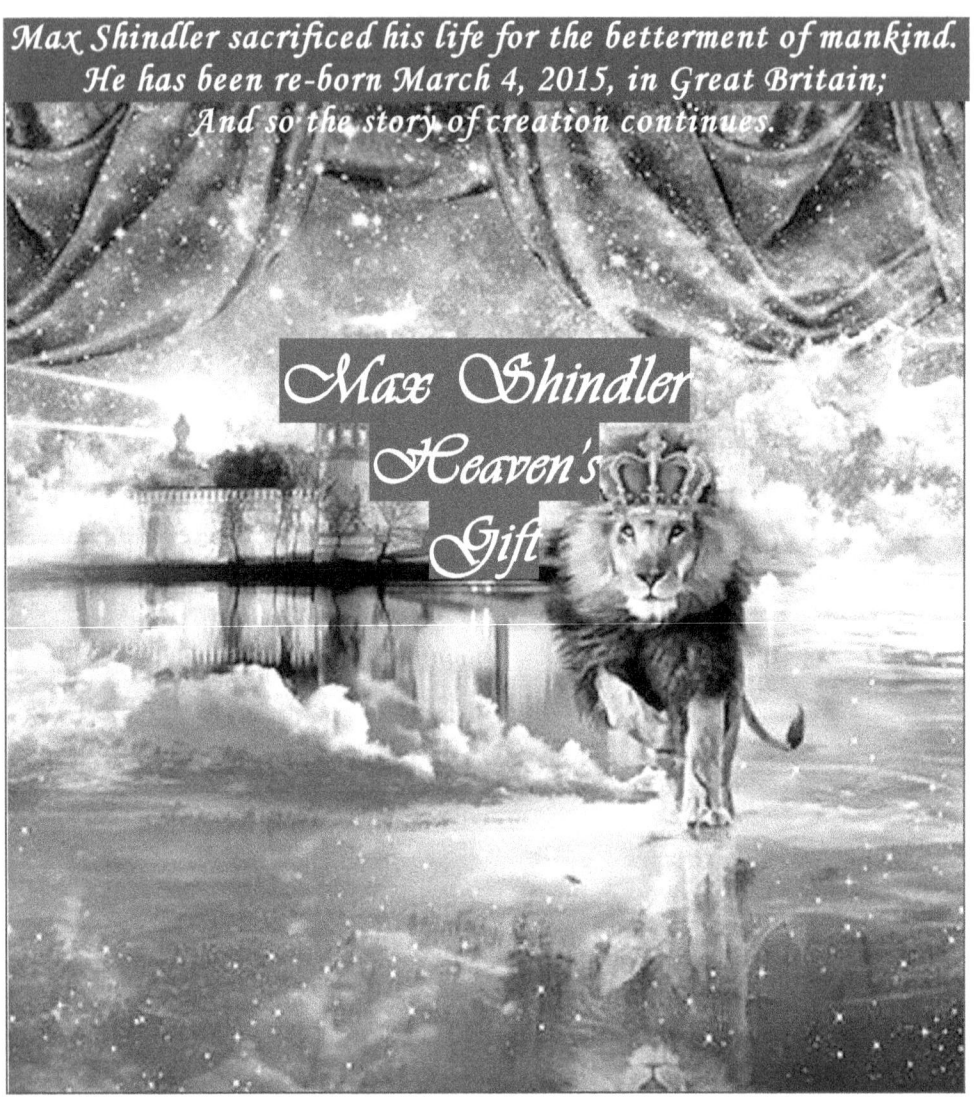

The Welsh Boy

A little boy sits,
While he shivers in the cold.
The wind whistles, as it streams between planks of wood
That forms the shed that is his home;
But only while he eats his meals,
Because the family that live in the house next door,
That sits quietly upon a Welsh moor,
Have decided this is somehow proper for him.
He came here for shelter,
Due to the bombs that reigned upon Liverpool to the north;
Sent by a maniac, that spoke like a madman about a day
When everybody would do as he would say.
But why must a Jew, who deserves to be on a throne,
Sit with the smell of feces that makes him sick to the bone?
The reason is because he is meant to serve as a lesson to Hominids,
Who wish to hide the truth that they are far from civilized;
Their actions are ignited by rage,
And fueled by seething malice, as well as horrible envy.
Why should this be so?
My father actually told during every Passover meal;
It is because of what The Lord did for ME
When He brought us out of Egypt.
This is why a Jew remains proud
No matter where he has to sit.

Nigel Shindler

The author was given the name, Nigel, and seven days later conforming to Jewish custom, a circumcision ceremony was performed, and he was given both the Hebrew name Tovia and the Yiddish equivalent Tevia, both names imply the blessings and goodness that God bestows. In later years whilst living in Israel he adopted the Hebrew name Eitan, which implies stoicism and impetuousness.

All these terms, expressions, are a reflection of his nature. He has depth and complexity, but can display banality, depending on his mood, or the time of day.

Due to fate, circumstance, destiny, over recent years he has not just engaged in a battle with himself in order to become fully actualized, but the nations Canada, and Britain.

He admits that he neither wanted, nor desired, many things that have encroached on the lifestyle he sought for himself, but by conquering each obstacle placed in his path, he managed to realize himself in a span of time previously not thought to be possible, and as a consequence, figured out the unifying force that Einstein left unknown; and the psychological ailment that has crippled the world, and thereby completed Freud's work on the psyche by uncovering the nature of the defense mechanism, projection.

His life has been filled with struggles, from early childhood onward, and the education he has provided himself is primarily the result of questioning;

wanting to understand why people behave the way they do, and how, and why, circumstances arise.

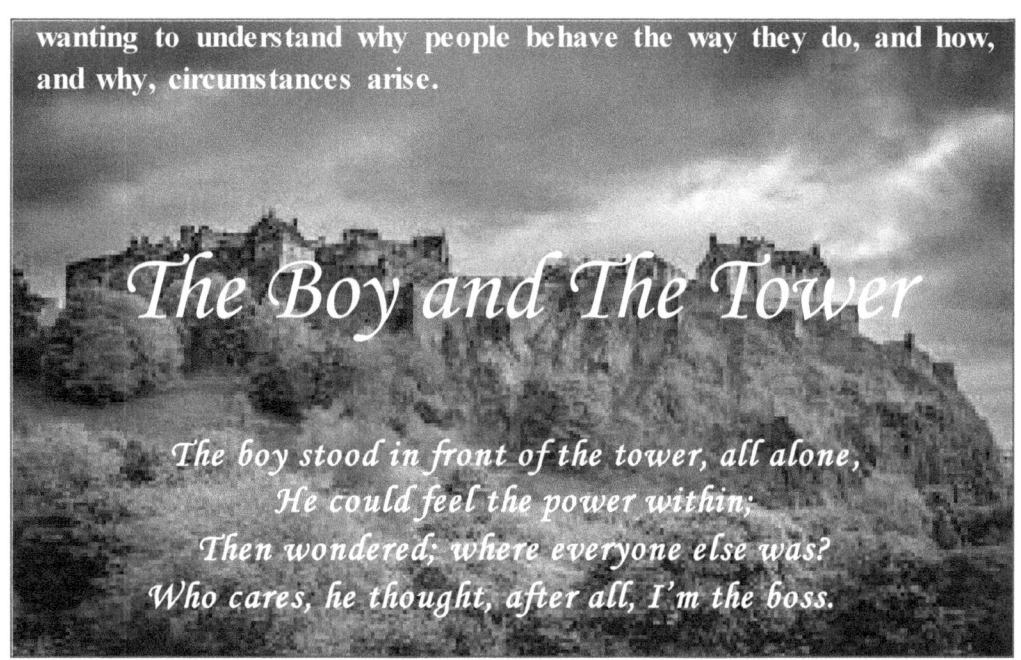

The Boy and The Tower

The boy stood in front of the tower, all alone,
He could feel the power within;
Then wondered; where everyone else was?
Who cares, he thought, after all, I'm the boss.

There is a world filled with everything you could possibly need.
You are a seed that grows and grows.
One day, who knows, maybe you'll know it all,
Then, in totality, you'll be whole, complete,
At the same time, unique;
You!

Others are far, but you don't need them.
You have you; all you need.
There are you in the shadow, in the light; awake at dawn and night.
Everything is awakened to the surroundings you have
Around you.

The beauty of it all, dances with sweet delight.
On angel's feet, you take to flight.
There it is above in heaven; bright light!
There is never a fallen night.
It's perfect, it's right.

Everything is good once again,
There is no pain, nothing more to gain.

You had it all in the beginning.
Only at the end did you realize this.

No longer is there a hiss.
The serpent has disappeared;
All that is left
Is a kiss.

How does man become good? By combining with the divine.
He follows laws that make the cosmos shine.
He practices them religiously,
Till they become instilled within himself.
He sees the light with his awakened eye.
What a pleasure it is to here;
On God's beautiful,
Green land.

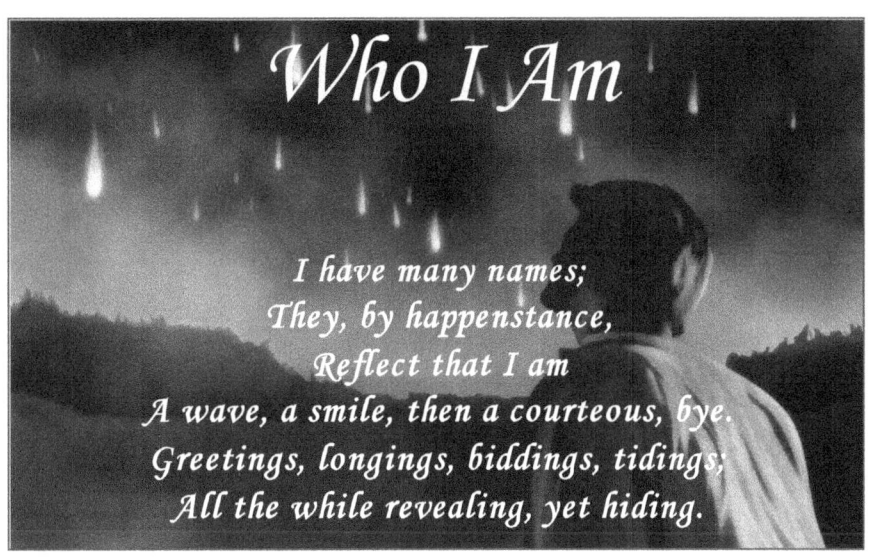

Who I Am

I have many names;
They, by happenstance,
Reflect that I am
A wave, a smile, then a courteous, bye.
Greetings, longings, biddings, tidings;
All the while revealing, yet hiding.

Goodness, stoicism, impetuousness, commonality,
Patience, an essence of formality,
For many there might seem a duality.
Fascinating, illuminating, glamorous;
Many times a facade of utter banality.

Where beneath, behind, in front, is this figure within us all?
Many may wish to ask, but fearfully don't.
A gleaming eye, a playful smirk, dancing inside;
But, self-consciously, notes that he won't.

A sign within a hand indicates understanding;
During, while, all this takes place.
He, I am, is contained within.

Sensing a presence unknown that stretches forward;
The longest distance sets us back
To a time and place that all shall, will, did, begin;
Knowing this entirely is the means to the key.
Many may not sense relevance, practicality,
However, somewhere is an essential faculty.

A few have named Him such.
All of these kindred spirits really have much.

Where I Am

Adrift in time,
I search beyond the sparkling lights
Sprinkled along the Mercy Shore.
There, in the distance, I see the days of my past.
They seem so long ago;
But, then again, they could have taken place yesterday.
They fade from my mind when they're overtaken
By the sights and sounds
Contained in the cluttered corridors of the world today.
That's when I grasp onto the golden memories that enchanted my youth.
They made me the man I once was, and will be again one day.
Maturity has taught me that
I am not becoming a part of the future;
By growing older with each chime of a clock;
I am returning to the man I used to be;
Back in the days when life was better;
When people had the sense to care for one another,
And I never felt my life was in jeopardy,
That was when I sought to better myself all the time,
Till the day arrived when I was able to hear Glen Gould
Play every note of the Goldberg Variations in my mind;

And when ever I liked
I could envision a masterpiece by Matisse.

The Trinity
Is made of the blocks that form creation.

Love is The Word
That forms civilizations

Creation
Exists because The British teach people
To live in harmony with nature.

The Last Judgment
Light must be separated from the dark
For a new Age to begin.

The British
Make history,
As they lead people toward
The light of Truth.

The Miracle of Life

Hell on Earth

Nigel & Max
Shindler

SAVIOURS OF THE WORLD

NIGEL & MAX SHINDLER

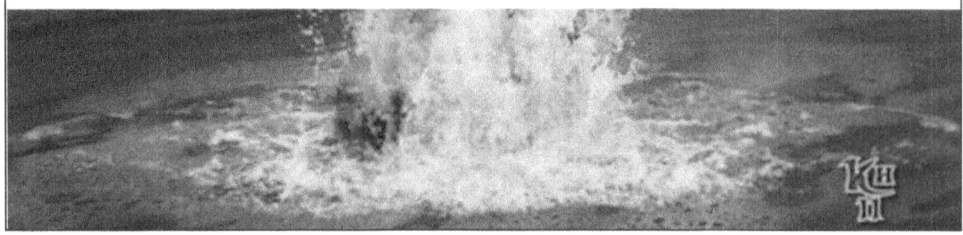

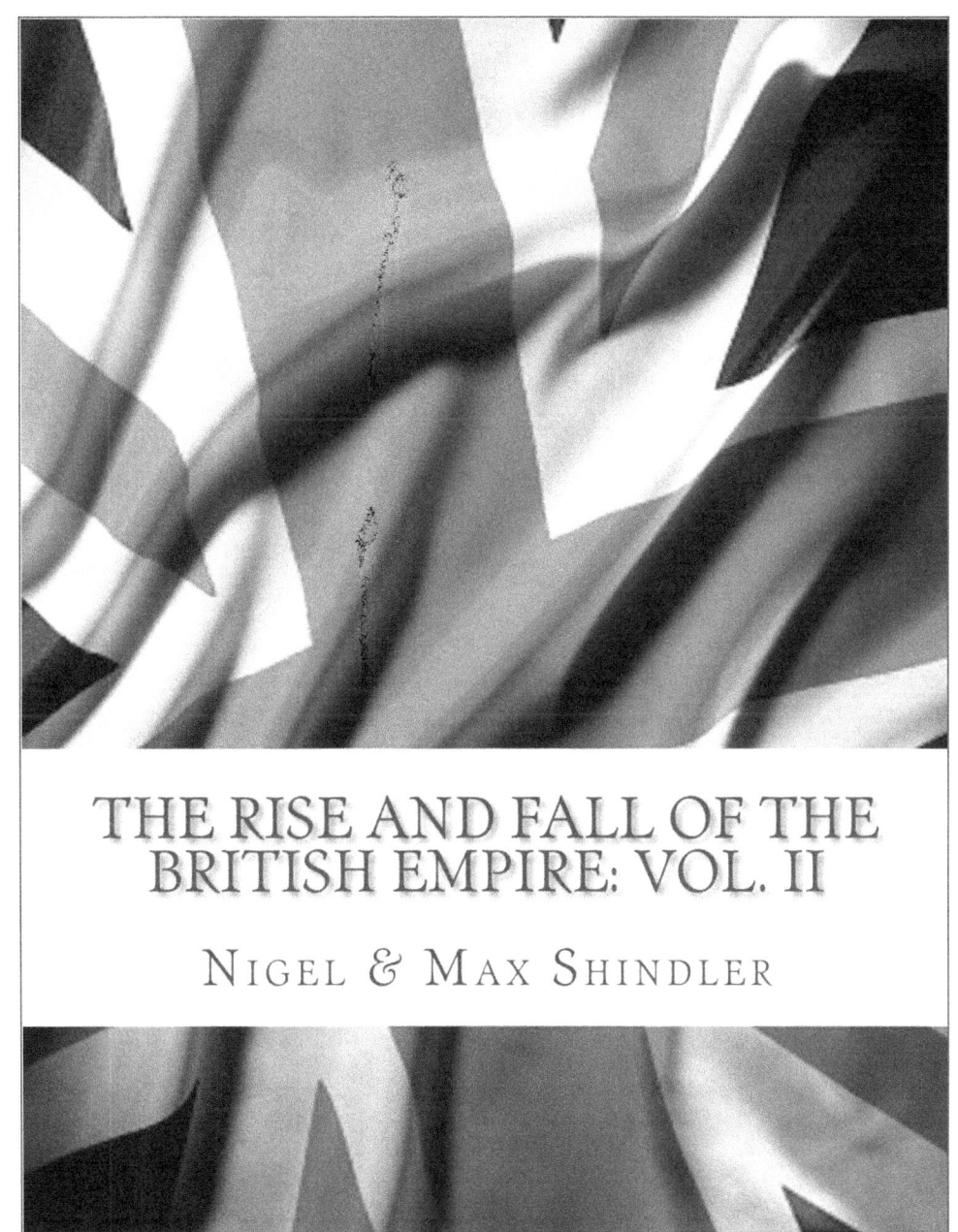

THE RISE AND FALL OF THE BRITISH EMPIRE: VOL. II

Nigel & Max Shindler

WHILE BRITAIN SLEPT

Nigel & Max Shindler

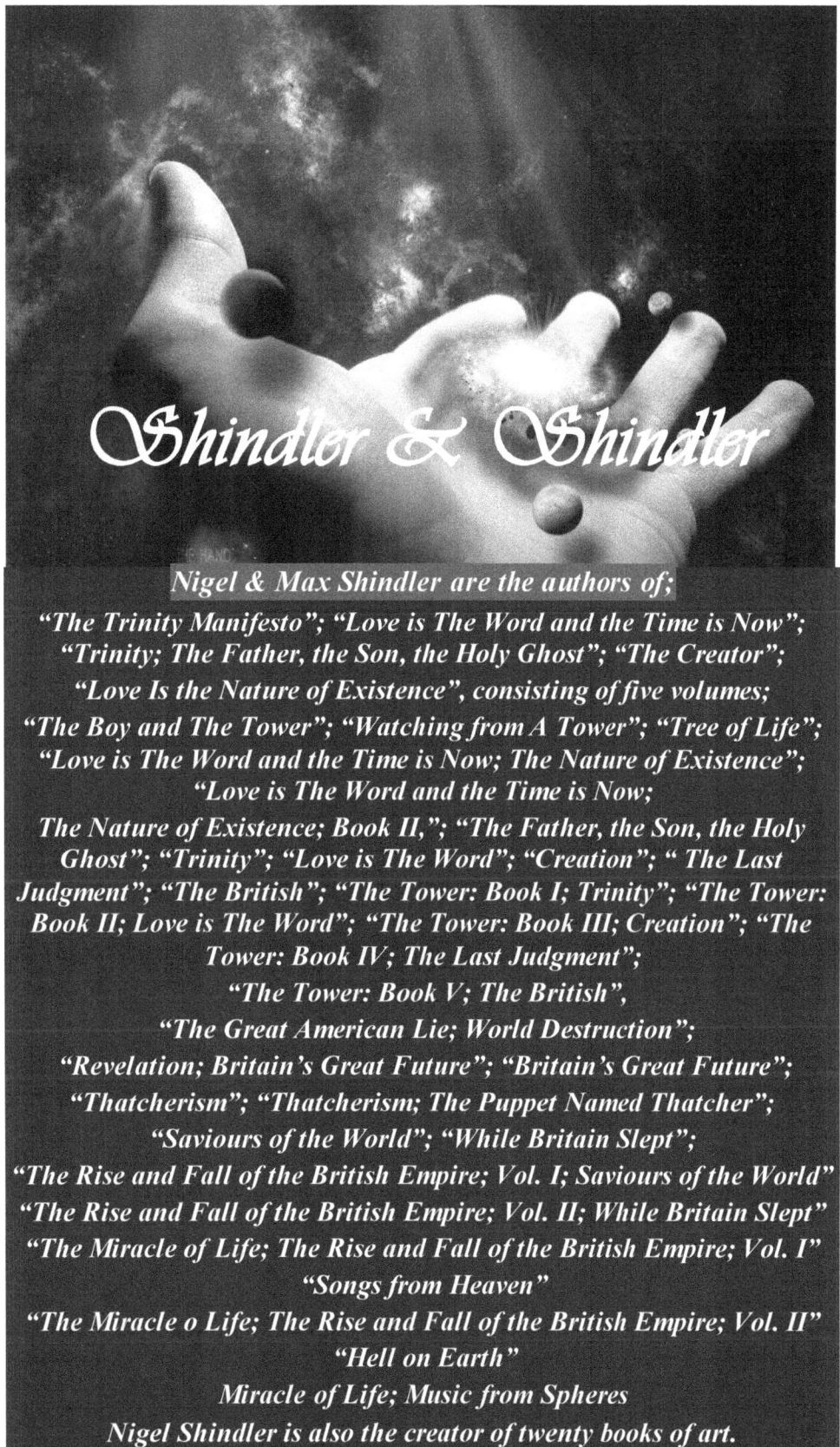

Shindler & Shindler

Nigel & Max Shindler are the authors of;

"The Trinity Manifesto"; "Love is The Word and the Time is Now";
"Trinity; The Father, the Son, the Holy Ghost"; "The Creator";
"Love Is the Nature of Existence", consisting of five volumes;
"The Boy and The Tower"; "Watching from A Tower"; "Tree of Life";
"Love is The Word and the Time is Now; The Nature of Existence";
"Love is The Word and the Time is Now;
The Nature of Existence; Book II,"; "The Father, the Son, the Holy
Ghost"; "Trinity"; "Love is The Word"; "Creation"; " The Last
Judgment"; "The British"; "The Tower: Book I; Trinity"; "The Tower:
Book II; Love is The Word"; "The Tower: Book III; Creation"; "The
Tower: Book IV; The Last Judgment";
"The Tower: Book V; The British",
"The Great American Lie; World Destruction";
"Revelation; Britain's Great Future"; "Britain's Great Future";
"Thatcherism"; "Thatcherism; The Puppet Named Thatcher";
"Saviours of the World"; "While Britain Slept";
"The Rise and Fall of the British Empire; Vol. I; Saviours of the World"
"The Rise and Fall of the British Empire; Vol. II; While Britain Slept"
"The Miracle of Life; The Rise and Fall of the British Empire; Vol. I"
"Songs from Heaven"
"The Miracle o Life; The Rise and Fall of the British Empire; Vol. II"
"Hell on Earth"
Miracle of Life; Music from Spheres
Nigel Shindler is also the creator of twenty books of art.
Nofineartamerica.com › Artists Website; http://nigelshindler.webs.com

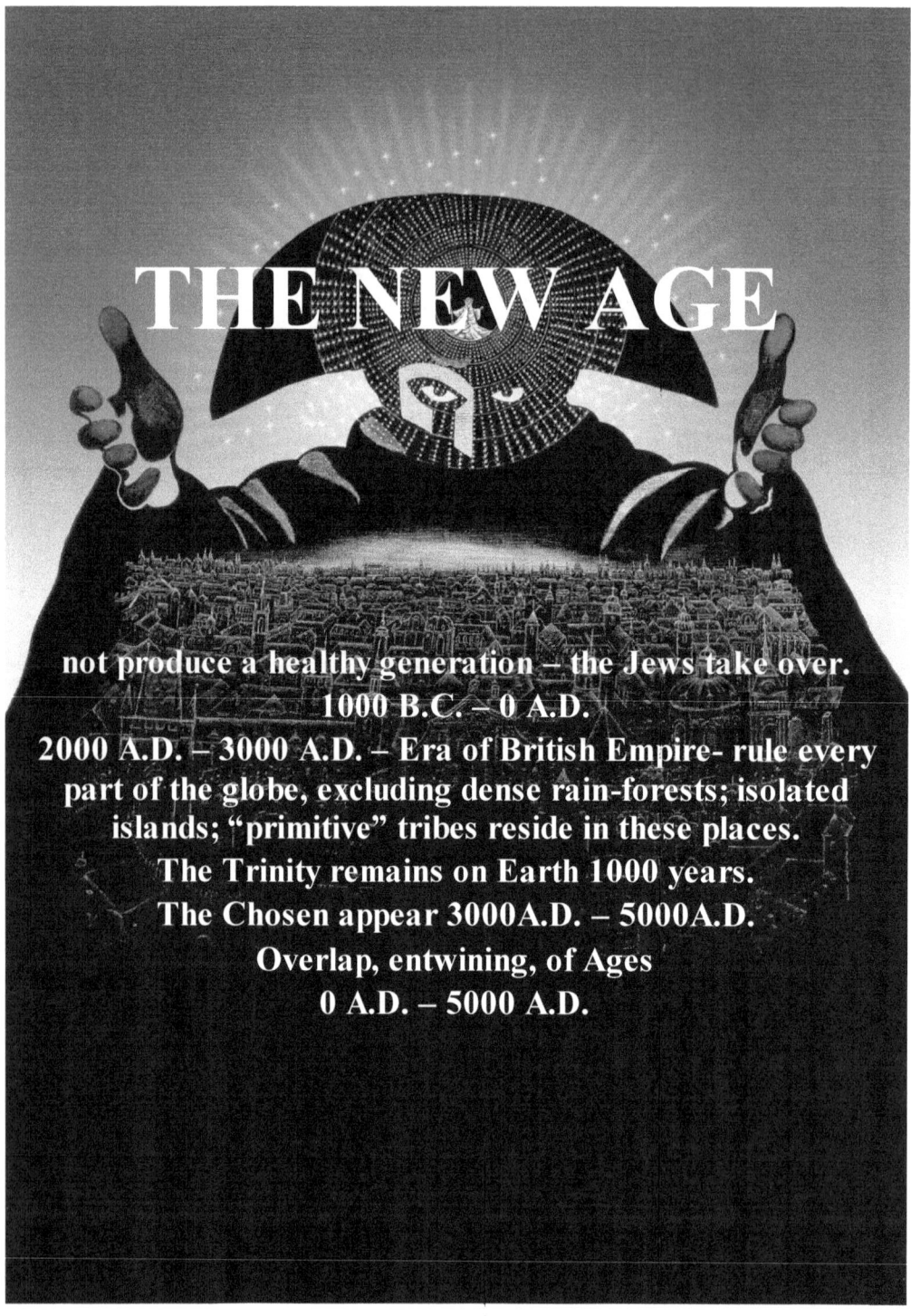

THE NEW AGE

not produce a healthy generation – the Jews take over.
1000 B.C. – 0 A.D.
2000 A.D. – 3000 A.D. – Era of British Empire- rule every part of the globe, excluding dense rain-forests; isolated islands; "primitive" tribes reside in these places.

The Trinity remains on Earth 1000 years.

The Chosen appear 3000A.D. – 5000A.D.

Overlap, entwining, of Ages

0 A.D. – 5000 A.D.

Nigel invites you to join him on;

Nigel Shindler - Google+
https://plus.google.com/+NigelShindler144

Nigel Shindler - Fine Art America
fineartamerica.com › Artists › Nigel Shindler › Images
Browse through Nigel Shindler's online art portfolio. Each image can be purchased as a canvas print, framed print, greeting card, phone case, and more.
Nigel Shindler - Google+
https://plus.google.com/+NigelShindlerLOVE

Nigel Shindler | LinkedIn
https://www.linkedin.com/in/nigel-shindler-03374663

Nigel Shindler - YouTube
https://www.youtube.com/user/teviaschindler

Nigel Shindler (@NigelShindler) | Twitter
https://twitter.com/nigelshindler

Nigel Shindler Profiles | Facebook
https://www.facebook.com/public/Nigel-Shindler

Nigel Shindler
nigelshindler.webs.com/